The Human Condition

The Human Condition

BY HANNAH ARENDT

Second Edition

with a New Foreword by
Danielle Allen

Introduction by
Margaret Canovan

THE UNIVERSITY OF CHICAGO PRESS

CHICAGO & LONDON

The University of Chicago Press, Chicago 60637
The University of Chicago Press, Ltd., London
© 1958, 1998 by The University of Chicago
Foreword by Danielle S. Allen © 2018 by Danielle Allen
Introduction by Margaret Canovan © 1988 by The University of Chicago
Published 2018
Printed in the United States of America

27 26 25 24 23 22 21   3 4 5

ISBN-13: 978-0-226-58660-1 (paper)
ISBN-13: 978-0-226-58674-8 (e-book)
DOI: https://doi.org/10.7208/chicago/9780226586748.001.0001

Library of Congress Cataloging-in-Publication Data

Names: Arendt, Hannah, 1906–1975, author. | Allen, Danielle S., 1971– writer of
     preface. | Canovan, Margaret, writer of introduction.
Title: The human condition / Hannah Arendt ; with a new foreword by Danielle Allen ;
     introduction by Margaret Canovan.
Other titles: Human condition
Description: Second edition. | Sixtieth anniversary edition. | Chicago ; London :
     University of Chicago Press, 2018. | Includes index.
Identifiers: LCCN 2018011989 | ISBN 9780226586601 (pbk. : alk. paper) |
     ISBN 9780226586748 (e-book)
Subjects: LCSH: Sociology. | Economics. | Technology. | Philosophical anthropology. |
     Act (Philosophy)
Classification: LCC HM211 .A7 2018 BD450 | DDC 301—dc23
LC record available at https://lccn.loc.gov/2018011989
♾ This paper meets the requirements of ANSI/NISO Z39.48-1992 (Permanence of Paper).

Table of Contents

Foreword
Danielle Allen

2018

The Soviet Union's satellite Sputnik, the first man-made object to orbit the earth, is the astrological sign under which Hannah Arendt sets forth her argument in *The Human Condition*. "In 1957," she writes, "an earth-born object made by man was launched into the universe." This event, she continues, was "second in importance to no other, not even to the splitting of the atom" (p. 1). The glistening metal sphere of about twenty-three inches in diameter thus takes precedence over even the terrifying darkness of the mushroom cloud. The first subject of *The Human Condition* is science—and the transformation of human experience by not only technology but also scientific ways of knowing. The second subject is how, once we can see science's temptations and dangers as well as its self-evident virtues and rewards, we can restore human authority over it by reviving the power of political thinking to operate alongside it.

In invoking the splitting of the atom so immediately after Sputnik, Arendt adds a second astrological sign to the first with which she opens the book. The argument of the book thus begins not only under the sign of science's temptations—the joyful prospect that we earthlings might liberate ourselves from the earth—but also with an alert to its dangers. The temptations take precedence over the dangers only because they pave the way for them. Science tempts us into thinking we can put an end to politics and transform the human condition into a series of technical problems amenable to definite solutions. The danger that flows from this temptation, Arendt finally says explicitly at the end of the book, is that "an unprecedented and promising outburst of human activity . . . may end in the deadliest, most sterile passivity history has ever known" (p. 322).

To some extent, Arendt takes her bearings from the philosophical tradition known as phenomenology. That is, she looks to appearances in ordinary life as a starting point for insight into "the human condition." She investigates the world around her and observes in ordinary experience a rising habit of reducing all human action to, say, utility functions, or to an accounting of consequences and a toting up of costs and benefits, or to algorithms that depend on the mechanical application of deductive rules to action. Importantly, in contrast to the main line of phenomenologists, figures like her teacher Martin Heidegger, Arendt does not see such reductionist treatments of appearances as a natural and necessary pathology of ordinary experience. For her, they are not the inevitable if painful disappearance of true "being" behind its appearances. Instead, she sees such reductionist treatments as reflecting the problem of "rule," in Patchen Markell's formulation[1]—that is, the problem of the domination over possible interpretations of human life by a hegemonic intellectual paradigm. The reductionist paradigm that she has in her sights is science and the modes of thought on which it relies—styles of thinking that anticipate definite and final "right" answers.

In Arendt's argument, two key features of scientific thinking transform human experience. The first is a focus on cause and effect—on how things work—that makes human reasoning substitutable by machines, which are inevitably faster calculators. She writes, "It is not only and not even primarily contemplation which has become an entirely meaningless experience. Thought itself, when it became 'reckoning with consequences,' became a function of the brain, with the result that electronic instruments are found to fulfil these functions much better than we ever could" (pp. 321–22). The questions of *what* ends we should pursue and

---

1. See Patchen Markell, "Hannah Arendt and the Architecture of *The Human Condition*" (book in progress), and Markell, "Arendt's Work: On the Architecture of *The Human Condition*," *College Literature* 38, no. 1 (Winter 2011): 15–44.

why receded as science developed greater power to explain how things work and to devise technical solutions for making this or that thing, which someone with resources or power happened to want—for instance, nuclear weapons. Indeed, Arendt says that what she would like people to gain from her book "is very simple: it is nothing more than to think what we are doing" (p. 5).

Contemporary examples of science's links to passivity are not hard to find. *The Growth of Incarceration in the United States: Exploring Causes and Consequences,* a report released in 2014 by the National Academy of Sciences, is eloquent on this subject. The report seeks to assess the causes of the rapid rise in incarceration in the American criminal-justice system over the past three decades and offers a meta-analysis of a vast number of preexisting studies, tracing crime rates, the impact of changes in policing on incarceration, themes of electoral politics, and the like. The authors come to a powerful and disturbing conclusion: "In the domain of justice, empirical evidence by itself cannot point the way to policy, *yet an explicit and transparent expression of normative principles has been notably missing as U.S. incarceration rates dramatically rose over the past four decades.* Normative principles have deep roots in jurisprudence and theories of governance and are needed to supplement empirical evidence to guide future policy and research" (emphasis added).[2] In short, the authors admit that the transformation of criminal-justice policy over the last few decades was driven by the empirical work of social scientists focused on the sorts of questions amenable to being answered through quantification and that "a broad discussion of principles has been notably absent from the nation's recent policy debates on the use of imprisonment."[3] Policy makers were debating not the principles of punishment but only causes and consequences: Does incarceration make crime go up or down? What does it do to recidivism rates? And so forth. In noting the absence of normative debate in this policy arena, the authors of the report acknowledge about criminal-justice policy

2. National Research Council of the National Academies, *The Growth of Incarceration in the United States: Exploring Causes and Consequences* (Washington, DC: The National Academies Press, 2014), p. 8 and p. 333.

3. Ibid., p. 7.

just what Arendt points to more broadly: that we have failed "to think what we are doing."

The second key feature of science that leads to depoliticization—or, better, to a failure to engage in "thinking what we are doing"—is science's reliance on math. Arendt writes that scientists (and she does use that broad category regularly in the book) "move in a world where speech has lost its power." She continues, "For the sciences today have been forced to adopt a 'language' of mathematical symbols which, though it was originally meant only as an abbreviation for spoken statements, now contains statements that in no way can be translated back into speech" (p. 4). Although mathematization is a language of its own, it nonetheless forces a rupture in the shared linguistic fabric that human beings depend on to narrate histories, choose courses of action, and give meaning to life. Here Arendt is channeling Aristotle. Human beings are political animals, and they are political *because* they are equipped with words. She writes,

> There may be truths beyond speech, and they may be of great relevance to man in the singular, that is, to man in so far as he is not a political being, whatever else he may be. Men in the plural, that is, men in so far as they live and move and act in this world, can experience meaningfulness only because they can talk with and make sense to each other and to themselves. (p. 4)

As it happens, science itself can show us that Arendt was right to predict that an increasing reliance on mathematical language would yield a decline in political engagement. In recent years, data reveal that an education emphasizing science correlates with the depoliticization of those so educated. There is a statistically significant difference between the rate of political participation among humanities graduates and the rate among STEM (science, technology, engineering, and mathematics) graduates, for instance.[4] Data from the US Department of Education reveal that among

4. For data in this and the subsequent paragraph, please see Danielle Allen, *Education and Equality* (Chicago: University of Chicago Press, 2016), pp. 43–46.

2008 college graduates 92.8 percent of humanities majors have voted at least once since finishing school. Among STEM majors, that number is 83.5 percent. And, within ten years of graduation, 44.1 percent of 1993 humanities graduates had written to public officials, compared to 30.1 percent of STEM graduates. As college graduates, the students are generally of similar socioeconomic backgrounds, suggesting that other distinctions must account for the difference in political engagement.

Of course, the self-selection of students into the humanities and STEM majors may mean that these data reflect only underlying features of the students rather than the effects of teaching they receive. Yet the same pattern appears in a study by political scientist Sunshine Hillygus that controls for students' preexisting levels of interest in politics. Hillygus also finds that the differences in political engagement among college graduates are mirrored in K–12 education. High SAT verbal scores correlate with increased likelihood of political participation, while high SAT math scores correlate with decreased likelihood of participation. Again, since socioeconomic effects on SAT scores move both verbal and math scores in the same direction, this difference between how high verbal scores and high math scores correlate to the likelihood of participation must be telling us something about the relationship between attainment in specific subject domains and preparedness for political engagement. To identify a correlation is not, of course, to identify, let alone prove, causation. But those with more sophisticated verbal skills are clearly more ready to participate in civic life. The decline in civic participation by those whose education emphasizes the sciences is surely one harbinger of that dangerous passivity Arendt saw in the offing.

A reader might wonder whether I offer this data ironically, given Arendt's critique of scientific modes of thought, mathematization, and mere "calculation." But this is to misunderstand that critique. The tools of scientific thinking are powerful and of great value. The point is not to abandon them but to integrate the knowledge they provide into a broader, richer conversation about what we human beings are doing and should be doing, and why. Science is not a guide but a support—a crutch, a tool, an accelerator—for human beings ready to think what they are doing.

Foreword

In rereading *The Human Condition* now, sixty years after its initial publication, I am struck most powerfully by Arendt's prescience. Science and a related phenomenon, technocracy, have grown only more powerful in the intervening six decades. By technocracy, I mean the now-common expectation that political questions should simply be decided by policy experts equipped with the relevant expertise—economics, climate science, genomics, and the like. Our inegalitarian economies and divisive politics are, I think, the bitter fruit that we are now harvesting from the depoliticization of much of human life. For almost two decades, the exceptionally prescient economist Dani Rodrik has powerfully argued for an "impossibility theorem" for the global economy, namely that "democracy, national sovereignty and global economic integration are mutually incompatible";[5] it is impossible, he argues, to have all three at the same time. Yet economic-policy makers have for two decades proceeded as if there were no tragic choices here, as if the question of growth were a purely technocratic one, to be managed through the right tax policies and market institutions. Rather than seeing the fundamental, underlying political questions and engaging with those, pursuing projects of educating and cultivating judgment within a citizenry, policy makers found themselves surprised by the Great Recession of 2008 and then, barely a decade later, surprised again by a tsunami of populist overturnings of what proved to be only seemingly stable political configurations.

Our painful struggle to deal with the politics of climate change is surely also a product of the strange standoff between science and political thinking. This standoff consists of a refusal of science to recognize its role as aide and support, not guide, to political thinking and a refusal of political thinking to recognize the value of science for itself. Arendt, I'm sure, would not have been surprised by our stasis or passivity on that issue. In contrast, technocracy and technocrats find themselves surprised by politics, over and over

5. This quotation, from Rodrik's blog post "The inescapable trilemma of the world economy" (*Dani Rodrik's weblog*, June 27, 2007, http://rodrik.typepad.com/dani_rodriks_weblog/2007/06/the-inescapable.html), summarizes the argument of Rodrik's book *The Globalization Paradox* (New York: Norton, 2012).

again. The fact that our policy makers are startled by eruptions of politics will be dangerous and damaging until we learn again to think what we are doing and equip ourselves to re-politicize our world. I say this not in the hopes of dethroning science, for we can and should celebrate its actualization of extraordinary human potentials. But the time has come, I think, to reunite science with political thinking, calculation and algorithmic intelligence with judgment. This is the work Arendt was calling us to many years ago, and the call has, I think, gotten only more urgent.

To suggest that depoliticization leads to divisiveness and inequality is surely counterintuitive. Indeed, the burden of *The Human Condition* is to help us see why politicization is a good thing.

In *The Human Condition*, Arendt famously expounds on the content and import of three core human activities: labor, work, and action. Labor is that which we undertake out of biological necessity; that which we do, in other words, to feed ourselves. It also encompasses sexual reproduction and the energies devoted to child-rearing. Work is that which we do out of creative effort, to build the things—whether physical or cultural—that shape our world and establish our social connections with others. Labor and work overlap with each other since our romantic relationships are products of our social art and, of course, create the context within which we may also pursue biological reproduction. Finally, action identifies the effort we make together as political creatures, struggling in conditions of pluralistic diversity, to come to collective decisions about our polity's course of action.

Arendt shows us how these three human "doings"—labor, work, and action—can lessen the space between us; give us the opportunity to build worlds together that turn our plurality, or difference, into productive engagements; and enable us to make ourselves visible and knowable to one another, not armed strangers in the four corners of cold and uninhabitable rooms. The third "doing," action, belongs to the political realm, and Arendt is often interpreted as if she focuses exclusively somehow on rescuing action from other human activities. But in fact, her rescue of the political and of politicization is more complicated than that. Labor and work are also categories that scientific analyses flatten and reduce—whether those are the analyses of economists or engineers or computer scientists. Just as we worry now that automation will soon leave us all without a sense of purpose, so Arendt's contemporaries also worried.

Politics is rescued, in Arendt's work, by means of a broader rescue of all three fundamental and interrelated human doings. Arendt's project is not to separate labor, work, and action from one another but, having distinguished them, to show their inter-actions and, more importantly, the articulations of each activity with the others. In so doing, Arendt overturns centuries of intel-lectual tradition that insisted on assigning action, work, and labor to different categories of people. Arendt instead insists that these three areas are pertinent to every human being; the project of politics is precisely to define the terms of how these three activ-ities articulate with one another in each of our lives. The revival of politics demands a renewal of the capacities of each and every democratic citizen or civic agent to diagnose the circumstances of human experience—what is possible for them with regard to opportunities for labor and work—and to pursue the political work of responding to those circumstances.

Life moves faster than science, whether natural or social. Facto-ries close. People find themselves out of work and smitten by depression. People die. People go to prison over the many years it takes the social scientist to hypothesize, collect data, test, confirm or disconfirm, and replicate. And on it goes. When we confront the hardest social problems—like mass incarceration, or economic disruption as occasioned by globalization, or climate change—we need to accelerate the pace of our acquisition of understanding. We have to use every available tool of thought to think what we are doing. I, as a sensible, cognizing, learning, processing, feeling being, can make a judgment even before the economic study gets completely rendered, and so can you, and then we can debate our differing judgments, wrangling our way to a decision about what should be done. This is what it means to think what we are doing.

Of course, where we can, we also want to draw on and vali-date our judgments through rigorous science or social-science research. But it is crucial to keep in mind that science will always remain limited by the fact that it can't catch up with reality, and it can never tell us what we should do. We must supplement science with other instruments for understanding current, lived reality. Arendt can tell us what we should do no more than science can. Instead, she seeks to give us back those instruments of thought and the doings that they shape: labor, work, and action. In proffering these gifts and modeling political thinking, *The Human Condition* offers a much-needed remedy for our times.

Introduction
Margaret Canovan

With the creation of man, the principle of beginning came
into the world. . . . It is in the nature of beginning that something new is started which cannot be expected. (Below, p. 177)

I

Hannah Arendt is preeminently the theorist of beginnings. All
her books are tales of the unexpected (whether concerned with
the novel horrors of totalitarianism or the new dawn of revolution), and reflections on the human capacity to start something
new pervade her thinking. When she published *The Human Condition* in 1958, she herself sent something unexpected out into
the world, and forty years later the book's originality is as striking
as ever. Belonging to no genre, it has had no successful imitators,
and its style and manner remain highly idiosyncratic. Although
Arendt never tried to gather disciples and found a school of
thought, she has been a great educator, opening her readers' eyes
to new ways of looking at the world and at human affairs. Often
the way she sheds light into neglected corners of experience is
by making new distinctions, many of them threefold, as if conventional dichotomies were too constricting for her intellectual
imagination. *The Human Condition* is crammed with distinctions:
between labor, work, and action; between power, violence, and
strength; between the earth and the world; between property and
wealth; and many more, often established through etymological
explorations. But these distinctions are linked to a more contro-
versial way of challenging contemporary truisms. For (in what is
surely the most unexpected feature of the book) she finds in ancient Greece an Archimedean point from which to cast a critical

Introduction

eye on ways of thinking and behaving that we take for granted. Indeed, her calm assumption that we may be able to learn important lessons from the experience of people who lived two and a half millennia ago itself challenges the modern belief in progress. Continual references to the Greeks have added to the sense of bewilderment experienced by many readers of *The Human Condition*, who have found it hard to understand what is actually going on in the book. Here is a long, complex piece of writing that conforms to no established pattern, crammed with unexpected insights but lacking a clearly apparent argumentative structure. The most urgent question to be addressed by way of introduction is, therefore, what is Arendt actually *doing*?

Both the book's difficulty and its enduring fascination arise from the fact that she is doing a great many things at once. There are more intertwined strands of thought than can possibly be followed at first reading, and even repeated readings are liable to bring surprises. But one thing she is clearly *not* doing is writing political philosophy as conventionally understood: that is to say, offering political prescriptions backed up by philosophical arguments. Readers accustomed to that genre have tried to find something like it in *The Human Condition*, usually by stressing Arendt's account of the human capacity for action. Since the book is laced with criticism of modern society, it is tempting to suppose that she intended to present a utopia of political action, a kind of New Athens. Nor is this caricature entirely without foundation. Arendt was certainly drawn to participatory democracy, and was an enthusiastic observer of outbreaks of civic activity ranging from American demonstrations against the Vietnam War to the formation of grassroots citizens' "councils" during the short-lived Hungarian Revolution of 1956. Reminding us that the capacity to act is present even in unlikely circumstances was certainly one of her purposes. But she emphatically denied that her role as a political thinker was to propose a blueprint for the future or to tell anyone what to do. Repudiating the title of "political philosopher," she argued that the mistake made by all political philosophers since Plato has been to ignore the fundamental condition of politics: that it goes on among *plural* human

Introduction

beings, each of whom can act and start something new. The results that emerge from such interaction are contingent and unpredictable, "matters of practical politics, subject to the agreement of many; they can never lie in theoretical considerations or the opinion of one person" (p. 5).

Not political philosophy, then; and, indeed, a good deal of the book does not on the face of it appear to be about politics at all. The long analyses of labor and work, and of the implications of modern science and economic growth, are concerned with the setting for politics rather than politics itself. Even the discussion of action is only partially related to specifically political acts. Shortly after the book's publication, Arendt herself described *The Human Condition* as "a kind of prolegomena" to a more systematic work of political theory which she planned (but never completed). Since "the central political activity is action," she explained, it had been necessary first to carry out a preliminary exercise in clarification "to separate action conceptually from other human activities with which it is usually confounded, such as labor and work."[1] And indeed the book's most obvious organizing principle lies in its phenomenological analysis of three forms of activity that are fundamental to the human condition: labor, which corresponds to the biological life of man as an animal; work, which corresponds to the artificial world of objects that human beings build upon the earth; and action, which corresponds to our plurality as distinct individuals. Arendt argues that these distinctions (and the hierarchy of activities implicit in them) have been ignored within an intellectual tradition shaped by philosophical and religious priorities. However, there is considerably more to the book than the phenomenological analysis, and more even than Arendt's critique of traditional political philosophy's misrepresentation of human activity. For those concerns are framed by her response to contemporary events. When she says in her prologue that she proposes "nothing more than

1. From a research proposal submitted to the Rockefeller Foundation after the publication of *The Human Condition*, probably in 1959. Correspondence with the Rockefeller Foundation, Library of Congress MSS Box 20, p. 013872.

to think what we are doing," she also makes clear that what she has in mind is not just a general analysis of human activity, but "a reconsideration of the human condition from the vantage point of our newest experiences and our most recent fears." *What* experiences and fears?

II

The prologue opens with reflections on one of those events that reveal the human capacity for making new beginnings: the launch of the first space satellite in 1957, which Arendt describes as an "event, second in importance to no other, not even to the splitting of the atom." Like the Hungarian Revolution of 1956, which also occurred while she was working on the book, this un-expected event led her to rearrange her ideas, but was at the same time a vindication of observations already made. For, noting that this amazing demonstration of human power was greeted on all sides not with pride or awe but rather as a sign that mankind might escape from the earth, she comments that this "rebellion against human existence as it has been given" had been under way for some time. By escaping from the earth into the skies, and through enterprises such as nuclear technology, human be-ings are successfully challenging natural limits, posing political questions made vastly more difficult by the inaccessibility of modern science to public discussion.

Arendt's prologue moves from this theme to "another no less threatening event" that seems at first sight strangely unconnec-ted: the advent of automation. While liberating us from the bur-den of hard labor, automation is causing unemployment in a "so-ciety of laborers" where all occupations are conceived of as ways of making a living. Over the course of the book, framing the phenomenological analysis of human activities, a dialectical con-trast between these two apparently unrelated topics is gradually developed. On the one hand, the dawn of the space age demon-strates that human beings literally transcend nature. As a result of modern science's "alienation from the *earth*," the human ca-pacity to start new things calls all natural limits into question, leaving the future alarmingly open. On the other hand, in a development Arendt traces to "alienation from the *world*," modern, automated societies engrossed by ever more efficient production and consumption encourage us to behave and think of ourselves simply as an animal species governed by natural laws.

Human animals unconscious of their capacities and responsibilities are not well fitted to take charge of earth-threatening powers. This conjunction echoes Arendt's earlier analysis of totalitarianism as a nihilistic process propelled by a paradoxical combination of convictions: on the one hand the belief that "everything is possible," and on the other that human beings are merely an animal species governed by laws of nature or history, in the service of which individuals are entirely dispensable. The echo is not surprising, for *The Human Condition* is organically linked to Arendt's work on totalitarianism, and the two together contain an original and striking diagnosis of the contemporary human predicament.

The book grew from the Charles R. Walgreen Foundation lectures which Arendt gave at the University of Chicago in April 1956, themselves an outgrowth of a much larger project on "Totalitarian Elements in Marxism." Arendt had embarked on this project after finishing *The Origins of Totalitarianism*, which contained a good deal about the antecedents of Nazi anti-Semitism and racism, but nothing about the Marxist background to Stalin's murderous version of class struggle. Her new enterprise was to consider what features of Marxist theory might have contributed to this disaster. In the event, her trawl brought up so rich and variegated a catch that the Marx book was never written, but many of the trains of thought involved found their way into *The Human Condition*, notably her conclusion that Marx had fatally misconceived political action in terms of a mixture of the other human activities she calls *work* and *labor.*

To understand political action as *making* something is in Arendt's view a dangerous mistake. Making—the activity she calls *work*—is something a craftsman does by forcing raw material to conform to his model. The raw material has no say in the process, and neither do human beings cast as raw material for an

Introduction

attempt to create a new society or make history.[2] Talk of "Man" making his own history is misleading, for (as Arendt continually reminds us) there is no such person: "men, not Man, live on the earth and inhabit the world." To conceive of politics as making is to ignore human plurality in theory and to coerce individuals in practice. Nonetheless, Arendt found that Marx had inherited this particular misconception of politics from the great tradition of Western political thought. Ever since Plato turned his back on the Athenian democracy and set out his scheme for an ideal city, political philosophers had been writing about politics in a way that systematically ignored the most salient political features of human beings—that they are plural, that each of them is capable of new perspectives and new actions, and that they will not fit a tidy, predictable model unless these political capacities are crushed. One of Arendt's main purposes in *The Human Condition* is therefore to challenge the entire tradition of political philosophy by recovering and bringing to light these neglected human capacities.

But this critique of political philosophy is not the only grand theme in the book that stems from her reflections on Marx. For although Marx spoke of *making*, using the terminology of craftsmanship, Arendt claims that he actually understood history in terms of processes of production and consumption much closer to animal life—labor, in fact. His vision of human history as a predictable process is a story not of unique, mortal individuals but of the collective life-process of a species. While he was in Arendt's view quite wrong to suppose that this process could lead through revolution to "the realm of freedom," she was struck by his picture of individuality submerged in the collective life of a human species, devoted to production and consumption and moving inexorably on its way. She found this a revealing representation of modern society, in which economic concerns have come to dominate both politics and human self-consciousness.

2. Arendt's point is illustrated by Mussolini's admiring comment on the Bolshevik revolution, "Lenin is an artist who has worked in men as others have worked in marble or metal," quoted by Alan Bullock in *Hitler and Stalin: Parallel Lives* (London: Fontana Press, 1993), page 374.

Introduction

A second grand theme interwoven with Arendt's phenomenology of human activities is therefore her account of the rise of a "laborers' society."

This theme of "the social" remains one of the most baffling and contentious aspects of the book. Many readers have taken offense at Arendt's derogatory references to social concerns, and have also assumed that in criticizing the conformist materialism of modern society, Arendt intends to recommend a life of heroic action. But that reading misses the book's complexity, for another of its central themes concerns the *dangers* of action, which sets off new processes beyond the actors' control, including the very processes that have given rise to modern society. At the heart of her analysis of the human condition is the vital importance for civilized existence of a durable human world, built upon the earth to shield us against natural processes and provide a stable setting for our mortal lives. Like a table around which people are gathered, that world "relates and separates men at the same time" (p. 52). Only the experience of sharing a common human world with others who look at it from different perspectives can enable us to see reality in the round and to develop a shared common sense. Without it, we are each driven back on our own subjective experience, in which only our feelings, wants, and desires have reality.

The main threat to the human world has for several centuries been the economic modernization that (as Marx pointed out) destroyed all stability and set everything in motion. Unlike Marx, for whom this change was part of an inevitable historical process, Arendt traces it to the unintended effects of contingent human actions, notably the massive expropriation of ecclesiastical and peasant property carried out in the course of the Reformation. For property (in the sense of rights to land passed down through the generations) had always been the chief bastion of the civilized world, giving owners an interest in maintaining its stability. The great change set in motion by the expropriations of the sixteenth century was twofold. For one thing, peasants with a stake in the stability of the world were turned into day laborers entirely absorbed in the struggle to satisfy their bodily needs. For another, stable property was converted into fluid wealth—capital, in fact—with the dynamic effects that Marx had described so well. Instead of inhabiting a stable world of objects made to last, human beings found themselves sucked into an accelerating process of production and consumption.

By the time that Arendt was reflecting on the implications of automation, this process of production and consumption had gone far beyond catering for natural needs; indeed the activities, methods, and consumer goods involved were all highly artificial. But she points out that this modern artificiality is quite unlike the stable worldly artifice inhabited by earlier civilizations. Objects, furniture, houses themselves have become items of consumption, while automatic production processes have taken on a quasi-natural rhythm to which human beings have had to adjust themselves. It is, she says, "as though we had forced open the distinguishing boundaries which protected the world, the human artifice, from nature, the biological process which goes on in its very midst as well as the natural cyclical processes which surround it, delivering and abandoning to them the always threatened stability of a human world" (p. 126). Elsewhere in *The Human Condition* she describes what has happened as an "unnatural growth of the natural" or a "liberation of the life process," for modernization has turned out to be extraordinarily good at increasing production, consumption, and procreation, giving rise to a vastly expanded human race which is producing and consuming more than ever before. Her contention is that since these economic concerns came to be the center of public attention and public policy (instead of being hidden away in the privacy of the household as in all previous civilizations), the costs have been devastation of the world and an ever-increasing tendency for human beings to conceive of themselves in terms of their desire to consume.

The implication of her argument is not, however, that all we need to do is to haul ourselves up out of our immersion in labor and take action. For this modern hegemony of laboring does not mean that human beings have ceased to act, to make new beginnings, or to start new processes—only that science and technology have become the arena for "action into nature." At the very same time when men were becoming more and more inclined to think of themselves as an animal species, their ability to transcend such limits was being dramatically revealed by scientific inventions. For the counterpart of the "world-alienation" suffered by laborers was "earth-alienation" among scientists. While Archimedes had declared long ago that he would be able to move the earth if he could find a place to stand, Arendt argues that (from the time of Galileo to contemporary space engineers and nuclear scientists) men *have* found ways of looking at the earth from a cosmic perspective, and (exercising the human privilege of making new beginnings) have challenged natural limits to the point of threatening the future of life itself. According to her diagnosis of the contemporary predicament, Promethean powers—releasing processes with unfathomable consequences—are being exercised in a society of beings too absorbed in consumption to take any responsibility for the human world or to understand their political capacities. She observes in her prologue that "thoughtlessness" (itself related to the loss of the common human world) is "among the outstanding characteristics of our time," and her object in thinking aloud was surely to encourage thought in others.

III

In so far as Arendt's purpose was to provoke thought and discussion, she has been resoundingly successful. Like many of her writings, *The Human Condition* has been the subject of intense debate ever since its appearance. Indeed, few other works of modern political theory have had such a mixed press, regarded by some as a work of genius and by others as beneath refutation. Many academics have taken exception to the book's unorthodox style and manner. Paying no attention to mainstream debates, Arendt sets out her own analysis without defining her terms or engaging in conventional argumentation. Political controversies have also raged about the book. Its treatment of the *animal laborans* and its analysis of social concerns made its author unpopular with many on the left, but her account of action brought a message of hope and encouragement to other radicals, including some in the Civil Rights movement and behind the Iron Curtain.

Introduction

During the students' movement of the 1960s *The Human Condition* was hailed as a textbook of participatory democracy, and association with that movement in turn alienated its critics.

In recent years, as Arendt's thought has attracted increased attention (partly for reasons she would not herself have welcomed, such as interest in her gender, her ethnicity, and her romantic relationship with Heidegger), the book's importance has come to be very widely recognized, but its meaning remains in dispute. Such is the complexity of its interwoven threads that there is scope for many different readings. Aristotelians, phenomenologists, Habermasians, postmodernists, feminists, and many others have found inspiration in different strands of its rich fabric, and the forty years since its publication are not nearly long enough to allow an assessment of its lasting significance. If we can extract a central theme from so complex a book, that theme must be its reminder of the vital importance of politics, and of properly understanding our political capacities and the dangers and opportunities they offer.

Arendt's account of the human condition reminds us that human beings are creatures who *act* in the sense of starting things and setting off trains of events. This is something we go on doing whether we understand the implications or not, with the result that both the human world and the earth itself have been devastated by our self-inflicted catastrophes. Looking at what she calls "the modern age" (from the seventeenth to the early twentieth century), she diagnoses a paradoxical situation in which radical economic processes were set off by human action, while those concerned increasingly thought of themselves as helpless flotsam on the currents of socioeconomic forces. Both trends, she believed, were linked with a new focusing of public attention on economic activities that had traditionally been private matters for the household. In her prologue, however, she observes that this "modern age" of which she writes has itself now passed away, for the advent of nuclear technology has begun a "new and yet unknown age" in the long interaction between human beings and their natural habitat. If she were alive today, she might point to a novel variation on the familiar theme of power and helplessness, again connected with the emergence into the public

Introduction

realm of a natural function hitherto cloaked in privacy. On the one hand, the advent of genetic engineering (with its power to set off new processes that burst the bonds of nature) strikingly confirms human transcendence and what she called "a rebellion against human existence as it has been given" (p. 2). On the other hand, our self-understanding as animals has deepened into an unprecedented stress not just on production but on reproduction. Matters of sex, allowed only recently into the public arena, seem rapidly to be elbowing other topics out of public discourse, while neo-Darwinian scientists encourage us to believe that everything about us is determined by our genes.

Since the gap between power and responsibility seems wider than ever, her reminder of the human capacity for action and her attempt "to think what we are doing" are particularly timely. However, we need to listen carefully to what she is saying, for we can easily misunderstand her message as a call for humanity to rise from its torpor, take charge of events, and consciously make our own future. The trouble with that quasi-Marxist scenario is that there is no "humanity" that *could* take responsibility in this way. Human beings are plural and mortal, and it is these features of the human condition that give politics both its miraculous openness and its desperate contingency.

The most heartening message of *The Human Condition* is its reminder of human natality and the miracle of beginning. In sharp contrast to Heidegger's stress on our mortality, Arendt argues that faith and hope in human affairs come from the fact that new people are continually coming into the world, each of them unique, each capable of new initiatives that may interrupt or divert the chains of events set in motion by previous actions. She speaks of action as "the one miracle-working faculty of man" (p. 246), pointing out that in human affairs it is actually quite reasonable to expect the unexpected, and that new beginnings cannot be ruled out even when society seems locked in stagnation or set on an inexorable course. Since the book's publication, her observations on the unpredictability of politics have been strikingly confirmed, not least by the collapse of communism. The revolutions of 1989 were notably Arendtian, illustrating her account of how power can spring up as if from nowhere when

Introduction

people begin to "act in concert," and can ebb away unexpectedly from apparently powerful regimes.

But if her analysis of action is a message of hope in dark times, it also carries warnings. For the other side of that miraculous unpredictability of action is lack of control over its effects. Action sets things in motion, and one cannot foresee even the effects of one's own initiatives, let alone control what happens when they are entangled with other people's initiatives in the public arena. Action is therefore deeply frustrating, for its results can turn out to be quite different from what the actor intended. It is because of this "haphazardness" of action amongst plural actors that political philosophers ever since Plato have tried to substitute for action a model of politics as *making* a work of art. Following the philosopher-king who sees the ideal model and molds his passive subjects to fit it, scheme after scheme has been elaborated for perfect societies in which everyone conforms to the author's blueprint. The curious sterility of utopias comes from the absence within them of any scope for initiative, any room for plurality. Although it is now forty years since Arendt made this point, mainstream political philosophy is still caught in the same trap, still unwilling to take action and plurality seriously, still searching for theoretical principles so rationally compelling that even generations yet unborn must accept them, thus making redundant the haphazard contingency of accommodations reached in actual political arenas.

Arendt observes that there are some remedies for the predicaments of action, but she stresses their limited reach. One is simply the permanent possibility of taking *further* action to interrupt apparently inexorable processes or set politics off on a different direction, but that in itself does nothing to cure the damage of the past or make safe the unpredictable future. Only the human capacities to forgive and to promise can deal with these problems, and then only in part. Faced (as so many contemporary polities are) with the wearisome sequence of revenge for past wrongs that only provokes further revenge, forgiveness *can* break that chain, and recent efforts at reconciliation between the races in South Africa offer an impressive illustration of Arendt's point. As she notes, however, no one can forgive himself:

only the unpredictable cooperation of others can do that, and some evils are beyond forgiveness. Furthermore, this way of breaking the chain of consequences set off by action works only for *human* consequences; there is no remedy through forgiveness for the "action into nature" that sets off nuclear reaction or causes the extinction of species.

Another way of coping with the unpredictable consequences of plural initiatives is the human capacity to make and keep promises. Promises made to oneself have no reliability, but when plural persons come together to bind themselves for the future, the covenants they create among themselves can throw "islands of predictability" into the "ocean of uncertainty," creating a new kind of assurance and enabling them to exercise power collectively. Contracts, treaties, and constitutions are all of this kind; they may be enormously strong and reliable, like the U.S. Constitution, or (like Hitler's Munich agreement) they may be not worth the paper they are written on. In other words they are utterly contingent, quite unlike the hypothetical agreements reached in philosophers' imaginations.

Arendt is well known for her celebration of action, particularly for the passages where she talks about the immortal fame earned by Athenian citizens when they engaged with their peers in the public realm. But *The Human Condition* is just as much concerned with action's dangers, and with the myriad processes set off by human initiative and now raging out of control. She reminds us, of course, that we are not helpless animals: we can engage in further action, take initiatives to interrupt such processes, and try to bring them under control through agreements. But apart from the physical difficulties of gaining control over processes thoughtlessly set off by action into nature, she also reminds us of the political problems caused by plurality itself. In principle, if we can all agree to work together we can exercise great power; but agreement between plural persons is hard to achieve, and never safe from the disruptive initiatives of further actors.

As we stand at the threshold of a new millennium, the one safe prediction we can make is that, despite the continuation of processes already in motion, the open future will become an arena for countless human initiatives that are beyond our present

Introduction

imagination. Perhaps it is not too rash to make another prediction: that future readers will find food for thought and scope for debate in *The Human Condition*, picking up and developing different strands and themes in this extraordinary book. That would have suited Arendt very well. As she said toward the end of her life,

> Each time you write something and you send it out into the world and it becomes public, obviously everybody is free to do with it what he pleases, and this is as it should be. I do not have any quarrel with this. You should not try to hold your hand now on whatever may happen to what you have been thinking for yourself. You should rather try to learn from what other people do with it.[3]

3. Remarks to the American Society of Christian Ethics, 1973. Library of Congress MSS Box 70, p. 011828.

*The Human
Condition*

Prologue

In 1957, an earth-born object made by man was launched into the universe, where for some weeks it circled the earth according to the same laws of gravitation that swing and keep in motion the celestial bodies—the sun, the moon, and the stars. To be sure, the man-made satellite was no moon or star, no heavenly body which could follow its circling path for a time span that to us mortals, bound by earthly time, lasts from eternity to eternity. Yet, for a time it managed to stay in the skies; it dwelt and moved in the proximity of the heavenly bodies as though it had been admitted tentatively to their sublime company.

This event, second in importance to no other, not even to the splitting of the atom, would have been greeted with unmitigated joy if it had not been for the uncomfortable military and political circumstances attending it. But, curiously enough, this joy was not triumphal; it was not pride or awe at the tremendousness of human power and mastery which filled the hearts of men, who now, when they looked up from the earth toward the skies, could behold there a thing of their own making. The immediate reaction, expressed on the spur of the moment, was relief about the first "step toward escape from men's imprisonment to the earth." And this strange statement, far from being the accidental slip of some American reporter, unwittingly echoed the extraordinary line which, more than twenty years ago, had been carved on the funeral obelisk for one of Russia's great scientists: "Mankind will not remain bound to the earth forever."

Such feelings have been commonplace for some time. They show that men everywhere are by no means slow to catch up and adjust to scientific discoveries and technical developments, but that, on the contrary, they have outsped them by decades. Here, as in other respects, science has realized and affirmed what men anticipated in dreams that were neither wild nor idle. What is new is only that one of this country's most respectable newspapers finally brought to its front page what up to then had been buried in the highly non-respectable literature of science fiction (to which, unfortunately, nobody yet has paid the attention it deserves as a vehicle of mass sentiments and mass desires). The banality of the statement should not make us overlook how extraordinary in fact it was; for although Christians have spoken of the earth as a vale of tears and philosophers have looked upon their body as a prison of mind or soul, nobody in the history of mankind has ever conceived of the earth as a prison for men's bodies or shown such eagerness to go literally from here to the moon. Should the emancipation and secularization of the modern age, which began with a turning-away, not necessarily from God, but from a god who was the Father of men in heaven, end with an even more fateful repudiation of an Earth who was the Mother of all living creatures under the sky?

The earth is the very quintessence of the human condition, and earthly nature, for all we know, may be unique in the universe in providing human beings with a habitat in which they can move and breathe without effort and without artifice. The human artifice of the world separates human existence from all mere animal environment, but life itself is outside this artificial world, and through life man remains related to all other living organisms. For some time now, a great many scientific endeavors have been directed toward making life also "artificial," toward cutting the last tie through which even man belongs among the children of nature. It is the same desire to escape from imprisonment to the earth that is manifest in the attempt to create life in the test tube, in the desire to mix "frozen germ plasm from people of demonstrated ability under the microscope to produce superior human beings" and "to alter [their] size, shape and function"; and the wish to escape the human condition, I suspect, also underlies the hope to extend man's life-span far beyond the hundred-year limit.

This future man, whom the scientists tell us they will produce in no more than a hundred years, seems to be possessed by a rebellion against human existence as it has been given, a free gift from nowhere (secularly speaking), which he wishes to exchange, as it were, for something he has made himself. There is no reason to doubt our abilities to accomplish such an exchange, just as there is no reason to doubt our present ability to destroy all organic life on earth. The question is only whether we wish to use our new scientific and technical knowledge in this direction, and this question cannot be decided by scientific means; it is a political question of the first order and therefore can hardly be left to the decision of professional scientists or professional politicians.

While such possibilities still may lie in a distant future, the first boomerang effects of science's great triumphs have made themselves felt in a crisis within the natural sciences themselves. The trouble concerns the fact that the "truths" of the modern scientific world view, though they can be demonstrated in mathematical formulas and proved technologically, will no longer lend themselves to normal expression in speech and thought. The moment these "truths" are spoken of conceptually and coherently, the resulting statements will be "not perhaps as meaningless as a 'triangular circle,' but much more so than a 'winged lion' " (Erwin Schrödinger). We do not yet know whether this situation is final. But it could be that we, who are earth-bound creatures and have begun to act as though we were dwellers of the universe, will forever be unable to understand, that is, to think and speak about the things which nevertheless we are able to do. In this case, it would be as though our brain, which constitutes the physical, material condition of our thoughts, were unable to follow what we do, so that from now on we would indeed need artificial machines to do our thinking and speaking. If it should turn out to be true that knowledge (in the modern sense of know-how) and thought have parted company for good, then we would indeed become the helpless slaves, not so much of our machines as of our know-how, thoughtless creatures at the mercy of every gadget which is technically possible, no matter how murderous it is.

However, even apart from these last and yet uncertain consequences, the situation created by the sciences is of great political significance. Wherever the relevance of speech is at stake, matters become political by definition, for speech is what makes man a political being. If we would follow the advice, so frequently urged upon us, to adjust our cultural attitudes to the present status of scientific achievement, we would in all earnest adopt a way of life in which speech is no longer meaningful. For the sciences today have been forced to adopt a "language" of mathematical symbols which, though it was originally meant only as an abbreviation for spoken statements, now contains statements that in no way can be translated back into speech. The reason why it may be wise to distrust the political judgment of scientists *qua* scientists is not primarily their lack of "character"—that they did not refuse to develop atomic weapons—or their naïveté—that they did not understand that once these weapons were developed they would be the last to be consulted about their use—but precisely the fact that they move in a world where speech has lost its power. And whatever men do or know or experience can make sense only to the extent that it can be spoken about. There may be truths beyond speech, and they may be of great relevance to man in the singular, that is, to man in so far as he is not a political being, whatever else he may be. Men in the plural, that is, men in so far as they live and move and act in this world, can experience meaningfulness only because they can talk with and make sense to each other and to themselves.

Closer at hand and perhaps equally decisive is another no less threatening event. This is the advent of automation, which in a few decades probably will empty the factories and liberate mankind from its oldest and most natural burden, the burden of laboring and the bondage to necessity. Here, too, a fundamental aspect of the human condition is at stake, but the rebellion against it, the wish to be liberated from labor's "toil and trouble," is not modern but as old as recorded history. Freedom from labor itself is not new; it once belonged among the most firmly established privileges of the few. In this instance, it seems as though scientific progress and technical developments had been only taken advantage of to achieve something about which all former ages dreamed but which none had been able to realize.

However, this is so only in appearance. The modern age has carried with it a theoretical glorification of labor and has resulted in a factual transformation of the whole of society into a laboring society. The fulfilment of the wish, therefore, like the fulfilment of wishes in fairy tales, comes at a moment when it can only be self-defeating. It is a society of laborers which is about to be liberated from the fetters of labor, and this society does no longer know of those other higher and more meaningful activities for the sake of which this freedom would deserve to be won. Within this society, which is egalitarian because this is labor's way of making men live together, there is no class left, no aristocracy of either a political or spiritual nature from which a restoration of the other capacities of man could start anew. Even presidents, kings, and prime ministers think of their offices in terms of a job necessary for the life of society, and among the intellectuals, only solitary individuals are left who consider what they are doing in terms of work and not in terms of making a living. What we are confronted with is the prospect of a society of laborers without labor, that is, without the only activity left to them. Surely, nothing could be worse.

To these preoccupations and perplexities, this book does not offer an answer. Such answers are given every day, and they are matters of practical politics, subject to the agreement of many; they can never lie in theoretical considerations or the opinion of one person, as though we dealt here with problems for which only one solution is possible. What I propose in the following is a reconsideration of the human condition from the vantage point of our newest experiences and our most recent fears. This, obviously, is a matter of thought, and thoughtlessness—the heedless recklessness or hopeless confusion or complacent repetition of "truths" which have become trivial and empty—seems to me among the outstanding characteristics of our time. What I propose, therefore, is very simple: it is nothing more than to think what we are doing.

"What we are doing" is indeed the central theme of this book. It deals only with the most elementary articulations of the human condition, with those activities that traditionally, as well as according to current opinion, are within the range of every human being. For this and other reasons, the highest and perhaps purest activity of which men are capable, the activity of thinking, is left out of these present considerations. Systematically, therefore, the book is limited to a discussion of labor, work, and action, which forms its three central chapters. Historically, I deal in a last chapter with the modern age, and throughout the book with the various constellations within the hierarchy of activities as we know them from Western history.

However, the modern age is not the same as the modern world. Scientifically, the modern age which began in the seventeenth century came to an end at the beginning of the twentieth century; politically, the modern world, in which we live today, was born with the first atomic explosions. I do not discuss this modern world, against whose background this book was written. I confine myself, on the one hand, to an analysis of those general human capacities which grow out of the human condition and are permanent, that is, which cannot be irretrievably lost so long as the human condition itself is not changed. The purpose of the historical analysis, on the other hand, is to trace back modern world alienation, its twofold flight from the earth into the universe and from the world into the self, to its origins, in order to arrive at an understanding of the nature of society as it had developed and presented itself at the very moment when it was overcome by the advent of a new and yet unknown age.

CHAPTER I

The Human Condition

I

Vita Activa AND THE HUMAN CONDITION

With the term *vita activa*, I propose to designate three fundamental human activities: labor, work, and action. They are fundamental because each corresponds to one of the basic conditions under which life on earth has been given to man.

Labor is the activity which corresponds to the biological process of the human body, whose spontaneous growth, metabolism, and eventual decay are bound to the vital necessities produced and fed into the life process by labor. The human condition of labor is life itself.

Work is the activity which corresponds to the unnaturalness of human existence, which is not imbedded in, and whose mortality is not compensated by, the species' ever-recurring life cycle. Work provides an "artificial" world of things, distinctly different from all natural surroundings. Within its borders each individual life is housed, while this world itself is meant to outlast and transcend them all. The human condition of work is worldliness.

Action, the only activity that goes on directly between men without the intermediary of things or matter, corresponds to the human condition of plurality, to the fact that men, not Man, live on the earth and inhabit the world. While all aspects of the human condition are somehow related to politics, this plurality is specifically *the* condition—not only the *conditio sine qua non*, but the *conditio per quam*—of all political life. Thus the language of the Romans, perhaps the most political people we have known, used the words "to live" and "to be among men" (*inter homines esse*) or "to die" and "to cease to be among men" (*inter homines esse desinere*) as synonyms. But in its most elementary form, the human condition of action is implicit even in Genesis ("Male and female created He *them*"), if we understand that this story of man's creation is distinguished in principle from the one according to which God originally created Man (*adam*), "him" and not "them," so that the multitude of human beings becomes the result of multiplication.[1] Action would be an unnecessary luxury, a capricious interference with general laws of behavior, if men were endlessly reproducible repetitions of the same model, whose nature or essence was the same for all and as predictable as the nature or essence of any other thing. Plurality is the condition of human action because we are all the same, that is, human, in such a way that nobody is ever the same as anyone else who ever lived, lives, or will live.

All three activities and their corresponding conditions are intimately connected with the most general condition of human existence: birth and death, natality and mortality. Labor assures nòt only individual survival, but the life of the species. Work and its product, the human artifact, bestow a measure of permanence and durability upon the futility of mortal life and the fleeting character of human time. Action, in so far as it engages in founding and pre-

---

1. In the analysis of postclassical political thought, it is often quite illuminating to find out which of the two biblical versions of the creation story is cited. Thus it is highly characteristic of the difference between the teaching of Jesus of Nazareth and of Paul that Jesus, discussing the relationship between man and wife, refers to Genesis 1:27: "Have ye not read, that he which made *them* at the beginning made them male and female" (Matt. 19:4), whereas Paul on a similar occasion insists that the woman was created "of the man" and hence "for the man," even though he then somewhat attenuates the dependence: "neither is the man without the woman, neither the woman without the man" (I Cor. 11:8–12). The difference indicates much more than a different attitude to the role of woman. For Jesus, faith was closely related to action (cf. § 33 below); for Paul, faith was primarily related to salvation. Especially interesting in this respect is Augustine (*De civitate Dei* xii. 21), who not only ignores Genesis 1:27 altogether but sees the difference between man and animal in that man was created *unum ac singulum*, whereas all animals were ordered "to come into being several at once" (*plura simul iussit exsistere*). To Augustine, the creation story offers a welcome opportunity to stress the species character of animal life as distinguished from the singularity of human existence.

serving political bodies, creates the condition for remembrance, that is, for history. Labor and work, as well as action, are also rooted in natality in so far as they have the task to provide and preserve the world for, to foresee and reckon with, the constant influx of newcomers who are born into the world as strangers. However, of the three, action has the closest connection with the human condition of natality; the new beginning inherent in birth can make itself felt in the world only because the newcomer possesses the capacity of beginning something anew, that is, of acting. In this sense of initiative, an element of action, and therefore of natality, is inherent in all human activities. Moreover, since action is the political activity par excellence, natality, and not mortality, may be the central category of political, as distinguished from metaphysical, thought.

The human condition comprehends more than the conditions under which life has been given to man. Men are conditioned beings because everything they come in contact with turns immediately into a condition of their existence. The world in which the *vita activa* spends itself consists of things produced by human activities; but the things that owe their existence exclusively to men nevertheless constantly condition their human makers. In addition to the conditions under which life is given to man on earth, and partly out of them, men constantly create their own, self-made conditions, which, their human origin and their variability notwithstanding, possess the same conditioning power as natural things. Whatever touches or enters into a sustained relationship with human life immediately assumes the character of a condition of human existence. This is why men, no matter what they do, are always conditioned beings. Whatever enters the human world of its own accord or is drawn into it by human effort becomes part of the human condition. The impact of the world's reality upon human existence is felt and received as a conditioning force. The objectivity of the world—its object- or thing-character—and the human condition supplement each other; because human existence is conditioned existence, it would be impossible without things, and things would be a heap of unrelated articles, a non-world, if they were not the conditioners of human existence.

To avoid misunderstanding: the human condition is not the

same as human nature, and the sum total of human activities and capabilities which correspond to the human condition does not constitute anything like human nature. For neither those we discuss here nor those we leave out, like thought and reason, and not even the most meticulous enumeration of them all, constitute essential characteristics of human existence in the sense that without them this existence would no longer be human. The most radical change in the human condition we can imagine would be an emigration of men from the earth to some other planet. Such an event, no longer totally impossible, would imply that man would have to live under man-made conditions, radically different from those the earth offers him. Neither labor nor work nor action nor, indeed, thought as we know it would then make sense any longer. Yet even these hypothetical wanderers from the earth would still be human; but the only statement we could make regarding their "nature" is that they still are conditioned beings, even though their condition is now self-made to a considerable extent.

The problem of human nature, the Augustinian *quaestio mihi factus sum* ("a question have I become for myself"), seems unanswerable in both its individual psychological sense and its general philosophical sense. It is highly unlikely that we, who can know, determine, and define the natural essences of all things surrounding us, which we are not, should ever be able to do the same for ourselves—this would be like jumping over our own shadows. Moreover, nothing entitles us to assume that man has a nature or essence in the same sense as other things. In other words, if we have a nature or essence, then surely only a god could know and define it, and the first prerequisite would be that he be able to speak about a "who" as though it were a "what."[2] The perplexity

2. Augustine, who is usually credited with having been the first to raise the so-called anthropological question in philosophy, knew this quite well. He distinguishes between the questions of "Who am I?" and "What am I?" the first being directed by man at himself ("And I directed myself at myself and said to me: You, who are you? And I answered: A man"—*tu, quis es?* [*Confessiones* x. 6]) and the second being addressed to God ("What then am I, my God? What is my nature?"—*Quid ergo sum, Deus meus? Quae natura sum?* [x. 17]). For in the "great mystery," the *grande profundum*, which man is (iv. 14), there is "something of man [*aliquid hominis*] which the spirit of man which is in him itself

is that the modes of human cognition applicable to things with "natural" qualities, including ourselves to the limited extent that we are specimens of the most highly developed species of organic life, fail us when we raise the question: And *who* are we? This is why attempts to define human nature almost invariably end with some construction of a deity, that is, with the god of the philosophers, who, since Plato, has revealed himself upon closer inspection to be a kind of Platonic idea of man. Of course, to demask such philosophic concepts of the divine as conceptualizations of human capabilities and qualities is not a demonstration of, not even an argument for, the non-existence of God; but the fact that attempts to define the nature of man lead so easily into an idea which definitely strikes us as "superhuman" and therefore is identified with the divine may cast suspicion upon the very concept of "human nature."

On the other hand, the conditions of human existence—life itself, natality and mortality, worldliness, plurality, and the earth—can never "explain" what we are or answer the question of who we are for the simple reason that they never condition us absolutely. This has always been the opinion of philosophy, in distinction from the sciences—anthropology, psychology, biology, etc.—which also concern themselves with man. But today we may almost say that we have demonstrated even scientifically that, though we live now, and probably always will, under the earth's conditions, we are not mere earth-bound creatures. Modern natural science owes its great triumphs to having looked upon and treated earth-bound nature from a truly universal viewpoint, that is, from an Archimedean standpoint taken, wilfully and explicitly, outside the earth.

---

knoweth not. But Thou, Lord, who has made him [*fecisti eum*] knowest everything of him [*eius omnia*]" (x. 5). Thus, the most familiar of these phrases which I quoted in the text, the *quaestio mihi factus sum*, is a question raised in the presence of God, "in whose eyes I have become a question for myself" (x. 33). In brief, the answer to the question "Who am I?" is simply: "You are a man—whatever that may be"; and the answer to the question "What am I?" can be given only by God who made man. The question about the nature of man is no less a theological question than the question about the nature of God; both can be settled only within the framework of a divinely revealed answer.

2

THE TERM *Vita Activa*

The term *vita activa* is loaded and overloaded with tradition. It is as old as (but not older than) our tradition of political thought. And this tradition, far from comprehending and conceptualizing all the political experiences of Western mankind, grew out of a specific historical constellation: the trial of Socrates and the conflict between the philosopher and the *polis*. It eliminated many experiences of an earlier past that were irrelevant to its immediate political purposes and proceeded until its end, in the work of Karl Marx, in a highly selective manner. The term itself, in medieval philosophy the standard translation of the Aristotelian *bios politikos*, already occurs in Augustine, where, as *vita negotiosa* or *actuosa*, it still reflects its original meaning: a life devoted to public-political matters.[3]

Aristotle distinguished three ways of life (*bioi*) which men might choose in freedom, that is, in full independence of the necessities of life and the relationships they originated. This prerequisite of freedom ruled out all ways of life chiefly devoted to keeping one's self alive—not only labor, which was the way of life of the slave, who was coerced by the necessity to stay alive and by the rule of his master, but also the working life of the free craftsman and the acquisitive life of the merchant. In short, it excluded everybody who involuntarily or voluntarily, for his whole life or temporarily, had lost the free disposition of his movements and activities.[4] The remaining three ways of life have in common that

3. See Augustine *De civitate Dei* xix. 2, 19.

4. William L. Westermann ("Between Slavery and Freedom," *American Historical Review*, Vol. L [1945]) holds that the "statement of Aristotle . . . that craftsmen live in a condition of limited slavery meant that the artisan, when he made a work contract, disposed of two of the four elements of his free status [viz., of freedom of economic activity and right of unrestricted movement], but by his own volition and for a temporary period"; evidence quoted by Westermann shows that freedom was then understood to consist of "status, personal inviolability, freedom of economic activity, right of unrestricted movement," and slavery consequently "was the lack of these four attributes." Aristotle, in his enumeration of "ways of life" in the *Nicomachean Ethics* (i. 5) and the *Eudemian Ethics* (1215a35 ff.), does not even mention a craftsman's way of life; to him it

they were concerned with the "beautiful," that is, with things neither necessary nor merely useful: the life of enjoying bodily pleasures in which the beautiful, as it is given, is consumed; the life devoted to the matters of the *polis*, in which excellence produces beautiful deeds; and the life of the philosopher devoted to inquiry into, and contemplation of, things eternal, whose everlasting beauty can neither be brought about through the producing interference of man nor be changed through his consumption of them.[5]

The chief difference between the Aristotelian and the later medieval use of the term is that the *bios politikos* denoted explicitly only the realm of human affairs, stressing the action, *praxis*, needed to establish and sustain it. Neither labor nor work was considered to possess sufficient dignity to constitute a *bios* at all, an autonomous and authentically human way of life; since they served and produced what was necessary and useful, they could not be free, independent of human needs and wants.[6] That the political way of life escaped this verdict is due to the Greek understanding of *polis* life, which to them denoted a very special and freely chosen form of political organization and by no means just any form of action necessary to keep men together in an orderly fashion. Not that the Greeks or Aristotle were ignorant of the fact that human life always demands some form of political organization and that ruling over subjects might constitute a distinct way of life; but the despot's way of life, because it was "merely" a necessity, could not be considered free and had no relationship with the *bios politikos*.[7]

---

is obvious that a *banausos* is not free (cf. *Politics* 1337b5). He mentions, however, "the life of money-making" and rejects it because it too is "undertaken under compulsion" (*Nic. Eth.* 1096a5). That the criterion is freedom is stressed in the *Eudemian Ethics:* he enumerates only those lives that are chosen *ep' exousian*.

5. For the opposition of the beautiful to the necessary and the useful see *Politics* 1333a30 ff., 1332b32.

6. For the opposition of the free to the necessary and the useful see *ibid*. 1332b2.

7. See *ibid*. 1277b8 for the distinction between despotic rule and politics. For the argument that the life of the despot is not equal to the life of a free man because the former is concerned with "necessary things," see *ibid*. 1325a24.

With the disappearance of the ancient city-state—Augustine seems to have been the last to know at least what it once meant to be a citizen—the term *vita activa* lost its specifically political meaning and denoted all kinds of active engagement in the things of this world. To be sure, it does not follow that work and labor had risen in the hierarchy of human activities and were now equal in dignity with a life devoted to politics.[8] It was, rather, the other way round: action was now also reckoned among the necessities of earthly life, so that contemplation (the *bios theōrētikos*, translated into the *vita contemplativa*) was left as the only truly free way of life.[9]

However, the enormous superiority of contemplation over activity of any kind, action not excluded, is not Christian in origin. We find it in Plato's political philosophy, where the whole utopian reorganization of *polis* life is not only directed by the superior insight of the philosopher but has no aim other than to make possible the philosopher's way of life. Aristotle's very articulation of the different ways of life, in whose order the life of pleasure plays a minor role, is clearly guided by the ideal of contemplation (*theōria*). To the ancient freedom from the necessities of life and from compulsion by others, the philosophers added freedom and surcease from political activity (*skholē*),[10] so that the later Christian claim to be free from entanglement in worldly affairs, from all the busi-

8. On the widespread opinion that the modern estimate of labor is Christian in origin, see below, § 44.

9. See Aquinas *Summa theologica* ii. 2. 179, esp. art. 2, where the *vita activa* arises out of the *necessitas vitae praesentis*, and *Expositio in Psalmos* 45.3, where the body politic is assigned the task of finding all that is necessary for life: *in civitate oportet invenire omnia necessaria ad vitam.*

10. The Greek word *skholē*, like the Latin *otium*, means primarily freedom from political activity and not simply leisure time, although both words are also used to indicate freedom from labor and life's necessities. In any event, they always indicate a condition free from worries and cares. An excellent description of the everyday life of an ordinary Athenian citizen, who enjoys full freedom from labor and work, can be found in Fustel de Coulanges, *The Ancient City* (Anchor ed.; 1956), pp. 334–36; it will convince everybody how time-consuming political activity was under the conditions of the city-state. One can easily guess how full of worry this ordinary political life was if one remembers that Athenian law did not permit remaining neutral and punished those who did not want to take sides in factional strife with loss of citizenship.

ness of this world, was preceded by and originated in the philosophic *apolitia* of late antiquity. What had been demanded only by the few was now considered to be a right of all.

The term *vita activa*, comprehending all human activities and defined from the viewpoint of the absolute quiet of contemplation, therefore corresponds more closely to the Greek *askholia* ("unquiet"), with which Aristotle designated all activity, than to the Greek *bios politikos*. As early as Aristotle the distinction between quiet and unquiet, between an almost breathless abstention from external physical movement and activity of every kind, is more decisive than the distinction between the political and the theoretical way of life, because it can eventually be found within each of the three ways of life. It is like the distinction between war and peace: just as war takes place for the sake of peace, thus every kind of activity, even the processes of mere thought, must culminate in the absolute quiet of contemplation.[11] Every movement, the movements of body and soul as well as of speech and reasoning, must cease before truth. Truth, be it the ancient truth of Being or the Christian truth of the living God, can reveal itself only in complete human stillness.[12]

Traditionally and up to the beginning of the modern age, the term *vita activa* never lost its negative connotation of "un-quiet," *nec-otium, a-skholia*. As such it remained intimately related to the even more fundamental Greek distinction between things that are by themselves whatever they are and things which owe their existence to man, between things that are *physei* and things that are *nomō*. The primacy of contemplation over activity rests on the conviction that no work of human hands can equal in beauty and truth the physical *kosmos*, which swings in itself in changeless eternity without any interference or assistance from outside, from man or god. This eternity discloses itself to mortal eyes only when all human movements and activities are at perfect rest. Compared with this attitude of quiet, all distinctions and articulations within

11. See Aristotle *Politics* 1333a30–33. Aquinas defines contemplation as *quies ab exterioribus motibus* (*Summa theologica* ii. 2. 179. 1).

12. Aquinas stresses the stillness of the soul and recommends the *vita activa* because it exhausts and therefore "quietens interior passions" and prepares for contemplation (*Summa theologica* ii. 2. 182. 3).

the *vita activa* disappear. Seen from the viewpoint of contemplation, it does not matter what disturbs the necessary quiet, as long as it is disturbed.

Traditionally, therefore, the term *vita activa* receives its meaning from the *vita contemplativa;* its very restricted dignity is bestowed upon it because it serves the needs and wants of contemplation in a living body.[13] Christianity, with its belief in a hereafter whose joys announce themselves in the delights of contemplation,[14] conferred a religious sanction upon the abasement of the *vita activa* to its derivative, secondary position; but the determination of the order itself coincided with the very discovery of contemplation (*theōria*) as a human faculty, distinctly different from thought and reasoning, which occurred in the Socratic school and from then on has ruled metaphysical and political thought throughout our tradition.[15] It seems unnecessary to my present purpose to discuss the reasons for this tradition. Obviously they are deeper than the historical occasion which gave rise to the conflict between the *polis* and the philosopher and thereby, almost incidentally, also led to the discovery of contemplation as the philosopher's way of life. They must lie in an altogether different aspect of the human condition, whose diversity is not exhausted in the various articulations of the *vita activa* and, we may suspect, would not be exhausted even if thought and the movement of reasoning were included in it.

If, therefore, the use of the term *vita activa,* as I propose it here,

13. Aquinas is quite explicit on the connection between the *vita activa* and the wants and needs of the human body which men and animals have in common (*Summa theologica* ii. 2. 182. 1).

14. Augustine speaks of the "burden" (*sarcina*) of active life imposed by the duty of charity, which would be unbearable without the "sweetness" (*suavitas*) and the "delight of truth" given in contemplation (*De civitate Dei* xix. 19).

15. The time-honored resentment of the philosopher against the human condition of having a body is not identical with the ancient contempt for the necessities of life; to be subject to necessity was only one aspect of bodily existence, and the body, once freed of this necessity, was capable of that pure appearance the Greeks called beauty. The philosophers since Plato added to the resentment of being forced by bodily wants the resentment of movement of any kind. It is because the philosopher lives in complete quiet that it is only his body which, according to Plato, inhabits the city. Here lies also the origin of the early reproach of busy-bodiness (*polypragmosynē*) leveled against those who spent their lives in politics.

is in manifest contradiction to the tradition, it is because I doubt not the validity of the experience underlying the distinction but rather the hierarchical order inherent in it from its inception. This does not mean that I wish to contest or even to discuss, for that matter, the traditional concept of truth as revelation and therefore something essentially given to man, or that I prefer the modern age's pragmatic assertion that man can know only what he makes himself. My contention is simply that the enormous weight of contemplation in the traditional hierarchy has blurred the distinctions and articulations within the *vita activa* itself and that, appearances notwithstanding, this condition has not been changed essentially by the modern break with the tradition and the eventual reversal of its hierarchical order in Marx and Nietzsche. It lies in the very nature of the famous "turning upside down" of philosophic systems or currently accepted values, that is, in the nature of the operation itself, that the conceptual framework is left more or less intact.

The modern reversal shares with the traditional hierarchy the assumption that the same central human preoccupation must prevail in all activities of men, since without one comprehensive principle no order could be established. This assumption is not a matter of course, and my use of the term *vita activa* presupposes that the concern underlying all its activities is not the same as and is neither superior nor inferior to the central concern of the *vita contemplativa*.

3

ETERNITY VERSUS IMMORTALITY

That the various modes of active engagement in the things of this world, on one side, and pure thought culminating in contemplation, on the other, might correspond to two altogether different central human concerns has in one way or another been manifest ever since "the men of thought and the men of action began to take different paths,"[16] that is, since the rise of political thought in the

16. See F. M. Cornford, "Plato's Commonwealth," in *Unwritten Philosophy* (1950), p. 54: "The death of Pericles and the Peloponnesian War mark the moment when the men of thought and the men of action began to take different paths, destined to diverge more and more widely till the Stoic sage ceased to be a citizen of his own country and became a citizen of the universe."

Socratic school. However, when the philosophers discovered—and it is probable, though unprovable, that this discovery was made by Socrates himself—that the political realm did not as a matter of course provide for all of man's higher activities, they assumed at once, not that they had found something different in addition to what was already known, but that they had found a higher principle to replace the principle that ruled the *polis*. The shortest, albeit somewhat superficial, way to indicate these two different and to an extent even conflicting principles is to recall the distinction between immortality and eternity.

Immortality means endurance in time, deathless life on this earth and in this world as it was given, according to Greek understanding, to nature and the Olympian gods. Against this background of nature's ever-recurring life and the gods' deathless and ageless lives stood mortal men, the only mortals in an immortal but not eternal universe, confronted with the immortal lives of their gods but not under the rule of an eternal God. If we trust Herodotus, the difference between the two seems to have been striking to Greek self-understanding prior to the conceptual articulation of the philosophers, and therefore prior to the specifically Greek experiences of the eternal which underlie this articulation. Herodotus, discussing Asiatic forms of worship and beliefs in an invisible God, mentions explicitly that compared with this transcendent God (as we would say today) who is beyond time and life and the universe, the Greek gods are *anthrōpophyeis*, have the same nature, not simply the same shape, as man.[17] The Greeks' concern with immortality grew out of their experience of an immortal nature and immortal gods which together surrounded the individual lives of mortal men. Imbedded in a cosmos where everything was immortal, mortality became the hallmark of human existence. Men are "the mortals," the only mortal things in existence, because unlike animals they do not exist only as members of a species

17. Herodotus (i. 131), after reporting that the Persians have "no images of the gods, no temples nor altars, but consider these doings to be foolish," goes on to explain that this shows that they "do not believe, as the Greeks do, that the gods are *anthrōpophyeis*, of human nature," or, we may add, that gods and men have the same nature. See also Pindar *Carmina Nemaea* vi.

whose immortal life is guaranteed through procreation.[18] The mortality of men lies in the fact that individual life, with a recognizable life-story from birth to death, rises out of biological life. This individual life is distinguished from all other things by the rectilinear course of its movement, which, so to speak, cuts through the circular movement of biological life. This is mortality: to move along a rectilinear line in a universe where everything, if it moves at all, moves in a cyclical order.

The task and potential greatness of mortals lie in their ability to produce things—works and deeds and words[19]— which would deserve to be and, at least to a degree, are at home in everlastingness, so that through them mortals could find their place in a cosmos where everything is immortal except themselves. By their capacity for the immortal deed, by their ability to leave nonperishable traces behind, men, their individual mortality notwithstanding, attain an immortality of their own and prove themselves to be of a "divine" nature. The distinction between man and animal runs right through the human species itself: only the best (*aristoi*), who constantly prove themselves to be the best (*aristeuein*, a verb for which there is no equivalent in any other language) and who "prefer immortal fame to mortal things," are really human; the others, content with whatever pleasures nature will yield them, live and die like animals. This was still the opinion of Heraclitus,[20] an opinion whose equivalent one will find in hardly any philosopher after Socrates.

18. See Ps. Aristotle *Economics* 1343b24: Nature guarantees to the species their being forever through recurrence (*periodos*), but cannot guarantee such being forever to the individual. The same thought, "For living things, life is being," appears in *On the Soul* 415b13.

19. The Greek language does not distinguish between "works" and "deeds," but calls both *erga* if they are durable enough to last and great enough to be remembered. It is only when the philosophers, or rather the Sophists, began to draw their "endless distinctions" and to distinguish between making and acting (*poiein* and *prattein*) that the nouns *poiēmata* and *pragmata* received wider currency (see Plato's *Charmides* 163). Homer does not yet know the word *pragmata*, which in Plato (*ta tōn anthrōpōn pragmata*) is best rendered by "human affairs" and has the connotations of trouble and futility. In Herodotus *pragmata* can have the same connotation (cf., for instance, i. 155).

20. Heraclitus, frag. B29 (Diels, *Fragmente der Vorsokratiker* [4th ed.; 1922]).

In our context it is of no great importance whether Socrates himself or Plato discovered the eternal as the true center of strictly metaphysical thought. It weighs heavily in favor of Socrates that he alone among the great thinkers—unique in this as in many other respects—never cared to write down his thoughts; for it is obvious that, no matter how concerned a thinker may be with eternity, the moment he sits down to write his thoughts he ceases to be concerned primarily with eternity and shifts his attention to leaving some trace of them. He has entered the *vita activa* and chosen its way of permanence and potential immortality. One thing is certain: it is only in Plato that concern with the eternal and the life of the philosopher are seen as inherently contradictory and in conflict with the striving for immortality, the way of life of the citizen, the *bios politikos*.

The philosopher's experience of the eternal, which to Plato was *arrhēton* ("unspeakable"), and to Aristotle *aneu logou* ("without word"), and which later was conceptualized in the paradoxical *nunc stans* ("the standing now"), can occur only outside the realm of human affairs and outside the plurality of men, as we know from the Cave parable in Plato's *Republic*, where the philosopher, having liberated himself from the fetters that bound him to his fellow men, leaves the cave in perfect "singularity," as it were, neither accompanied nor followed by others. Politically speaking, if to die is the same as "to cease to be among men," experience of the eternal is a kind of death, and the only thing that separates it from real death is that it is not final because no living creature can endure it for any length of time. And this is precisely what separates the *vita contemplativa* from the *vita activa* in medieval thought.[21] Yet it is decisive that the experience of the eternal, in contradistinction to that of the immortal, has no correspondence with and cannot be transformed into any activity whatsoever, since even the activity of thought, which goes on within one's self by means of words, is obviously not only inadequate to render it but would interrupt and ruin the experience itself.

Theōria, or "contemplation," is the word given to the experience of the eternal, as distinguished from all other attitudes, which at

21. *In vita activa fixi permanere possumus; in contemplativa autem intenta mente manere nullo modo valemus* (Aquinas *Summa theologica* ii. 2. 181. 4).

most may pertain to immortality. It may be that the philosophers' discovery of the eternal was helped by their very justified doubt of the chances of the *polis* for immortality or even permanence, and it may be that the shock of this discovery was so overwhelming that they could not but look down upon all striving for immortality as vanity and vainglory, certainly placing themselves thereby into open opposition to the ancient city-state and the religion which inspired it. However, the eventual victory of the concern with eternity over all kinds of aspirations toward immortality is not due to philosophic thought. The fall of the Roman Empire plainly demonstrated that no work of mortal hands can be immortal, and it was accompanied by the rise of the Christian gospel of an everlasting individual life to its position as the exclusive religion of Western mankind. Both together made any striving for an earthly immortality futile and unnecessary. And they succeeded so well in making the *vita activa* and the *bios politikos* the handmaidens of contemplation that not even the rise of the secular in the modern age and the concomitant reversal of the traditional hierarchy between action and contemplation sufficed to save from oblivion the striving for immortality which originally had been the spring and center of the *vita activa*.

CHAPTER II

The Public
and the Private Realm

4

MAN: A SOCIAL OR A
POLITICAL ANIMAL

The *vita activa*, human life in so far as it is actively engaged in doing something, is always rooted in a world of men and of man-made things which it never leaves or altogether transcends. Things and men form the environment for each of man's activities, which would be pointless without such location; yet this environment, the world into which we are born, would not exist without the human activity which produced it, as in the case of fabricated things; which takes care of it, as in the case of cultivated land; or which established it through organization, as in the case of the body politic. No human life, not even the life of the hermit in nature's wilderness, is possible without a world which directly or indirectly testifies to the presence of other human beings.

All human activities are conditioned by the fact that men live together, but it is only action that cannot even be imagined outside the society of men. The activity of labor does not need the presence of others, though a being laboring in complete solitude would not be human but an *animal laborans* in the word's most literal significance. Man working and fabricating and building a world inhabited only by himself would still be a fabricator, though not *homo faber*: he would have lost his specifically human quality and, rather, be a god—not, to be sure, the Creator, but a divine demiurge as Plato described him in one of his myths. Action alone is the exclusive prerogative of man; neither a beast nor a god is capable of it,[1] and only action is entirely dependent upon the constant presence of others.

This special relationship between action and being together seems fully to justify the early translation of Aristotle's *zōon politikon* by *animal socialis*, already found in Seneca, which then became the standard translation through Thomas Aquinas: *homo est naturaliter politicus, id est, socialis* ("man is by nature political, that is, social").[2] More than any elaborate theory, this unconscious substitution of the social for the political betrays the extent to which the original Greek understanding of politics had been lost. For this, it is significant but not decisive that the word "social" is Roman in origin and has no equivalent in Greek language or thought. Yet the Latin usage of the word *societas* also originally had a clear, though limited, political meaning; it indicated an alliance between people for a specific purpose, as when men organize in order to rule others or to commit a crime.[3] It is only with the later

1. It seems quite striking that the Homeric gods act only with respect to men, ruling them from afar or interfering in their affairs. Conflicts and strife between the gods also seem to arise chiefly from their part in human affairs or their conflicting partiality with respect to mortals. What then appears is a story in which men and gods act together, but the scene is set by the mortals, even when the decision is arrived at in the assembly of gods on Olympus. I think such a "co-operation" is indicated in the Homeric *erg' andrōn te theōn te* (*Odyssey* i. 338): the bard sings the deeds of gods and men, not stories of the gods and stories of men. Similarly, Hesiod's *Theogony* deals not with the deeds of gods but with the genesis of the world (116); it therefore tells how things came into being through begetting and giving birth (constantly recurring). The singer, servant of the Muses, sings "the glorious deeds of men of old and the blessed gods" (97 ff.), but nowhere, as far as I can see, the glorious deeds of the gods.

2. The quotation is from the Index Rerum to the Taurinian edition of Aquinas (1922). The word "politicus" does not occur in the text, but the Index summarizes Thomas' meaning correctly, as can be seen from *Summa theologica* i. 96. 4; ii. 2. 109. 3.

3. *Societas regni* in Livius, *societas sceleris* in Cornelius Nepos. Such an alliance could also be concluded for business purposes, and Aquinas still holds that a "true *societas*" between businessmen exists only "where the investor himself shares in the risk," that is, where the partnership is truly an alliance (see W. J. Ashley, *An Introduction to English Economic History and Theory* [1931], p. 419).

concept of a *societas generis humani*, a "society of man-kind,"[4] that the term "social" begins to acquire the general meaning of a fundamental human condition. It is not that Plato or Aristotle was ignorant of, or unconcerned with, the fact that man cannot live outside the company of men, but they did not count this condition among the specifically human characteristics; on the contrary, it was something human life had in common with animal life, and for this reason alone it could not be fundamentally human. The natural, merely social companionship of the human species was considered to be a limitation imposed upon us by the needs of biological life, which are the same for the human animal as for other forms of animal life.

According to Greek thought, the human capacity for political organization is not only different from but stands in direct opposition to that natural association whose center is the home (*oikia*) and the family. The rise of the city-state meant that man received "besides his private life a sort of second life, his *bios politikos*. Now every citizen belongs to two orders of existence; and there is a sharp distinction in his life between what is his own (*idion*) and what is communal (*koinon*)."[5] It was not just an opinion or theory of Aristotle but a simple historical fact that the foundation of the *polis* was preceded by the destruction of all organized units resting on kinship, such as the *phratria* and the *phylē*.[6] Of all the activities

4. I use here and in the following the word "man-kind" to designate the human species, as distinguished from "mankind," which indicates the sum total of human beings.

5. Werner Jaeger, *Paideia* (1945), III, 111.

6. Although Fustel de Coulanges' chief thesis, according to the Introduction to *The Ancient City* (Anchor ed.; 1956), consists of demonstrating that "the same religion" formed the ancient family organization and the ancient city-state, he brings numerous references to the fact that the regime of the *gens* based on the religion of the family and the regime of the city "were in reality two antagonistic forms of government. . . . Either the city could not last, or it must in the course of time break up the family" (p. 252). The reason for the contradiction in this great book seems to me to be in Coulanges' attempt to treat Rome and the Greek city-states together; for his evidence and categories he relies chiefly on Roman institutional and political sentiment, although he recognizes that the Vesta cult "became weakened in Greece at a very early date . . . but it never became enfeebled at Rome" (p. 146). Not only was the gulf between household and city much deeper in Greece than in Rome, but only in Greece

The Public and the Private Realm

necessary and present in human communities, only two were deemed to be political and to constitute what Aristotle called the *bios politikos*, namely action (*praxis*) and speech (*lexis*), out of which rises the realm of human affairs (*ta tōn anthrōpōn pragmata*, as Plato used to call it) from which everything merely necessary or useful is strictly excluded.

However, while certainly only the foundation of the city-state enabled men to spend their whole lives in the political realm, in action and speech, the conviction that these two human capacities belonged together and are the highest of all seems to have preceded the *polis* and was already present in pre-Socratic thought. The stature of the Homeric Achilles can be understood only if one sees him as "the doer of great deeds and the speaker of great words."[7] In distinction from modern understanding, such words were not considered to be great because they expressed great thoughts; on the contrary, as we know from the last lines of *Antigone*, it may be the capacity for "great words" (*megaloi logoi*) with which to reply to striking blows that will eventually teach thought in old age.[8] Thought was secondary to speech, but

was the Olympian religion, the religion of Homer and the city-state, separate from and superior to the older religion of family and household. While Vesta, the goddess of the hearth, became the protectress of a "city hearth" and part of the official, political cult after the unification and second foundation of Rome, her Greek colleague, Hestia, is mentioned for the first time by Hesiod, the only Greek poet who, in conscious opposition to Homer, praises the life of the hearth and the household; in the official religion of the *polis*, she had to cede her place in the assembly of the twelve Olympian gods to Dionysos (see Mommsen, *Römische Geschichte* [5th ed.], Book I, ch. 12, and Robert Graves, *The Greek Myths* [1955], 27. k).

7. The passage occurs in Phoenix' speech, *Iliad* ix. 443. It clearly refers to education for war and *agora*, the public meeting, in which men can distinguish themselves. The literal translation is: "[your father] charged me to teach you all this, to be a speaker of words and a doer of deeds" (*mythōn te rhētēr' emenai prēktēra te ergōn*).

8. The literal translation of the last lines of *Antigone* (1350–54) is as follows: "But great words, counteracting [or paying back] the great blows of the overproud, teach understanding in old age." The content of these lines is so puzzling to modern understanding that one rarely finds a translator who dares to give the bare sense. An exception is Hölderlin's translation: "Grosse Blicke aber, / Grosse Streiche der hohen Schultern / Vergeltend, / Sie haben im Alter

speech and action were considered to be coeval and coequal, of the same rank and the same kind; and this originally meant not only that most political action, in so far as it remains outside the sphere of violence, is indeed transacted in words, but more fundamentally that finding the right words at the right moment, quite apart from the information or communication they may convey, is action. Only sheer violence is mute, and for this reason violence alone can never be great. Even when, relatively late in antiquity, the arts of war and speech (*rhetoric*) emerged as the two principal political subjects of education, the development was still inspired by this older pre-*polis* experience and tradition and remained subject to it.

In the experience of the *polis*, which not without justification has been called the most talkative of all bodies politic, and even more in the political philosophy which sprang from it, action and speech separated and became more and more independent activities. The emphasis shifted from action to speech, and to speech as a means of persuasion rather than the specifically human way of answering, talking back and measuring up to whatever happened or was done.[9] To be political, to live in a *polis*, meant that everything was decided through words and persuasion and not through force and violence. In Greek self-understanding, to force people

---

gelehrt, zu denken." An anecdote, reported by Plutarch, may illustrate the connection between acting and speaking on a much lower level. A man once approached Demosthenes and related how terribly he had been beaten. "But you," said Demosthenes, "suffered nothing of what you tell me." Whereupon the other raised his voice and cried out: "I suffered nothing?" "Now," said Demosthenes, "I hear the voice of somebody who was injured and who suffered" (*Lives*, "Demosthenes"). A last remnant of this ancient connection of speech and thought, from which our notion of expressing thought through words is absent, may be found in the current Ciceronian phrase of *ratio et oratio*.

9. It is characteristic for this development that every politician was called a "rhetor" and that rhetoric, the art of public speaking, as distinguished from dialectic, the art of philosophic speech, is defined by Aristotle as the art of persuasion (see *Rhetoric* 1354a12 ff., 1355b26 ff.). (The distinction itself is derived from Plato, *Gorgias* 448.) It is in this sense that we must understand the Greek opinion of the decline of Thebes, which was ascribed to Theban neglect of rhetoric in favor of military exercise (see Jacob Burckhardt, *Griechische Kulturgeschichte*, ed. Kroener, III, 190).

by violence, to command rather than persuade, were prepolitical ways to deal with people characteristic of life outside the *polis*, of home and family life, where the household head ruled with uncontested, despotic powers, or of life in the barbarian empires of Asia, whose despotism was frequently likened to the organization of the household.

Aristotle's definition of man as *zōon politikon* was not only unrelated and even opposed to the natural association experienced in household life; it can be fully understood only if one adds his second famous definition of man as a *zōon logon ekhon* ("a living being capable of speech"). The Latin translation of this term into *animal rationale* rests on no less fundamental a misunderstanding than the term "social animal." Aristotle meant neither to define man in general nor to indicate man's highest capacity, which to him was not *logos*, that is, not speech or reason, but *nous*, the capacity of contemplation, whose chief characteristic is that its content cannot be rendered in speech.[10] In his two most famous definitions, Aristotle only formulated the current opinion of the *polis* about man and the political way of life, and according to this opinion, everybody outside the *polis*—slaves and barbarians—was *aneu logou*, deprived, of course, not of the faculty of speech, but of a way of life in which speech and only speech made sense and where the central concern of all citizens was to talk with each other.

The profound misunderstanding expressed in the Latin translation of "political" as "social" is perhaps nowhere clearer than in a discussion in which Thomas Aquinas compares the nature of household rule with political rule: the head of the household, he finds, has some similarity to the head of the kingdom, but, he adds, his power is not so "perfect" as that of the king.[11] Not only in Greece and the *polis* but throughout the whole of occidental antiquity, it would indeed have been self-evident that even the power of the tyrant was less great, less "perfect" than the power with which the *paterfamilias*, the *dominus*, ruled over his household of slaves and family. And this was not because the power of the city's

10. *Nicomachean Ethics* 1142a25 and 1178a6 ff.

11. Aquinas *op. cit.* ii. 2. 50. 3.

ruler was matched and checked by the combined powers of household heads, but because absolute, uncontested rule and a political realm properly speaking were mutually exclusive.[12]

5

Although misunderstanding and equating the political and social realms is as old as the translation of Greek terms into Latin and their adaption to Roman-Christian thought, it has become even more confusing in modern usage and modern understanding of society. The distinction between a private and a public sphere of life corresponds to the household and the political realms, which have existed as distinct, separate entities at least since the rise of the ancient city-state; but the emergence of the social realm, which is neither private nor public, strictly speaking, is a relatively new phenomenon whose origin coincided with the emergence of the modern age and which found its political form in the nation-state.

What concerns us in this context is the extraordinary difficulty with which we, because of this development, understand the decisive division between the public and private realms, between the sphere of the *polis* and the sphere of household and family, and, finally, between activities related to a common world and those related to the maintenance of life, a division upon which all ancient political thought rested as self-evident and axiomatic. In our understanding, the dividing line is entirely blurred, because we see the body of peoples and political communities in the image of a family whose everyday affairs have to be taken care of by a gigantic, nation-wide administration of housekeeping. The scientific thought that corresponds to this development is no longer political science but "national economy" or "social economy" or *Volkswirtschaft*, all of which indicate a kind of "collective house-

12. The terms *dominus* and *paterfamilias* therefore were synonymous, like the terms *servus* and *familiaris: Dominum patrem familiae appellaverunt; servos . . . familiares* (Seneca *Epistolae* 47. 12). The old Roman liberty of the citizen disappeared when the Roman emperors adopted the title *dominus*, "ce nom, qu'Auguste et que Tibère encore, repoussaient comme une malédiction et une injure" (H. Wallon, *Histoire de l'esclavage dans l'antiquité* [1847], III, 21).

keeping";[13] the collective of families economically organized into the facsimile of one super-human family is what we call "society," and its political form of organization is called "nation."[14] We therefore find it difficult to realize that according to ancient thought on these matters, the very term "political economy" would have been a contradiction in terms: whatever was "economic," related to the life of the individual and the survival of the species, was a non-political, household affair by definition.[15]

Historically, it is very likely that the rise of the city-state and the public realm occurred at the expense of the private realm of family and household.[16] Yet the old sanctity of the hearth, though much less pronounced in classical Greece than in ancient Rome, was never entirely lost. What prevented the *polis* from violating the private lives of its citizens and made it hold sacred the boundaries surrounding each property was not respect for private property as we understand it, but the fact that without owning a house

13. According to Gunnar Myrdal (*The Political Element in the Development of Economic Theory* [1953], p. xl), the "idea of Social Economy or collective housekeeping (*Volkswirtschaft*)" is one of the "three main foci" around which "the political speculation which has permeated economics from the very beginning is found to be crystallized."

14. This is not to deny that the nation-state and its society grew out of the medieval kingdom and feudalism, in whose framework the family and household unit have an importance unequalled in classical antiquity. The difference, however, is marked. Within the feudal framework, families and households were mutually almost independent, so that the royal household, representing a given territorial region and ruling the feudal lords as *primus inter pares*, did not pretend, like an absolute ruler, to be the head of one family. The medieval "nation" was a conglomeration of families; its members did not think of themselves as members of one family comprehending the whole nation.

15. The distinction is very clear in the first paragraphs of the Ps. Aristotelian *Economics*, because it opposes the despotic one-man rule (*mon-archia*) of the household organization to the altogether different organization of the *polis*.

16. In Athens, one may see the turning point in Solon's legislation. Coulanges rightly sees in the Athenian law that made it a filial duty to support parents the proof of the loss of paternal power (*op. cit.*, pp. 315–16). However, paternal power was limited only if it conflicted with the interest of the city and never for the sake of the individual family member. Thus the sale of children and the exposure of infants lasted throughout antiquity (see R. H. Barrow, *Slavery in the Roman Empire* [1928], p. 8: "Other rights in the *patria potestas* had become obsolete; but the right of exposure remained unforbidden till A.D. 374").

a man could not participate in the affairs of the world because he had no location in it which was properly his own.[17] Even Plato, whose political plans foresaw the abolition of private property and an extension of the public sphere to the point of annihilating private life altogether, still speaks with great reverence of Zeus Herkeios, the protector of border lines, and calls the *horoi*, the boundaries between one estate and another, divine, without seeing any contradiction.[18]

The distinctive trait of the household sphere was that in it men lived together because they were driven by their wants and needs. The driving force was life itself—the penates, the household gods, were, according to Plutarch, "the gods who make us live and nourish our body"[19]—which, for its individual maintenance and its survival as the life of the species needs the company of others. That individual maintenance should be the task of the man and species survival the task of the woman was obvious, and both of these natural functions, the labor of man to provide nourishment and the labor of the woman in giving birth, were subject to the same urgency of life. Natural community in the household therefore was born of necessity, and necessity ruled over all activities performed in it.

The realm of the *polis*, on the contrary, was the sphere of freedom, and if there was a relationship between these two spheres, it was a matter of course that the mastering of the necessities of life

17. It is interesting for this distinction that there were Greek cities where citizens were obliged by law to share their harvest and consume it in common, whereas each of them had the absolute uncontested property of his soil. See Coulanges (*op. cit.*, p. 61), who calls this law "a singular contradiction"; it is no contradiction, because these two types of property had nothing in common in ancient understanding.

18. See *Laws* 842.

19. Quoted from Coulanges, *op. cit.*, p. 96; the reference to Plutarch is *Quaestiones Romanae* 51. It seems strange that Coulanges' one-sided emphasis on the underworld deities in Greek and Roman religion should have overlooked that these gods were not mere gods of the dead and the cult not merely a "death cult," but that this early earth-bound religion served life and death as two aspects of the same process. Life rises out of the earth and returns to it; birth and death are but two different stages of the same biological life over which the subterranean gods hold sway.

in the household was the condition for freedom of the *polis*. Under no circumstances could politics be only a means to protect society —a society of the faithful, as in the Middle Ages, or a society of property-owners, as in Locke, or a society relentlessly engaged in a process of acquisition, as in Hobbes, or a society of producers, as in Marx, or a society of jobholders, as in our own society, or a society of laborers, as in socialist and communist countries. In all these cases, it is the freedom (and in some instances so-called freedom) of society which requires and justifies the restraint of political authority. Freedom is located in the realm of the social, and force or violence becomes the monopoly of government.

What all Greek philosophers, no matter how opposed to *polis* life, took for granted is that freedom is exclusively located in the political realm, that necessity is primarily a prepolitical phenomenon, characteristic of the private household organization, and that force and violence are justified in this sphere because they are the only means to master necessity—for instance, by ruling over slaves—and to become free. Because all human beings are subject to necessity, they are entitled to violence toward others; violence is the prepolitical act of liberating oneself from the necessity of life for the freedom of world. This freedom is the essential condition of what the Greeks called felicity, *eudaimonia*, which was an objective status depending first of all upon wealth and health. To be poor or to be in ill health meant to be subject to physical necessity, and to be a slave meant to be subject, in addition, to manmade violence. This twofold and doubled "unhappiness" of slavery is quite independent of the actual subjective well-being of the slave. Thus, a poor free man preferred the insecurity of a daily-changing labor market to regular assured work, which, because it restricted his freedom to do as he pleased every day, was already felt to be servitude (*douleia*), and even harsh, painful labor was preferred to the easy life of many household slaves.[20]

20. The discussion between Socrates and Eutherus in Xenophon's *Memorabilia* (ii. 8) is quite interesting: Eutherus is forced by necessity to labor with his body and is sure that his body will not be able to stand this kind of life for very long and also that in his old age he will be destitute. Still, he thinks that to labor is better than to beg. Whereupon Socrates proposes that he look for somebody "who is better off and needs an assistant." Eutherus replies that he could not bear servitude (*douleia*).

The prepolitical force, however, with which the head of the household ruled over the family and its slaves and which was felt to be necessary because man is a "social" before he is a "political animal," has nothing in common with the chaotic "state of nature" from whose violence, according to seventeenth-century political thought, men could escape only by establishing a government that, through a monopoly of power and of violence, would abolish the "war of all against all" by "keeping them all in awe."[21] On the contrary, the whole concept of rule and being ruled, of government and power in the sense in which we understand them as well as the regulated order attending them, was felt to be prepolitical and to belong in the private rather than the public sphere.

The *polis* was distinguished from the household in that it knew only "equals," whereas the household was the center of the strictest inequality. To be free meant both not to be subject to the necessity of life or to the command of another *and* not to be in command oneself. It meant neither to rule nor to be ruled.[22] Thus within the realm of the household, freedom did not exist, for the household head, its ruler, was considered to be free only in so far as he had the power to leave the household and enter the political realm, where all were equals. To be sure, this equality of the political realm has very little in common with our concept of equality: it meant to live among and to have to deal only with one's peers, and it presupposed the existence of "unequals" who, as a matter of fact, were always the majority of the population in a city-state.[23] Equality, therefore, far from being connected with

21. The reference is to Hobbes, *Leviathan*, Part I, ch. 13.

22. The most famous and the most beautiful reference is the discussion of the different forms of government in Herodotus (iii. 80–83), where Otanes, the defender of Greek equality (*isonomiē*), states that he "wishes neither to rule nor to be ruled." But it is the same spirit in which Aristotle states that the life of a free man is better than that of a despot, denying freedom to the despot as a matter of course (*Politics* 1325a24). According to Coulanges, all Greek and Latin words which express some rulership over others, such as *rex, pater, anax, basileus*, refer originally to household relationships and were names the slaves gave to their master (*op. cit.*, pp. 89 ff., 228).

23. The proportion varied and is certainly exaggerated in Xenophon's report from Sparta, where among four thousand people in the market place, a foreigner counted no more than sixty citizens (*Hellenica* iii. 35).

justice, as in modern times, was the very essence of freedom: to be free meant to be free from the inequality present in rulership and to move in a sphere where neither rule nor being ruled existed.

However, the possibility of describing the profound difference between the modern and the ancient understanding of politics in terms of a clear-cut opposition ends here. In the modern world, the social and the political realms are much less distinct. That politics is nothing but a function of society, that action, speech, and thought are primarily superstructures upon social interest, is not a discovery of Karl Marx but on the contrary is among the axiomatic assumptions Marx accepted uncritically from the political economists of the modern age. This functionalization makes it impossible to perceive any serious gulf between the two realms; and this is not a matter of a theory or an ideology, since with the rise of society, that is, the rise of the "household" (*oikia*) or of economic activities to the public realm, housekeeping and all matters pertaining formerly to the private sphere of the family have become a "collective" concern.[24] In the modern world, the two realms indeed constantly flow into each other like waves in the never-resting stream of the life process itself.

The disappearance of the gulf that the ancients had to cross daily to transcend the narrow realm of the household and "rise" into the realm of politics is an essentially modern phenomenon. Such a gulf between the private and the public still existed somehow in the Middle Ages, though it had lost much of its significance

24. See Myrdal, *op. cit.*: "The notion that society, like the head of a family, keeps house for its members, is deeply rooted in economic terminology. . . . In German *Volkswirtschaftslehre* suggests . . . that there is a collective subject of economic activity . . . with a common purpose and common values. In English, . . . 'theory of wealth' or 'theory of welfare' express similar ideas" (p. 140). "What is meant by a social economy whose function is social housekeeping? In the first place, it implies or suggests an analogy between the individual who runs his own or his family household and society. Adam Smith and James Mill elaborated this analogy explicitly. After J. S. Mill's criticism, and with the wider recognition of the distinction between practical and theoretical political economy, the analogy was generally less emphasized" (p. 143). The fact that the analogy was no longer used may also be due to a development in which society devoured the family unit until it became a full-fledged substitute for it.

and changed its location entirely. It has been rightly remarked that after the downfall of the Roman Empire, it was the Catholic Church that offered men a substitute for the citizenship which had formerly been the prerogative of municipal government.[25] The medieval tension between the darkness of everyday life and the grandiose splendor attending everything sacred, with the concomitant rise from the secular to the religious, corresponds in many respects to the rise from the private to the public in antiquity. The difference is of course very marked, for no matter how "worldly" the Church became, it was always essentially an otherworldly concern which kept the community of believers together. While one can equate the public with the religious only with some difficulty, the secular realm under the rule of feudalism was indeed in its entirety what the private realm had been in antiquity. Its hallmark was the absorption of all activities into the household sphere, where they had only private significance, and consequently the very absence of a public realm.[26]

It is characteristic of this growth of the private realm, and incidentally of the difference between the ancient household head and the feudal lord, that the feudal lord could render justice within the limits of his rule, whereas the ancient household head, while he might exert a milder or harsher rule, knew neither of laws nor justice outside the political realm.[27] The bringing of all human

25. R. H. Barrow, *The Romans* (1953), p. 194.

26. The characteristics which E. Levasseur (*Histoire des classes ouvrières et le de l'industrie en France avant 1789* [1900]) finds for the feudal organization of labor are true for the whole of feudal communities: "Chacun vivait chez soi et vivait de soi-même, le noble sur sa seigneurie, le vilain sur sa culture, le citadin dans sa ville" (p. 229).

27. The fair treatment of slaves which Plato recommends in the *Laws* (777) has little to do with justice and is not recommended "out of regard for the [slaves], but more out of respect to ourselves." For the coexistence of two laws, the political law of justice and the household law of rule, see Wallon, *op. cit.*, II, 200: "La loi, pendant bien longtemps, donc . . . s'abstenait de pénétrer dans la famille, où elle reconnaissait l'empire d'une autre loi." Ancient, especially Roman, jurisdiction with respect to household matters, treatment of slaves, family relationships, etc., was essentially designed to restrain the otherwise unrestricted power of the household head; that there could be a rule of justice within the entirely "private" society of the slaves themselves was unthink-

activities into the private realm and the modeling of all human relationships upon the example of the household reached far into the specifically medieval professional organizations in the cities themselves, the guilds, *confrèries*, and *compagnons*, and even into the early business companies, where "the original joint household would seem to be indicated by the very word 'company' (*companis*) . . . [and] such phrases as 'men who eat one bread,' 'men who have one bread and one wine.' "[28] The medieval concept of the "common good," far from indicating the existence of a political realm, recognizes only that private individuals have interests in common, material and spiritual, and that they can retain their privacy and attend to their own business only if one of them takes it upon himself to look out for this common interest. What distinguishes this essentially Christian attitude toward politics from the modern reality is not so much the recognition of a "common good" as the exclusivity of the private sphere and the absence of that curiously hybrid realm where private interests assume public significance that we call "society."

It is therefore not surprising that medieval political thought, concerned exclusively with the secular realm, remained unaware of the gulf between the sheltered life in the household and the merciless exposure of the *polis* and, consequently, of the virtue of courage as one of the most elemental political attitudes. What remains surprising is that the only postclassical political theorist who, in an extraordinary effort to restore its old dignity to politics, perceived the gulf and understood something of the courage needed to cross it was Machiavelli, who described it in the rise "of the Condottiere from low condition to high rank," from privacy to princedom, that is, from circumstances common to all men to the shining glory of great deeds.[29]

---

able—they were by definition outside the realm of the law and subject to the rule of their master. Only the master himself, in so far as he was also a citizen, was subject to the rules of laws, which for the sake of the city eventually even curtailed his powers in the household.

28. W. J. Ashley, *op. cit.*, p. 415.

29. This "rise" from one realm or rank to a higher is a recurrent theme in Machiavelli (see esp. *Prince*, ch. 6 about Hiero of Syracuse and ch. 7; and *Discourses*, Book II, ch. 13).

To leave the household, originally in order to embark upon some adventure and glorious enterprise and later simply to devote one's life to the affairs of the city, demanded courage because only in the household was one primarily concerned with one's own life and survival. Whoever entered the political realm had first to be ready to risk his life, and too great a love for life obstructed freedom, was a sure sign of slavishness.[30] Courage therefore became the political virtue par excellence, and only those men who possessed it could be admitted to a fellowship that was political in content and purpose and thereby transcended the mere togetherness imposed on all—slaves, barbarians, and Greeks alike— through the urgencies of life.[31] The "good life," as Aristotle called the life of the citizen, therefore was not merely better, more carefree or nobler than ordinary life, but of an altogether different

30. "By Solon's time slavery had come to be looked on as worse than death" (Robert Schlaifer, "Greek Theories of Slavery from Homer to Aristotle," *Harvard Studies in Classical Philology* [1936], XLVII). Since then, *philopsychia* ("love of life") and cowardice became identified with slavishness. Thus, Plato could believe he had demonstrated the natural slavishness of slaves by the fact that they had not preferred death to enslavement (*Republic* 386A). A late echo of this might still be found in Seneca's answer to the complaints of slaves: "Is freedom so close at hand, yet is there any one a slave?" (*Ep.* 77. 14) or in his *vita si moriendi virtus abest, servitus est*—"life is slavery without the virtue which knows how to die" (77. 13). To understand the ancient attitude toward slavery, it is not immaterial to remember that the majority of slaves were defeated enemies and that generally only a small percentage were born slaves. And while under the Roman Republic slaves were, on the whole, drawn from outside the limits of Roman rule, Greek slaves usually were of the same nationality as their masters; they had proved their slavish nature by not committing suicide, and since courage was the political virtue par excellence, they had thereby shown their "natural" unworthiness, their unfitness to be citizens. The attitude toward slaves changed in the Roman Empire, not only because of the influence of Stoicism but because a much greater portion of the slave population were slaves by birth. But even in Rome, *labos* is considered to be closely connected with unglorious death by Vergil (*Aeneis* vi).

31. That the free man distinguishes himself from the slave through courage seems to have been the theme of a poem by the Cretan poet Hybrias: "My riches are spear and sword and the beautiful shield. . . . But those who do not dare to bear spear and sword and the beautiful shield that protects the body fall all down unto their knees with awe and address me as Lord and great King" (quoted from Eduard Meyer, *Die Sklaverei im Altertum* [1898], p. 22).

quality. It was "good" to the extent that by having mastered the necessities of sheer life, by being freed from labor and work, and by overcoming the innate urge of all living creatures for their own survival, it was no longer bound to the biological life process.

At the root of Greek political consciousness we find an unequaled clarity and articulateness in drawing this distinction. No activity that served only the purpose of making a living, of sustaining only the life process, was permitted to enter the political realm, and this at the grave risk of abandoning trade and manufacture to the industriousness of slaves and foreigners, so that Athens indeed became the "pensionopolis" with a "proletariat of consumers" which Max Weber so vividly described.[32] The true character of this *polis* is still quite manifest in Plato's and Aristotle's political philosophies, even if the borderline between household and *polis* is occasionally blurred, especially in Plato who, probably following Socrates, began to draw his examples and illustrations for the *polis* from everyday experiences in private life, but also in Aristotle when he, following Plato, tentatively assumed that at least the historical origin of the *polis* must be connected with the necessities of life and that only its content or inherent aim (*telos*) transcends life in the "good life."

These aspects of the teachings of the Socratic school, which soon were to become axiomatic to the point of banality, were then the newest and most revolutionary of all and sprang not from actual experience in political life but from the desire to be freed from its burden, a desire which in their own understanding the philosophers could justify only by demonstrating that even this freest of all ways of life was still connected with and subject to necessity. But the background of actual political experience, at least in Plato and Aristotle, remained so strong that the distinction between the spheres of household and political life was never doubted. Without mastering the necessities of life in the household, neither life nor the "good life" is possible, but politics is never for the sake of life. As far as the members of the *polis* are concerned, household life exists for the sake of the "good life" in the *polis*.

32. Max Weber, "Agrarverhältnisse im Altertum," *Gesammelte Aufsätze zur Sozial- und Wirtschaftsgeschichte* (1924), p. 147.

6

THE RISE OF THE SOCIAL

The emergence of society—the rise of housekeeping, its activities, problems, and organizational devices—from the shadowy interior of the household into the light of the public sphere, has not only blurred the old borderline between private and political, it has also changed almost beyond recognition the meaning of the two terms and their significance for the life of the individual and the citizen. Not only would we not agree with the Greeks that a life spent in the privacy of "one's own" (*idion*), outside the world of the common, is "idiotic" by definition, or with the Romans to whom privacy offered but a temporary refuge from the business of the *res publica*; we call private today a sphere of intimacy whose beginnings we may be able to trace back to late Roman, though hardly to any period of Greek antiquity, but whose peculiar manifoldness and variety were certainly unknown to any period prior to the modern age.

This is not merely a matter of shifted emphasis. In ancient feeling the privative trait of privacy, indicated in the word itself, was all-important; it meant literally a state of being deprived of something, and even of the highest and most human of man's capacities. A man who lived only a private life, who like the slave was not permitted to enter the public realm, or like the barbarian had chosen not to establish such a realm, was not fully human. We no longer think primarily of deprivation when we use the word "privacy," and this is partly due to the enormous enrichment of the private sphere through modern individualism. However, it seems even more important that modern privacy is at least as sharply opposed to the social realm—unknown to the ancients who considered its content a private matter—as it is to the political, properly speaking. The decisive historical fact is that modern privacy in its most relevant function, to shelter the intimate, was discovered as the opposite not of the political sphere but of the social, to which it is therefore more closely and authentically related.

The first articulate explorer and to an extent even theorist of intimacy was Jean-Jacques Rousseau who, characteristically enough, is the only great author still frequently cited by his first name alone. He arrived at his discovery through a rebellion not against the oppression of the state but against society's unbearable perversion of the human heart, its intrusion upon an innermost region in man which until then had needed no special protection. The intimacy of the heart, unlike the private household, has no objective tangible place in the world, nor can the society against which it protests and asserts itself be localized with the same certainty as the public space. To Rousseau, both the intimate and the social were, rather, subjective modes of human existence, and in his case, it was as though Jean-Jacques rebelled against a man called Rousseau. The modern individual and his endless conflicts, his inability either to be at home in society or to live outside it altogether, his ever-changing moods and the radical subjectivism of his emotional life, was born in this rebellion of the heart. The authenticity of Rousseau's discovery is beyond doubt, no matter how doubtful the authenticity of the individual who was Rousseau. The astonishing flowering of poetry and music from the middle of the eighteenth century until almost the last third of the nineteenth, accompanied by the rise of the novel, the only entirely social art form, coinciding with a no less striking decline of all the more public arts, especially architecture, is sufficient testimony to a close relationship between the social and the intimate.

The rebellious reaction against society during which Rousseau and the Romanticists discovered intimacy was directed first of all against the leveling demands of the social, against what we would call today the conformism inherent in every society. It is important to remember that this rebellion took place before the principle of equality, upon which we have blamed conformism since Tocqueville, had had the time to assert itself in either the social or the political realm. Whether a nation consists of equals or non-equals is of no great importance in this respect, for society always demands that its members act as though they were members of one enormous family which has only one opinion and one interest. Before the modern disintegration of the family, this common interest and single opinion was represented by the household head who ruled in accordance with it and prevented possible dis-

unity among the family members.[33] The striking coincidence of
the rise of society with the decline of the family indicates clearly
that what actually took place was the absorption of the family
unit into corresponding social groups. The equality of the members of these groups, far from being an equality among peers, resembles nothing so much as the equality of household members
before the despotic power of the household head, except that in
society, where the natural strength of one common interest and
one unanimous opinion is tremendously enforced by sheer number, actual rule exerted by one man, representing the common
interest and the right opinion, could eventually be dispensed with.
The phenomenon of conformism is characteristic of the last stage
of this modern development.

It is true that one-man, monarchical rule, which the ancients
stated to be the organizational device of the household, is transformed in society—as we know it today, when the peak of the
social order is no longer formed by the royal household of an absolute ruler—into a kind of no-man rule. But this nobody, the
assumed one interest of society as a whole in economics as well
as the assumed one opinion of polite society in the salon, does not
cease to rule for having lost its personality. As we know from the
most social form of government, that is, from bureaucracy (the
last stage of government in the nation-state just as one-man rule
in benevolent despotism and absolutism was its first), the rule by
nobody is not necessarily no-rule; it may indeed, under certain
circumstances, even turn out to be one of its cruelest and most
tyrannical versions.

It is decisive that society, on all its levels, excludes the possibility of action, which formerly was excluded from the household. Instead, society expects from each of its members a certain
kind of behavior, imposing innumerable and various rules, all of
which tend to "normalize" its members, to make them behave,
to exclude spontaneous action or outstanding achievement. With

33. This is well illustrated by a remark of Seneca, who, discussing the usefulness of highly educated slaves (who know all the classics by heart) to an assumedly rather ignorant master, comments: "What the household knows the
master knows" (*Ep.* 27. 6, quoted from Barrow, *Slavery in the Roman Empire*,
p. 61).

The Public and the Private Realm

Rousseau, we find these demands in the salons of high society, whose conventions always equate the individual with his rank within the social framework. What matters is this equation with social status, and it is immaterial whether the framework happens to be actual rank in the half-feudal society of the eighteenth century, title in the class society of the nineteenth, or mere function in the mass society of today. The rise of mass society, on the contrary, only indicates that the various social groups have suffered the same absorption into one society that the family units had suffered earlier; with the emergence of mass society, the realm of the social has finally, after several centuries of development, reached the point where it embraces and controls all members of a given community equally and with equal strength. But society equalizes under all circumstances, and the victory of equality in the modern world is only the political and legal recognition of the fact that society has conquered the public realm, and that distinction and difference have become private matters of the individual.

This modern equality, based on the conformism inherent in society and possible only because behavior has replaced action as the foremost mode of human relationship, is in every respect different from equality in antiquity, and notably in the Greek city-states. To belong to the few "equals" (*homoioi*) meant to be permitted to live among one's peers; but the public realm itself, the *polis*, was permeated by a fiercely agonal spirit, where everybody had constantly to distinguish himself from all others, to show through unique deeds or achievements that he was the best of all (*aien aristeuein*).[34] The public realm, in other words, was reserved for individuality; it was the only place where men could show who they really and inexchangeably were. It was for the sake of this chance, and out of love for a body politic that made it possible to them all, that each was more or less willing to share in the burden of jurisdiction, defense, and administration of public affairs.

It is the same conformism, the assumption that men behave and

34. *Aien aristeuein kai hypeirochon emmenai allōn* ("always to be the best and to rise above others") is the central concern of Homer's heroes (*Iliad* vi. 208), and Homer was "the educator of Hellas."

do not act with respect to each other, that lies at the root of the modern science of economics, whose birth coincided with the rise of society and which, together with its chief technical tool, statistics, became the social science par excellence. Economics—until the modern age a not too important part of ethics and politics and based on the assumption that men act with respect to their economic activities as they act in every other respect[35]—could achieve a scientific character only when men had become social beings and unanimously followed certain patterns of behavior, so that those who did not keep the rules could be considered to be asocial or abnormal.

The laws of statistics are valid only where large numbers or long periods are involved, and acts or events can statistically appear only as deviations or fluctuations. The justification of statistics is that deeds and events are rare occurrences in everyday life and in history. Yet the meaningfulness of everyday relationships is disclosed not in everyday life but in rare deeds, just as the significance of a historical period shows itself only in the few events that illuminate it. The application of the law of large numbers and long periods to politics or history signifies nothing less than the wilful obliteration of their very subject matter, and it is a hopeless enterprise to search for meaning in politics or signifi-

---

35. "The conception of political economy as primarily a 'science' dates only from Adam Smith" and was unknown not only to antiquity and the Middle Ages, but also to canonist doctrine, the first "complete and economic doctrine" which "differed from modern economics in being an 'art' rather than a 'science'" (W. J. Ashley, *op. cit.*, pp. 379 ff.). Classical economics assumed that man, in so far as he is an active being, acts exclusively from self-interest and is driven by only one desire, the desire for acquisition. Adam Smith's introduction of an "invisible hand to promote an end which was no part of [anybody's] intention" proves that even this minimum of action with its uniform motivation still contains too much unpredictable initiative for the establishment of a science. Marx developed classical economics further by substituting group or class interests for individual and personal interests and by reducing these class interests to two major classes, capitalists and workers, so that he was left with one conflict, where classical economics had seen a multitude of contradictory conflicts. The reason why the Marxian economic system is more consistent and coherent, and therefore apparently so much more "scientific" than those of his predecessors, lies primarily in the construction of "socialized man," who is even less an acting being than the "economic man" of liberal economics.

cance in history when everything that is not everyday behavior or automatic trends has been ruled out as immaterial.

However, since the laws of statistics are perfectly valid where we deal with large numbers, it is obvious that every increase in population means an increased validity and a marked decrease of "deviation." Politically, this means that the larger the population in any given body politic, the more likely it will be the social rather than the political that constitutes the public realm. The Greeks, whose city-state was the most individualistic and least conformable body politic known to us, were quite aware of the fact that the *polis*, with its emphasis on action and speech, could survive only if the number of citizens remained restricted. Large numbers of people, crowded together, develop an almost irresistible inclination toward despotism, be this the despotism of a person or of majority rule; and although statistics, that is, the mathematical treatment of reality, was unknown prior to the modern age, the social phenomena which make such treatment possible—great numbers, accounting for conformism, behaviorism, and automatism in human affairs—were precisely those traits which, in Greek self-understanding, distinguished the Persian civilization from their own.

The unfortunate truth about behaviorism and the validity of its "laws" is that the more people there are, the more likely they are to behave and the less likely to tolerate non-behavior. Statistically, this will be shown in the leveling out of fluctuation. In reality, deeds will have less and less chance to stem the tide of behavior, and events will more and more lose their significance, that is, their capacity to illuminate historical time. Statistical uniformity is by no means a harmless scientific ideal; it is the no longer secret political ideal of a society which, entirely submerged in the routine of everyday living, is at peace with the scientific outlook inherent in its very existence.

The uniform behavior that lends itself to statistical determination, and therefore to scientifically correct prediction, can hardly be explained by the liberal hypothesis of a natural "harmony of interests," the foundation of "classical" economics; it was not Karl Marx but the liberal economists themselves who had to introduce the "communistic fiction," that is, to assume that there is one interest of society as a whole which with "an invisible hand" guides the behavior of men and produces the harmony of their conflicting interests.[36] The difference between Marx and his forerunners was only that he took the reality of conflict, as it presented itself in the society of his time, as seriously as the hypothetical fiction of harmony; he was right in concluding that the "socialization of man" would produce automatically a harmony of all interests, and was only more courageous than his liberal teachers when he proposed to establish in reality the "communistic fiction" underlying all economic theories. What Marx did not— and, at his time, could not—understand was that the germs of communistic society were present in the reality of a national household, and that their full development was not hindered by any class-interest as such, but only by the already obsolete monarchical structure of the nation-state. Obviously, what prevented society from smooth functioning was only certain traditional remnants that interfered and still influenced the behavior of "backward" classes. From the viewpoint of society, these were merely disturbing factors in the way of a full development of "social forces"; they no longer corresponded to reality and were therefore, in a sense, much more "fictitious" than the scientific "fiction" of one interest.

A complete victory of society will always produce some sort of "communistic fiction," whose outstanding political characteristic is that it is indeed ruled by an "invisible hand," namely, by

36. That liberal utilitarianism, and not socialism, is "forced into an untenable 'communistic fiction' about the unity of society" and that "the communist fiction [is] implicit in most writings on economics" constitutes one of the chief theses of Myrdal's brilliant work (*op. cit.*, pp. 54 and 150). He shows conclusively that economics can be a science only if one assumes that one interest pervades society as a whole. Behind the "harmony of interests" stands always the "communistic fiction" of one interest, which may then be called welfare or commonwealth. Liberal economists consequently were always guided by a "communistic" ideal, namely, by "interest of society as a whole" (pp. 194–95). The crux of the argument is that this "amounts to the assertion that society must be conceived as a single subject. This, however, is precisely what cannot be conceived. If we tried, we would be attempting to abstract from the essential fact that social activity is the result of the intentions of several individuals" (p. 154).

nobody. What we traditionally call state and government gives place here to pure administration—a state of affairs which Marx rightly predicted as the "withering away of the state," though he was wrong in assuming that only a revolution could bring it about, and even more wrong when he believed that this complete victory of society would mean the eventual emergence of the "realm of freedom."[37]

To gauge the extent of society's victory in the modern age, its early substitution of behavior for action and its eventual substitution of bureaucracy, the rule of nobody, for personal rulership, it may be well to recall that its initial science of economics, which substitutes patterns of behavior only in this rather limited field of human activity, was finally followed by the all-comprehensive pretension of the social sciences which, as "behavioral sciences," aim to reduce man as a whole, in all his activities, to the level of a conditioned and behaving animal. If economics is the science of society in its early stages, when it could impose its rules of behavior only on sections of the population and on parts of their activities, the rise of the "behavioral sciences" indicates clearly the final stage of this development, when mass society has devoured all strata of the nation and "social behavior" has become the standard for all regions of life.

Since the rise of society, since the admission of household and housekeeping activities to the public realm, an irresistible tendency to grow, to devour the older realms of the political and private as well as the more recently established sphere of intimacy, has been one of the outstanding characteristics of the new realm. This constant growth, whose no less constant acceleration we can observe over at least three centuries, derives its strength from the fact that through society it is the life process itself which in one form or another has been channeled into the public realm. The private realm of the household was the sphere where the necessities of life, of individual survival as well as of continuity of the species, were taken care of and guaranteed. One of the character-

37. For a brilliant exposition of this usually neglected aspect of Marx's relevance for modern society, see Siegfried Landshut, "Die Gegenwart im Lichte der Marxschen Lehre," *Hamburger Jahrbuch für Wirtschafts- und Gesellschaftspolitik*, Vol. I (1956).

istics of privacy, prior to the discovery of the intimate, was that man existed in this sphere not as a truly human being but only as a specimen of the animal species man-kind. This, precisely, was the ultimate reason for the tremendous contempt held for it by antiquity. The emergence of society has changed the estimate of this whole sphere but has hardly transformed its nature. The monolithic character of every type of society, its conformism which allows for only one interest and one opinion, is ultimately rooted in the one-ness of man-kind. It is because this one-ness of man-kind is not fantasy and not even merely a scientific hypothesis, as in the "communistic fiction" of classical economics, that mass society, where man as a social animal rules supreme and where apparently the survival of the species could be guaranteed on a world-wide scale, can at the same time threaten humanity with extinction.

Perhaps the clearest indication that society constitutes the public organization of the life process itself may be found in the fact that in a relatively short time the new social realm transformed all modern communities into societies of laborers and jobholders; in other words, they became at once centered around the one activity necessary to sustain life. (To have a society of laborers, it is of course not necessary that every member actually be a laborer or worker—not even the emancipation of the working class and the enormous potential power which majority rule accords to it are decisive here—but only that all members consider whatever they do primarily as a way to sustain their own lives and those of their families.) Society is the form in which the fact of mutual dependence for the sake of life and nothing else assumes public significance and where the activities connected with sheer survival are permitted to appear in public.

Whether an activity is performed in private or in public is by no means a matter of indifference. Obviously, the character of the public realm must change in accordance with the activities admitted into it, but to a large extent the activity itself changes its own nature too. The laboring activity, though under all circumstances connected with the life process in its most elementary, biological sense, remained stationary for thousands of years, imprisoned in the eternal recurrence of the life process to which it was tied. The

admission of labor to public stature, far from eliminating its character as a process—which one might have expected, remembering
that bodies politic have always been designed for permanence and
their laws always understood as limitations imposed upon movement—has, on the contrary, liberated this process from its circular, monotonous recurrence and transformed it into a swiftly
progressing development whose results have in a few centuries
totally changed the whole inhabited world.

The moment laboring was liberated from the restrictions imposed by its banishment into the private realm—and this emancipation of labor was not a consequence of the emancipation of the
working class, but preceded it—it was as though the growth element inherent in all organic life had completely overcome and
overgrown the processes of decay by which organic life is checked
and balanced in nature's household. The social realm, where the
life process has established its own public domain, has let loose
an unnatural growth, so to speak, of the natural; and it is against
this growth, not merely against society but against a constantly
growing social realm, that the private and intimate, on the one
hand, and the political (in the narrower sense of the word), on the
other, have proved incapable of defending themselves.

What we described as the unnatural growth of the natural is
usually considered to be the constantly accelerated increase in the
productivity of labor. The greatest single factor in this constant
increase since its inception has been the organization of laboring,
visible in the so-called division of labor, which preceded the industrial revolution; even the mechanization of labor processes,
the second greatest factor in labor's productivity, is based upon it.
Inasmuch as the organizational principle itself clearly derives from
the public rather than the private realm, division of labor is precisely what happens to the laboring activity under conditions of
the public realm and what could never have happened in the privacy
of the household.[38] In no other sphere of life do we appear to have

38. Here and later I apply the term "division of labor" only to modern labor
conditions where one activity is divided and atomized into innumerable minute
manipulations, and not to the "division of labor" given in professional specialization. The latter can be so classified only under the assumption that society must
be conceived as one single subject, the fulfilment of whose needs are then sub-

attained such excellence as in the revolutionary transformation of laboring, and this to the point where the verbal significance of the word itself (which always had been connected with hardly bearable "toil and trouble," with effort and pain and, consequently, with a deformation of the human body, so that only extreme misery and poverty could be its source), has begun to lose its meaning for us.[39] While dire necessity made labor indispensable to sustain life, excellence would have been the last thing to expect from it.

Excellence itself, *aretē* as the Greeks, *virtus* as the Romans

---

divided by "an invisible hand" among its members. The same holds true, *mutatis mutandis*, for the odd notion of a division of labor between the sexes, which is even considered by some writers to be the most original one. It presumes as its single subject man-kind, the human species, which has divided its labors among men and women. Where the same argument is used in antiquity (see, for instance, Xenophon *Oeconomicus* vii. 22), emphasis and meaning are quite different. The main division is between a life spent indoors, in the household, and a life spent outside, in the world. Only the latter is a life fully worthy of man, and the notion of equality between man and woman, which is a necessary assumption for the idea of division of labor, is of course entirely absent (cf. n. 81). Antiquity seems to have known only professional specialization, which assumedly was predetermined by natural qualities and gifts. Thus work in the gold mines, which occupied several thousand workers, was distributed according to strength and skill. See J.-P. Vernant, "Travail et nature dans la Grèce ancienne," *Journal de psychologie normale et pathologique*, Vol. LII, No. 1 (January–March, 1955).

39. All the European words for "labor," the Latin and English *labor*, the Greek *ponos*, the French *travail*, the German *Arbeit*, signify pain and effort and are also used for the pangs of birth. *Labor* has the same etymological root as *labare* ("to stumble under a burden"); *ponos* and *Arbeit* have the same etymological roots as "poverty" (*penia* in Greek and *Armut* in German). Even Hesiod, currently counted among the few defenders of labor in antiquity, put *ponon alginoenta* ("painful labor") as first of the evils plaguing man (*Theogony* 226). For the Greek usage, see G. Herzog-Hauser, "*Ponos*," in Pauly-Wissowa. The German *Arbeit* and *arm* are both derived from the Germanic *arbma-*, meaning lonely and neglected, abandoned. See Kluge/Götze, *Etymologisches Wörterbuch* (1951). In medieval German, the word is used to translate *labor, tribulatio, persecutio, adversitas, malum* (see Klara Vontobel, *Das Arbeitsethos des deutschen Protestantismus* [Dissertation, Bern, 1946]).

would have called it, has always been assigned to the public realm where one could excel, could distinguish oneself from all others. Every activity performed in public can attain an excellence never matched in privacy; for excellence, by definition, the presence of others is always required, and this presence needs the formality of the public, constituted by one's peers, it cannot be the casual, familiar presence of one's equals or inferiors.[40] Not even the social realm—though it made excellence anonymous, emphasized the progress of mankind rather than the achievements of men, and changed the content of the public realm beyond recognition—has been able altogether to annihilate the connection between public performance and excellence. While we have become excellent in the laboring we perform in public, our capacity for action and speech has lost much of its former quality since the rise of the social realm banished these into the sphere of the intimate and the private. This curious discrepancy has not escaped public notice, where it is usually blamed upon an assumed time lag between our technical capacities and our general humanistic development or between the physical sciences, which change and control nature, and the social sciences, which do not yet know how to change and control society. Quite apart from other fallacies of the argument which have been pointed out so frequently that we need not repeat them, this criticism concerns only a possible change in the psychology of human beings—their so-called behavior patterns—not a change of the world they move in. And this psychological interpretation, for which the absence or presence of a public realm is as irrelevant as any tangible, worldly reality, seems rather doubtful in view of the fact that no activity can become excellent if the world does not provide a proper space for its exercise. Neither education nor ingenuity nor talent can replace the constituent elements of the public realm, which make it the proper place for human excellence.

40. Homer's much quoted thought that Zeus takes away half of a man's excellence (*aretē*) when the day of slavery catches him (*Odyssey* xvii. 320 ff.) is put into the mouth of Eumaios, a slave himself, and meant as an objective statement, not a criticism or a moral judgment. The slave lost excellence because he lost admission to the public realm, where excellence can show.

7

THE PUBLIC REALM: THE COMMON

The term "public" signifies two closely interrelated but not altogether identical phenomena:

It means, first, that everything that appears in public can be seen and heard by everybody and has the widest possible publicity. For us, appearance—something that is being seen and heard by others as well as by ourselves—constitutes reality. Compared with the reality which comes from being seen and heard, even the greatest forces of intimate life—the passions of the heart, the thoughts of the mind, the delights of the senses—lead an uncertain, shadowy kind of existence unless and until they are transformed, deprivatized and deindividualized, as it were, into a shape to fit them for public appearance.[41] The most current of such transformations occurs in storytelling and generally in artistic transposition of individual experiences. But we do not need the form of the artist to witness this transfiguration. Each time we talk about things that can be experienced only in privacy or intimacy, we bring them out into a sphere where they will assume a kind of reality which, their intensity notwithstanding, they never could have had before. The presence of others who see what we see and hear what we hear assures us of the reality of the world and ourselves, and while the intimacy of a fully developed private life, such as had never been known before the rise of the modern age and the concomitant decline of the public realm, will always greatly intensify and enrich the whole scale of subjective emotions and private feelings, this intensification will always come to pass at the expense of the assurance of the reality of the world and men.

Indeed, the most intense feeling we know of, intense to the point of blotting out all other experiences, namely, the experience of great bodily pain, is at the same time the most private and least

41. This is also the reason why it is impossible "to write a character sketch of any slave who lived. . . . Until they emerge into freedom and notoriety, they remain shadowy types rather than persons" (Barrow, *Slavery in the Roman Empire*, p. 156).

communicable of all. Not only is it perhaps the only experience which we are unable to transform into a shape fit for public appearance, it actually deprives us of our feeling for reality to such an extent that we can forget it more quickly and easily than anything else. There seems to be no bridge from the most radical subjectivity, in which I am no longer "recognizable," to the outer world of life.[42] Pain, in other words, truly a borderline experience between life as "being among men" (*inter homines esse*) and death, is so subjective and removed from the world of things and men that it cannot assume an appearance at all.[43]

Since our feeling for reality depends utterly upon appearance and therefore upon the existence of a public realm into which things can appear out of the darkness of sheltered existence, even the twilight which illuminates our private and intimate lives is ultimately derived from the much harsher light of the public realm. Yet there are a great many things which cannot withstand the implacable, bright light of the constant presence of others on the public scene; there, only what is considered to be relevant, worthy of being seen or heard, can be tolerated, so that the irrelevant becomes automatically a private matter. This, to be sure, does not mean that private concerns are generally irrelevant; on the contrary, we shall see that there are very relevant matters which can survive only in the realm of the private. For instance, love, in distinction from friendship, is killed, or rather extinguished, the moment it is displayed in public. ("Never seek to tell

42. I use here a little-known poem on pain from Rilke's deathbed: The first lines of the untitled poem are: "Komm du, du letzter, den ich anerkenne, / heilloser Schmerz im leiblichen Geweb"; and it concludes as follows: "Bin ich es noch, der da unkenntlich brennt? / Erinnerungen reiss ich nicht herein. / O Leben, Leben: Draussensein. / Und ich in Lohe. Niemand, der mich kennt."

43. On the subjectivity of pain and its relevance for all variations of hedonism and sensualism, see §§ 15 and 43. For the living, death is primarily dis-appearance. But unlike pain, there is one aspect of death in which it is as though death appeared among the living, and that is in old age. Goethe once remarked that growing old is "gradually receding from appearance" (*stufenweises Zurücktreten aus der Erscheinung*); the truth of this remark as well as the actual appearance of this process of disappearing becomes quite tangible in the old-age self-portraits of the great masters—Rembrandt, Leonardo, etc.—in which the intensity of the eyes seems to illuminate and preside over the receding flesh.

thy love / Love that never told can be.") Because of its inherent worldlessness, love can only become false and perverted when it is used for political purposes such as the change or salvation of the world.

What the public realm considers irrelevant can have such an extraordinary and infectious charm that a whole people may adopt it as their way of life, without for that reason changing its essentially private character. Modern enchantment with "small things," though preached by early twentieth-century poetry in almost all European tongues, has found its classical presentation in the *petit bonheur* of the French people. Since the decay of their once great and glorious public realm, the French have become masters in the art of being happy among "small things," within the space of their own four walls, between chest and bed, table and chair, dog and cat and flowerpot, extending to these things a care and tenderness which, in a world where rapid industrialization constantly kills off the things of yesterday to produce today's objects, may even appear to be the world's last, purely humane corner. This enlargement of the private, the enchantment, as it were, of a whole people, does not make it public, does not constitute a public realm, but, on the contrary, means only that the public realm has almost completely receded, so that greatness has given way to charm everywhere; for while the public realm may be great, it cannot be charming precisely because it is unable to harbor the irrelevant.

Second, the term "public" signifies the world itself, in so far as it is common to all of us and distinguished from our privately owned place in it. This world, however, is not identical with the earth or with nature, as the limited space for the movement of men and the general condition of organic life. It is related, rather, to the human artifact, the fabrication of human hands, as well as to affairs which go on among those who inhabit the man-made world together. To live together in the world means essentially that a world of things is between those who have it in common, as a table is located between those who sit around it; the world, like every in-between, relates and separates men at the same time.

The public realm, as the common world, gathers us together and yet prevents our falling over each other, so to speak. What makes mass society so difficult to bear is not the number of people

involved, or at least not primarily, but the fact that the world between them has lost its power to gather them together, to relate
and to separate them. The weirdness of this situation resembles a
spiritualistic séance where a number of people gathered around a
table might suddenly, through some magic trick, see the table
vanish from their midst, so that two persons sitting opposite each
other were no longer separated but also would be entirely unrelated to each other by anything tangible.

Historically, we know of only one principle that was ever devised to keep a community of people together who had lost their
interest in the common world and felt themselves no longer related
and separated by it. To find a bond between people strong enough
to replace the world was the main political task of early Christian
philosophy, and it was Augustine who proposed to found not only
the Christian "brotherhood" but all human relationships on charity. But this charity, though its worldlessness clearly corresponds
to the general human experience of love, is at the same time clearly
distinguished from it in being something which, like the world, is
between men: "Even robbers have between them [*inter se*] what
they call charity."[44] This surprising illustration of the Christian
political principle is in fact very well chosen, because the bond of
charity between people, while it is incapable of founding a public
realm of its own, is quite adequate to the main Christian principle
of worldlessness and is admirably fit to carry a group of essentially
worldless people through the world, a group of saints or a group
of criminals, provided only it is understood that the world itself
is doomed and that every activity in it is undertaken with the pro-
viso *quamdiu mundus durat* ("as long as the world lasts").[45] The
unpolitical, non-public character of the Christian community was
early defined in the demand that it should form a *corpus*, a "body,"
whose members were to be related to each other like brothers of
the same family.[46] The structure of communal life was modeled

44. *Contra Faustum Manichaeum* v. 5.

45. This is of course still the presupposition even of Aquinas' political philosophy (see *op. cit.* ii. 2. 181. 4).

46. The term *corpus rei publicae* is current in pre-Christian Latin, but has the
connotation of the population inhabiting a *res publica*, a given political realm. The
corresponding Greek term *sōma* is never used in pre-Christian Greek in a political

on the relationships between the members of a family because these were known to be non-political and even antipolitical. A public realm had never come into being between the members of a family, and it was therefore not likely to develop from Christian community life if this life was ruled by the principle of charity and nothing else. Even then, as we know from the history and the rules of the monastic orders—the only communities in which the principle of charity as a political device was ever tried—the danger that the activities undertaken under "the necessity of present life" (*necessitas vitae praesentis*)[47] would lead by themselves, because they were performed in the presence of others, to the establishment of a kind of counterworld, a public realm within the orders themselves, was great enough to require additional rules and regulations, the most relevant one in our context being the prohibition of excellence and its subsequent pride.[48]

Worldlessness as a political phenomenon is possible only on the assumption that the world will not last; on this assumption, however, it is almost inevitable that worldlessness, in one form or another, will begin to dominate the political scene. This happened after the downfall of the Roman Empire and, albeit for quite other reasons and in very different, perhaps even more disconsolate forms, it seems to happen again in our own days. The Christian abstention from worldly things is by no means the only conclusion one can draw from the conviction that the human artifice, a product of mortal hands, is as mortal as its makers. This, on the contrary, may also intensify the enjoyment and consump-

sense. The metaphor seems to occur for the first time in Paul (I Cor. 12: 12–27) and is current in all early Christian writers (see, for instance, Tertullian *Apologeticus* 39, or Ambrosius *De officiis ministrorum* iii. 3. 17). It became of the greatest importance for medieval political theory, which unanimously assumed that all men were *quasi unum corpus* (Aquinas *op. cit.* ii. 1. 81. 1). But while the early writers stressed the equality of the members, which are all equally necessary for the well-being of the body as a whole, the emphasis later shifted to the difference between the head and the members, to the duty of the head to rule and of the members to obey. (For the Middle Ages, see Anton-Hermann Chroust, "The Corporate Idea in the Middle Ages," *Review of Politics*, Vol. VIII [1947].)

47. Aquinas *op. cit.* ii. 2. 179. 2.

48. See Article 57 of the Benedictine rule, in Levasseur, *op. cit.*, p. 187: If one of the monks became proud of his work, he had to give it up.

tion of the things of the world, all manners of intercourse in which the world is not primarily understood to be the *koinon*, that which is common to all. Only the existence of a public realm and the world's subsequent transformation into a community of things which gathers men together and relates them to each other depends entirely on permanence. If the world is to contain a public space, it cannot be erected for one generation and planned for the living only; it must transcend the life-span of mortal men.

Without this transcendence into a potential earthly immortality, no politics, strictly speaking, no common world and no public realm, is possible. For unlike the common good as Christianity understood it—the salvation of one's soul as a concern common to all—the common world is what we enter when we are born and what we leave behind when we die. It transcends our life-span into past and future alike; it was there before we came and will outlast our brief sojourn in it. It is what we have in common not only with those who live with us, but also with those who were here before and with those who will come after us. But such a common world can survive the coming and going of the generations only to the extent that it appears in public. It is the publicity of the public realm which can absorb and make shine through the centuries whatever men may want to save from the natural ruin of time. Through many ages before us—but now not any more— men entered the public realm because they wanted something of their own or something they had in common with others to be more permanent than their earthly lives. (Thus, the curse of slavery consisted not only in being deprived of freedom and of visibility, but also in the fear of these obscure people themselves "that from being obscure they should pass away leaving no trace that they have existed.")[49] There is perhaps no clearer testimony to the loss of the public realm in the modern age than the almost complete loss of authentic concern with immortality, a loss somewhat overshadowed by the simultaneous loss of the metaphysical concern with eternity. The latter, being the concern of the philosophers

49. Barrow (*Slavery in the Roman Empire*, p. 168), in an illuminating discussion of the membership of slaves in the Roman colleges, which provided, besides "good fellowship in life and the certainty of a decent burial . . . the crowning glory of an epitaph; and in this last the slave found a melancholy pleasure."

and the *vita contemplativa*, must remain outside our present considerations. But the former is testified to by the current classification of striving for immortality with the private vice of vanity. Under modern conditions, it is indeed so unlikely that anybody should earnestly aspire to an earthly immortality that we probably are justified in thinking it is nothing but vanity.

The famous passage in Aristotle, "Considering human affairs, one must not . . . consider man as he is and not consider what is mortal in mortal things, but think about them [only] to the extent that they have the possibility of immortalizing," occurs very properly in his political writings.[50] For the *polis* was for the Greeks, as the *res publica* was for the Romans, first of all their guarantee against the futility of individual life, the space protected against this futility and reserved for the relative permanence, if not immortality, of mortals.

What the modern age thought of the public realm, after the spectacular rise of society to public prominence, was expressed by Adam Smith when, with disarming sincerity, he mentions "that unprosperous race of men commonly called men of letters" for whom "public admiration . . . makes always a part of their reward . . . , a considerable part . . . in the profession of physic; a still greater perhaps in that of law; in poetry and philosophy it makes almost the whole."[51] Here it is self-evident that public admiration and monetary reward are of the same nature and can become substitutes for each other. Public admiration, too, is something to be used and consumed, and status, as we would say today, fulfils one need as food fulfils another: public admiration is consumed by individual vanity as food is consumed by hunger. Obviously, from this viewpoint the test of reality does not lie in the public presence of others, but rather in the greater or lesser urgency of needs to whose existence or non-existence nobody can ever testify except the one who happens to suffer them. And since the need for food has its demonstrable basis of reality in the life process itself, it is also obvious that the entirely subjective pangs of hunger are more real than "vainglory," as Hobbes used

50. *Nicomachean Ethics* 1177b31.

51. *Wealth of Nations*, Book I, ch. 10 (pp. 120 and 95 of Vol. I of Everyman's ed.).

to call the need for public admiration. Yet, even if these needs, through some miracle of sympathy, were shared by others, their very futility would prevent their ever establishing anything so solid and durable as a common world. The point then is not that there is a lack of public admiration for poetry and philosophy in the modern world, but that such admiration does not constitute a space in which things are saved from destruction by time. The futility of public admiration, which daily is consumed in ever greater quantities, on the contrary, is such that monetary reward, one of the most futile things there is, can become more "objective" and more real.

As distinguished from this "objectivity," whose only basis is money as a common denominator for the fulfilment of all needs, the reality of the public realm relies on the simultaneous presence of innumerable perspectives and aspects in which the common world presents itself and for which no common measurement or denominator can ever be devised. For though the common world is the common meeting ground of all, those who are present have different locations in it, and the location of one can no more coincide with the location of another than the location of two objects. Being seen and being heard by others derive their significance from the fact that everybody sees and hears from a different position. This is the meaning of public life, compared to which even the richest and most satisfying family life can offer only the prolongation or multiplication of one's own position with its attending aspects and perspectives. The subjectivity of privacy can be prolonged and multiplied in a family, it can even become so strong that its weight is felt in the public realm; but this family "world" can never replace the reality rising out of the sum total of aspects presented by one object to a multitude of spectators. Only where things can be seen by many in a variety of aspects without changing their identity, so that those who are gathered around them know they see sameness in utter diversity, can worldly reality truly and reliably appear.

Under the conditions of a common world, reality is not guaranteed primarily by the "common nature" of all men who constitute it, but rather by the fact that, differences of position and the resulting variety of perspectives notwithstanding, everybody is always concerned with the same object. If the sameness of the object can no longer be discerned, no common nature of men, least of all the unnatural conformism of a mass society, can prevent the destruction of the common world, which is usually preceded by the destruction of the many aspects in which it presents itself to human plurality. This can happen under conditions of radical isolation, where nobody can any longer agree with anybody else, as is usually the case in tyrannies. But it may also happen under conditions of mass society or mass hysteria, where we see all people suddenly behave as though they were members of one family, each multiplying and prolonging the perspective of his neighbor. In both instances, men have become entirely private, that is, they have been deprived of seeing and hearing others, of being seen and being heard by them. They are all imprisoned in the subjectivity of their own singular experience, which does not cease to be singular if the same experience is multiplied innumerable times. The end of the common world has come when it is seen only under one aspect and is permitted to present itself in only one perspective.

8

THE PRIVATE REALM: PROPERTY

It is with respect to this multiple significance of the public realm that the term "private," in its original privative sense, has meaning. To live an entirely private life means above all to be deprived of things essential to a truly human life: to be deprived of the reality that comes from being seen and heard by others, to be deprived of an "objective" relationship with them that comes from being related to and separated from them through the intermediary of a common world of things, to be deprived of the possibility of achieving something more permanent than life itself. The privation of privacy lies in the absence of others; as far as they are concerned, private man does not appear, and therefore it is as though he did not exist. Whatever he does remains without significance and consequence to others, and what matters to him is without interest to other people.

Under modern circumstances, this deprivation of "objective"

relationships to others and of a reality guaranteed through them has become the mass phenomenon of loneliness, where it has assumed its most extreme and most antihuman form.[52] The reason for this extremity is that mass society not only destroys the public realm but the private as well, deprives men not only of their place in the world but of their private home, where they once felt sheltered against the world and where, at any rate, even those excluded from the world could find a substitute in the warmth of the hearth and the limited reality of family life. The full development of the life of hearth and family into an inner and private space we owe to the extraordinary political sense of the Roman people who, unlike the Greeks, never sacrificed the private to the public, but on the contrary understood that these two realms could exist only in the form of coexistence. And although the conditions of slaves probably were hardly better in Rome than in Athens, it is quite characteristic that a Roman writer should have believed that to slaves the household of the master was what the *res publica* was to citizens.[53] Yet no matter how bearable private life in the family might have been, it could obviously never be more than a substitute, even though the private realm in Rome as in Athens offered plenty of room for activities which we today class higher than political activity, such as the accumulation of wealth in Greece or the devotion to art and science in Rome. This "liberal" attitude, which could under certain circumstances result in very prosperous and highly educated slaves, meant only that to be prosperous had no reality in the Greek *polis* and to be a philosopher was without much consequence in the Roman republic.[54]

52. For modern loneliness as a mass phenomenon see David Riesman, *The Lonely Crowd* (1950).

53. So Plinius Junior, quoted in W. L. Westermann, "Sklaverei," in Pauly-Wissowa, Suppl. VI, p. 1045.

54. There is plenty of evidence for this different estimation of wealth and culture in Rome and Greece. But it is interesting to note how consistently this estimate coincided with the position of slaves. Roman slaves played a much greater role in Roman culture than in Greece, where, on the other hand, their role in economic life was much more important (see Westermann, in Pauly-Wissowa, p. 984).

It is a matter of course that the privative trait of privacy, the consciousness of being deprived of something essential in a life spent exclusively in the restricted sphere of the household, should have been weakened almost to the point of extinction by the rise of Christianity. Christian morality, as distinguished from its fundamental religious precepts, has always insisted that everybody should mind his own business and that political responsibility constituted first of all a burden, undertaken exclusively for the sake of the well-being and salvation of those it freed from worry about public affairs.[55] It is surprising that this attitude should have survived into the secular modern age to such an extent that Karl Marx, who in this as in other respects only summed up, conceptualized, and transformed into a program the underlying assumptions of two hundred years of modernity, could eventually predict and hope for the "withering away" of the whole public realm. The difference between the Christian and socialist viewpoints in this respect, the one viewing government as a necessary evil because of man's sinfulness and the other hoping to abolish it eventually, is not a difference in estimate of the public sphere itself, but of human nature. What is impossible to perceive from either point of view is that Marx's "withering away of the state" had been preceded by a withering away of the public realm, or rather by its transformation into a very restricted sphere of government; in Marx's day, this government had already begun to wither further, that is, to be transformed into a nation-wide "housekeeping," until in our own day it has begun to disappear altogether into the even more restricted, impersonal sphere of administration.

It seems to be in the nature of the relationship between the public and private realms that the final stage of the disappearance

55. Augustine (*De civitate Dei* xix. 19) sees in the duty of *caritas* toward the *utilitas proximi* ("the interest of one's neighbor") the limitation of *otium* and contemplation. But "in active life, it is not the honors or power of this life we should covet, . . . but the welfare of those who are under us [*salutem subditorum*]." Obviously, this kind of responsibility resembles the responsibility of the household head for his family more than political responsibility, properly speaking. The Christian precept to mind one's own business is derived from I Thess. 4: 11: "that ye study to be quiet and to do your own business" (*prattein ta idia*, whereby *ta idia* is understood as opposed to *ta koina* ["public common affairs"]).

of the public realm should be accompanied by the threatened liquidation of the private realm as well. Nor is it an accident that the whole discussion has eventually turned into an argument about the desirability or undesirability of privately owned property. For the word "private" in connection with property, even in terms of ancient political thought, immediately loses its privative character and much of its opposition to the public realm in general; property apparently possesses certain qualifications which, though lying in the private realm, were always thought to be of utmost importance to the political body.

The profound connection between private and public, manifest on its most elementary level in the question of private property, is likely to be misunderstood today because of the modern equation of property and wealth on one side and propertylessness and poverty on the other. This misunderstanding is all the more annoying as both, property as well as wealth, are historically of greater relevance to the public realm than any other private matter or concern and have played, at least formally, more or less the same role as the chief condition for admission to the public realm and full-fledged citizenship. It is therefore easy to forget that wealth and property, far from being the same, are of an entirely different nature. The present emergence everywhere of actually or potentially very wealthy societies which at the same time are essentially propertyless, because the wealth of any single individual consists of his share in the annual income of society as a whole, clearly shows how little these two things are connected.

Prior to the modern age, which began with the expropriation of the poor and then proceeded to emancipate the new propertyless classes, all civilizations have rested upon the sacredness of private property. Wealth, on the contrary, whether privately owned or publicly distributed, had never been sacred before. Originally, property meant no more or less than to have one's location in a particular part of the world and therefore to belong to the body politic, that is, to be the head of one of the families which together constituted the public realm. This piece of privately owned world was so completely identical with the family who owned it[56] that

56. Coulanges (*op. cit.*) holds: "The true signification of *familia* is property; it designates the field, the house, money, and slaves" (p. 107). Yet, this "prop-

the expulsion of a citizen could mean not merely the confiscation of his estate but the actual destruction of the building itself.[57] The wealth of a foreigner or a slave was under no circumstances a substitute for this property,[58] and poverty did not deprive the head of a family of this location in the world and the citizenship resulting from it. In early times, if he happened to lose his location, he almost automatically lost his citizenship and the protection of the law as well.[59] The sacredness of this privacy was like the sacredness of the hidden, namely, of birth and death, the beginning and end of the mortals who, like all living creatures, grow out of and return to the darkness of an underworld.[60] The non-privative trait of the household realm originally lay in its being the realm of birth and death which must be hidden from the public realm because it harbors the things hidden from human eyes and

erty" is not seen as attached to the family; on the contrary, "the family is attached to the hearth, the hearth is attached to the soil" (p. 62). The point is: "The fortune is immovable like the hearth and the tomb to which it is attached. It is the man who passes away" (p. 74).

57. Levasseur (*op. cit.*) relates the medieval foundation of a community and the conditions of admission to it: "Il ne suffisait pas d'habiter la ville pour avoir droit à cette admission. Il fallait . . . posséder une maison. . . ." Furthermore: "Toute injure proférée en public contre la commune entraînait la démolition de la maison et le bannissement du coupable" (p. 240, including n. 3).

58. The distinction is most obvious in the case of slaves who, though without property in the ancient understanding (that is, without a place of their own), were by no means propertyless in the modern sense. The *peculium* (the "private possession of a slave") could amount to considerable sums and even contain slaves of his own (*vicarii*). Barrow speaks of "the property which the humblest of his class possessed" (*Slavery in the Roman Empire*, p. 122; this work is the best report on the role of the *peculium*).

59. Coulanges reports a remark of Aristotle that in ancient times the son could not be a citizen during the lifetime of his father; upon his death, only the eldest son enjoyed political rights (*op. cit.*, p. 228). Coulanges holds that the Roman *plebs* originally consisted of people without home and hearth, that it therefore was clearly distinct from the *populus Romanus* (pp. 229 ff.).

60. "The whole of this religion was inclosed within the walls of each house. . . . All these gods, the Hearth, the Lares, and the Manes, were called the hidden gods, or gods of the interior. To all the acts of this religion secrecy was necessary, *sacrificia occulta*, as Cicero said (De arusp. respl. 17)" (Coulanges, *op. cit.*, p. 37).

impenetrable to human knowledge.[61] It is hidden because man does not know where he comes from when he is born and where he goes when he dies.

Not the interior of this realm, which remains hidden and of no public significance, but its exterior appearance is important for the city as well, and it appears in the realm of the city through the boundaries between one household and the other. The law originally was identified with this boundary line,[62] which in ancient times was still actually a space, a kind of no man's land[63] between the private and the public, sheltering and protecting both realms while, at the same time, separating them from each other. The law of the *polis*, to be sure, transcended this ancient understanding from which, however, it retained its original spatial significance. The law of the city-state was neither the content of political action (the idea that political activity is primarily legislating, though Roman in origin, is essentially modern and found its greatest expession in Kant's political philosophy) nor was it a catalogue of prohibitions, resting, as all modern laws still do, upon the Thou Shalt Nots of the Decalogue. It was quite literally a

61. It seems as though the Eleusinian Mysteries provided for a common and quasi-public experience of this whole realm, which, because of its very nature and even though it was common to all, needed to be hidden, kept secret from the public realm: Everybody could participate in them, but nobody was permitted to talk about them. The mysteries concerned the unspeakable, and experiences beyond speech were non-political and perhaps antipolitical by definition (see Karl Kerenyi, *Die Geburt der Helena* [1943–45], pp. 48 ff.). That they concerned the secret of birth and death seems proved by a fragment of Pindar: *oide men biou teleutan, oiden de diosdoton archan* (frag. 137*a*), where the initiated is said to know "the end of life and the Zeus-given beginning."

62. The Greek word for law, *nomos*, derives from *nemein*, which means to distribute, to possess (what has been distributed), and to dwell. The combination of law and hedge in the word *nomos* is quite manifest in a fragment of Heraclitus: *machesthai chrē ton dēmon hyper tou nomou hokōsper teicheos* ("the people should fight for the law as for a wall"). The Roman word for law, *lex*, has an entirely different meaning; it indicates a formal relationship between people rather than the wall that separates them from others. But the boundary and its god, Terminus, who separated the *agrum publicum a privato* (Livius) was more highly revered than the corresponding *theoi horoi* in Greece.

63. Coulanges reports an ancient Greek law according to which two buildings never were permitted to touch (*op. cit.*, p. 63).

wall, without which there might have been an agglomeration of houses, a town (*asty*), but not a city, a political community. This wall-like law was sacred, but only the inclosure was political.[64] Without it a public realm could no more exist than a piece of property without a fence to hedge it in; the one harbored and inclosed political life as the other sheltered and protected the biological life process of the family.[65]

It is therefore not really accurate to say that private property, prior to the modern age, was thought to be a self-evident condition for admission to the public realm; it is much more than that. Privacy was like the other, the dark and hidden side of the public realm, and while to be political meant to attain the highest possibility of human existence, to have no private place of one's own (like a slave) meant to be no longer human.

Of an altogether different and historically later origin is the political significance of private wealth from which one draws the means of one's livelihood. We mentioned earlier the ancient identification of necessity with the private realm of the household, where each had to master the necessities of life for himself. The free man, who disposed of his own privacy and was not, like a slave, at the disposition of a master, could still be "forced" by poverty. Poverty forces the free man to act like a slave.[66] Private wealth, therefore, became a condition for admission to public life not because its owner was engaged in accumulating it but, on the contrary, because it assured with reasonable certainty that its owner would not have to engage in providing for himself the

64. The word *polis* originally connoted something like a "ring-wall," and it seems the Latin *urbs* also expressed the notion of a "circle" and was derived from the same root as *orbis*. We find the same connection in our word "town," which originally, like the German *Zaun*, meant a surrounding fence (see R. B. Onians, *The Origins of European Thought* [1954], p. 444, n. 1).

65. The legislator therefore did not need to be a citizen and frequently was called in from the outside. His work was not political; political life, however, could begin only after he had finished his legislation.

66. Demosthenes *Orationes* 57. 45: "Poverty forces the free to do many slavish and base things" (*polla doulika kai tapeina pragmata tous eleutherous hē penia biazetai poiein*).

means of use and consumption and was free for public activity.[67] Public life, obviously, was possible only after the much more urgent needs of life itself had been taken care of. The means to take care of them was labor, and the wealth of a person therefore was frequently counted in terms of the number of laborers, that is, slaves, he owned.[68] To own property meant here to be master over one's own necessities of life and therefore potentially to be a free person, free to transcend his own life and enter the world all have in common.

Only with the emergence of such a common world in concrete tangibility, that is, with the rise of the city-state, could this kind of private ownership acquire its eminent political significance, and it is therefore almost a matter of course that the famous "disdain for menial occupations" is not yet to be found in the Homeric world. If the property-owner chose to enlarge his property instead of using it up in leading a political life, it was as though he willingly sacrificed his freedom and became voluntarily what the slave was against his own will, a servant of necessity.[69]

67. This condition for admission to the public realm was still in existence in the earlier Middle Ages. The English "Books of Customs" still drew "a sharp distinction between the craftsman and the freeman, *franke homme*, of the town. . . . If a craftsman became so rich that he wished to become a freeman, he must first foreswear his craft and get rid of all his tools from his house" (W. J. Ashley, *op. cit.*, p. 83). It was only under the rule of Edward III that the craftsmen became so rich that "instead of the craftsmen being incapable of citizenship, citizenship came to be bound up with membership of one of the companies" (p. 89).

68. Coulanges, in distinction from other authors, stresses the time- and strength-consuming activities demanded from an ancient citizen, rather than his "leisure," and sees rightly that Aristotle's statement that no man who had to work for his livelihood could be a citizen is a simple statement of fact rather than the expression of a prejudice (*op. cit.*, pp. 335 ff.). It is characteristic of the modern development that riches as such, regardless of the occupation of their owner, became a qualification for citizenship: only now was it a mere privilege to be a citizen, unconnected with any specifically political activities.

69. This seems to me to be the solution of the "well-known puzzle in the study of the economic history of the ancient world that industry developed up to a certain point, but stopped short of making progress which might have been expected . . . [in view of the fact that] thoroughness and capacity for organization on a large scale is shown by the Romans in other departments, in the public services and the army" (Barrow, *Slavery in the Roman Empire*, pp. 109–10). It

Up to the beginning of the modern age, this kind of property had never been held to be sacred, and only where wealth as the source of income coincided with the piece of land on which a family was located, that is, in an essentially agricultural society, could these two types of property coincide to such an extent that all property assumed the character of sacredness. Modern advocates of private property, at any rate, who unanimously understand it as privately owned wealth and nothing else, have little cause to appeal to a tradition according to which there could be no free public realm without a proper establishment and protection of privacy. For the enormous and still proceeding accumulation of wealth in modern society, which was started by expropriation—the expropriation of the peasant classes which in turn was the almost accidental consequence of the expropriation of Church and monastic property after the Reformation[70]—has never shown

---

seems a prejudice due to modern conditions to expect the same capacity for organization in private as in "public services." Max Weber, in his remarkable essay (*op. cit.*) had already insisted on the fact that ancient cities were rather "centers of consumption than of production" and that the ancient slave owner was a *"rentier* and not a capitalist [*Unternehmer*]" (pp. 13, 22 ff., and 144). The very indifference of ancient writers to economic questions, and the lack of documents in this respect, give additional weight to Weber's argument.

70. All histories of the working class, that is, a class of people who are without any property and live only from the work of their hands, suffer from the naïve assumption that there has always been such a class. Yet, as we saw, even slaves were not without property in antiquity, and the so-called free labor in antiquity usually turns out to consist of "free shopkeepers, traders and craftsmen" (Barrow, *Slavery in the Roman Empire*, p. 126). M. E. Park (*The Plebs Urbana in Cicero's Day* [1921]), therefore, comes to the conclusion that there was no free labor, since the free man always appears to be an owner of some sort. W. J. Ashley sums up the situation in the Middle Ages up to the fifteenth century: "There was as yet no large class of wage laborers, no 'working class' in the modern sense of the term. By 'working men,' we mean a number of men, from among whom individuals may indeed rise to become masters, but the majority of whom cannot hope ever to rise to a higher position. But in the fourteenth century a few years' work as a journeyman was but a stage through which the poorer men had to pass, while the majority probably set up for themselves as master craftsmen as soon as apprenticeship was over" (*op. cit.*, pp. 93–94).

Thus, the working class in antiquity was neither free nor without property; if, through manumission, the slave was given (in Rome) or had bought (in

much consideration for private property but has sacrificed it whenever it came into conflict with the accumulation of wealth. Proudhon's dictum that property is theft has a solid basis of truth in the origins of modern capitalism; it is all the more significant that even Proudhon hesitated to accept the doubtful remedy of general expropriation, because he knew quite well that the abolition of private property, while it might cure the evil of poverty, was only too likely to invite the greater evil of tyranny.[71] Since he did not distinguish between property and wealth, his two insights appear in his work like contradictions, which in fact they are not. Individual appropriation of wealth will in the long run respect private property no more than socialization of the accumulation process. It is not an invention of Karl Marx but actually in the very nature of this society itself that privacy in every sense can only hinder the development of social "productivity" and that considerations of private ownership therefore should be overruled in favor of the ever-increasing process of social wealth.[72]

Athens) his freedom, he did not become a free laborer but instantly became an independent businessman or craftsman. ("Most slaves seem to have taken into freedom some capital of their own" to set up in trade and industry [Barrow, *Slavery in the Roman Empire*, p. 103]). And in the Middle Ages, to be a worker in the modern sense of the term was a temporary stage in one's life, a preparation for mastership and manhood. Hired labor in the Middle Ages was an exception, and the German day laborers (the *Tagelöhner* in Luther's Bible translation) or the French *manœuvres* lived outside the settled communities and were identical with the poor, the "labouring poor" in England (see Pierre Brizon, *Histoire du travail et des travailleurs* [1926], p. 40). Moreover, the fact that no code of law before the *Code Napoléon* offers any treatment of free labor (see W. Endemann, *Die Behandlung der Arbeit im Privatrecht* [1896], pp. 49, 53) shows conclusively how recent the existence of a working class is.

71. See the ingenious comment on "property is theft" which occurs in Proudhon's posthumously published *Théorie de la propriété*, pp. 209–10, where he presents property in its "egoist, satanic nature" as the "most efficient means to resist despotism without overthrowing the state."

72. I must confess that I fail to see on what grounds in present-day society liberal economists (who today call themselves conservatives) can justify their optimism that the private appropriation of wealth will suffice to guard individual liberties—that is, will fulfil the same role as private property. In a jobholding

9

THE SOCIAL AND THE PRIVATE

What we called earlier the rise of the social coincided historically with the transformation of the private care for private property into a public concern. Society, when it first entered the public realm, assumed the disguise of an organization of property-owners who, instead of claiming access to the public realm because of their wealth, demanded protection from it for the accumulation of more wealth. In the words of Bodin, government belonged to kings and property to subjects, so that it was the duty of the kings to rule in the interest of their subjects' property. "The commonwealth," as has recently been pointed out, "largely existed for the common *wealth*."[73]

When this common wealth, the result of activities formerly banished to the privacy of the households, was permitted to take over the public realm, private possessions—which are essentially much less permanent and much more vulnerable to the mortality of their owners than the common world, which always grows out of the past and is intended to last for future generations—began to undermine the durability of the world. It is true that wealth can be accumulated to a point where no individual life-span can use it up, so that the family rather than the individual becomes its owner. Yet wealth remains something to be used and consumed no matter how many individual life-spans it may sustain. Only when wealth became capital, whose chief function was to generate more capital, did private property equal or come close to the permanence inherent in the commonly shared world.[74] How-

---

society, these liberties are safe only as long as they are guaranteed by the state, and even now they are constantly threatened, not by the state, but by society, which distributes the jobs and determines the share of individual appropriation.

73. R. W. K. Hinton, "Was Charles I a Tyrant?" *Review of Politics*, Vol. XVIII (January, 1956).

74. For the history of the word "capital" deriving from the Latin *caput*, which in Roman law was employed for the principal of a debt, see W. J. Ashley, *op. cit.*, pp. 429 and 433, n. 183. Only eighteenth-century writers began to use the word in the modern sense as "wealth invested in such a way as to bring gain."

ever, this permanence is of a different nature; it is the permanence of a process rather than the permanence of a stable structure. Without the process of accumulation, wealth would at once fall back into the opposite process of disintegration through use and consumption.

Common wealth, therefore, can never become common in the sense we speak of a common world; it remained, or rather was intended to remain, strictly private. Only the government, appointed to shield the private owners from each other in the competitive struggle for more wealth, was common. The obvious contradiction in this modern concept of government, where the only thing people have in common is their private interests, need no longer bother us as it still bothered Marx, since we know that the contradiction between private and public, typical of the initial stages of the modern age, has been a temporary phenomenon which introduced the utter extinction of the very difference between the private and public realms, the submersion of both in the sphere of the social. By the same token, we are in a far better position to realize the consequences for human existence when both the public and private spheres of life are gone, the public because it has become a function of the private and the private because it has become the only common concern left.

Seen from this viewpoint, the modern discovery of intimacy seems a flight from the whole outer world into the inner subjectivity of the individual, which formerly had been sheltered and protected by the private realm. The dissolution of this realm into the social may most conveniently be watched in the progressing transformation of immobile into mobile property until eventually the distinction between property and wealth, between the *fungibiles* and the *consumptibiles* of Roman law, loses all significance because every tangible, "fungible" thing has become an object of "consumption"; it lost its private use value which was determined by its location and acquired an exclusively social value determined through its ever-changing exchangeability whose fluctuation could itself be fixed only temporarily by relating it to the common denominator of money.[75] Closely connected with this social evapora-

75. Medieval economic theory did not yet conceive of money as a common denominator and yardstick but counted it among the *consumptibiles*.

tion of the tangible was the most revolutionary modern contribution to the concept of property, according to which property was not a fixed and firmly located part of the world acquired by its owner in one way or another but, on the contrary, had its source in man himself, in his possession of a body and his indisputable ownership of the strength of this body, which Marx called "labor-power."

Thus modern property lost its worldly character and was located in the person himself, that is, in what an individual could lose only along with his life. Historically, Locke's assumption that the labor of one's body is the origin of property is more than doubtful; but in view of the fact that we already live under conditions where our only reliable property is our skill and our labor power, it is more than likely that it will become true. For wealth, after it became a public concern, has grown to such proportions that it is almost unmanageable by private ownership. It is as though the public realm had taken its revenge against those who tried to use it for their private interests. The greatest threat here, however, is not the abolition of private ownership of wealth but the abolition of private property in the sense of a tangible, worldly place of one's own.

In order to understand the danger to human existence from the elimination of the private realm, for which the intimate is not a very reliable substitute, it may be best to consider those non-privative traits of privacy which are older than, and independent of, the discovery of intimacy. The difference between what we have in common and what we own privately is first that our private possessions, which we use and consume daily, are much more urgently needed than any part of the common world; without property, as Locke pointed out, "the common is of no use."[76] The same necessity that, from the standpoint of the public realm, shows only its negative aspect as a deprivation of freedom possesses a driving force whose urgency is unmatched by the so-called higher desires and aspirations of man; not only will it always be the first among man's needs and worries, it will also prevent the apathy and disappearance of initiative which so obvi-

76. *Second Treatise of Civil Government*, sec. 27.

ously threatens all overly wealthy communities.[77] Necessity and life are so intimately related and connected that life itself is threatened where necessity is altogether eliminated. For the elimination of necessity, far from resulting automatically in the establishment of freedom, only blurs the distinguishing line between freedom and necessity. (Modern discussions of freedom, where freedom is never understood as an objective state of human existence but either presents an unsolvable problem of subjectivity, of an entirely undetermined or determined will, or develops out of necessity, all point to the fact that the objective, tangible difference between being free and being forced by necessity is no longer perceived.)

The second outstanding non-privative characteristic of privacy is that the four walls of one's private property offer the only reliable hiding place from the common public world, not only from everything that goes on in it but also from its very publicity, from being seen and being heard. A life spent entirely in public, in the presence of others, becomes, as we would say, shallow. While it retains its visibility, it loses the quality of rising into sight from some darker ground which must remain hidden if it is not to lose its depth in a very real, non-subjective sense. The only efficient way to guarantee the darkness of what needs to be hidden against the light of publicity is private property, a privately owned place to hide in.[78]

While it is only natural that the non-privative traits of privacy should appear most clearly when men are threatened with deprivation of it, the practical treatment of private property by premodern political bodies indicates clearly that men have always been conscious of their existence and importance. This, however, did not make them protect the activities in the private realm directly, but rather the boundaries separating the privately owned from other parts of the world, most of all from the common world itself. The distinguishing mark of modern political and economic theory,

77. The relatively few instances of ancient authors praising labor and poverty are inspired by this danger (for references see G. Herzog-Hauser, *op. cit.*).

78. The Greek and Latin words for the interior of the house, *megaron* and *atrium*, have a strong connotation of darkness and blackness (see Mommsen, *op. cit.*, pp. 22 and 236).

on the other hand, in so far as it regards private property as a crucial issue, has been its stress upon the private activities of property-owners and their need of government protection for the sake of accumulation of wealth at the expense of the tangible property itself. What is important to the public realm, however, is not the more or less enterprising spirit of private businessmen but the fences around the houses and gardens of citizens. The invasion of privacy by society, the "socialization of man" (Marx), is most efficiently carried through by means of expropriation, but this is not the only way. Here, as in other respects, the revolutionary measures of socialism or communism can very well be replaced by a slower and no less certain "withering away" of the private realm in general and of private property in particular.

The distinction between the private and public realms, seen from the viewpoint of privacy rather than of the body politic, equals the distinction between things that should be shown and things that should be hidden. Only the modern age, in its rebellion against society, has discovered how rich and manifold the realm of the hidden can be under the conditions of intimacy; but it is striking that from the beginning of history to our own time it has always been the bodily part of human existence that needed to be hidden in privacy, all things connected with the necessity of the life process itself, which prior to the modern age comprehended all activities serving the subsistence of the individual and the survival of the species. Hidden away were the laborers who "with their bodies minister to the [bodily] needs of life,"[79] and the women who with their bodies guarantee the physical survival of the species. Women and slaves belonged to the same category and were hidden away not only because they were somebody else's property but because their life was "laborious," devoted to bodily functions.[80] In the beginning of the modern age, when "free"

79. Aristotle *Politics* 1254b25.

80. The life of a woman is called *ponētikos* by Aristotle, *On the Generation of Animals* 775a33. That women and slaves belonged and lived together, that no woman, not even the wife of the household head, lived among her equals—other free women—so that rank depended much less on birth than on "occupation" or function, is very well presented by Wallon (*op. cit.*, I, 77 ff.), who speaks of a "confusion des rangs, ce partage de toutes les fonctions domestiques": "Les

labor had lost its hiding place in the privacy of the household, the laborers were hidden away and segregated from the community like criminals behind high walls and under constant supervision.[81] The fact that the modern age emancipated the working classes and the women at nearly the same historical moment must certainly be counted among the characteristics of an age which no longer believes that bodily functions and material concerns should be hidden. It is all the more symptomatic of the nature of these phenomena that the few remnants of strict privacy even in our own civilization relate to "necessities" in the original sense of being necessitated by having a body.

10

THE LOCATION OF HUMAN ACTIVITIES

Although the distinction between private and public coincides with the opposition of necessity and freedom, of futility and permanence, and, finally, of shame and honor, it is by no means true that only the necessary, the futile, and the shameful have their proper place in the private realm. The most elementary meaning of the two realms indicates that there are things that need to be hidden and others that need to be displayed publicly if they are to exist at all. If we look at these things, regardless of where we find them in any given civilization, we shall see that each human activity points to its proper location in the world. This is true for the chief activities of the *vita activa*, labor, work, and action; but there is one, admittedly extreme, example of this phenomenon, whose advantage for illustration is that it played a considerable role in political theory.

Goodness in an absolute sense, as distinguished from the "good-for" or the "excellent" in Greek and Roman antiquity, became known in our civilization only with the rise of Christianity. Since

femmes . . . se confondaient avec leurs esclaves dans les soins habituels de la vie intérieure. De quelque rang qu'elles fussent, le travail était leur apanage, comme aux hommes la guerre."

81. See Pierre Brizon, *Histoire du travail et des travailleurs* (4th ed.; 1926), p. 184, concerning the conditions of factory work in the seventeenth century.

then, we know of good works as one important variety of possible human action. The well-known antagonism between early Christianity and the *res publica*, so admirably summed up in Tertullian's formula: *nec ulla magis res aliena quam publica* ("no matter is more alien to us than what matters publicly"),[82] is usually and rightly understood as a consequence of early eschatological expectations that lost their immediate significance only after experience had taught that even the downfall of the Roman Empire did not mean the end of the world.[83] Yet the otherworldliness of Christianity has still another root, perhaps even more intimately related to the teachings of Jesus of Nazareth, and at any rate so independent of the belief in the perishability of the world that one is tempted to see in it the true inner reason why Christian alienation from the world could so easily survive the obvious non-fulfilment of its eschatological hopes.

The one activity taught by Jesus in word and deed is the activity of goodness, and goodness obviously harbors a tendency to hide from being seen or heard. Christian hostility toward the public realm, the tendency at least of early Christians to lead a life as far removed from the public realm as possible, can also be understood as a self-evident consequence of devotion to good works, independent of all beliefs and expectations. For it is manifest that the moment a good work becomes known and public, it loses its specific character of goodness, of being done for nothing but goodness' sake. When goodness appears openly, it is no longer goodness, though it may still be useful as organized charity or an act of solidarity. Therefore: "Take heed that ye do not your alms before men, to be seen of them." Goodness can exist only when it is not perceived, not even by its author; whoever sees himself performing a good work is no longer good, but at best a useful member of society or a dutiful member of a church. Therefore: "Let not thy left hand know what thy right hand doeth."

It may be this curious negative quality of goodness, the lack of outward phenomenal manifestation, that makes Jesus of Naza-

82. Tertullian *op. cit.* 38.

83. This difference of experience may partly explain the difference between the great sanity of Augustine and the horrible concreteness of Tertullian's views on politics. Both were Romans and profoundly shaped by Roman political life.

reth's appearance in history such a profoundly paradoxical event; it certainly seems to be the reason why he thought and taught that no man can be good: "Why callest thou me good? none is good, save one, that is, God."[84] The same conviction finds its expression in the talmudic story of the thirty-six righteous men, for the sake of whom God saves the world and who also are known to nobody, least of all to themselves. We are reminded of Socrates' great insight that no man can be wise, out of which love for wisdom, or philo-sophy, was born; the whole life story of Jesus seems to testify how love for goodness arises out of the insight that no man can be good.

Love of wisdom and love of goodness, if they resolve themselves into the activities of philosophizing and doing good works, have in common that they come to an immediate end, cancel themselves, so to speak, whenever it is assumed that man can *be* wise or *be* good. Attempts to bring into being that which can never survive the fleeting moment of the deed itself have never been lacking and have always led into absurdity. The philosophers of late antiquity who demanded of themselves to *be* wise were absurd when they claimed to be happy when roasted alive in the famous Phaleric Bull. And no less absurd is the Christian demand to *be* good and to turn the other cheek, when not taken metaphorically but tried as a real way of life.

But the similarity between the activities springing from love of goodness and love of wisdom ends here. Both, it is true, stand in a certain opposition to the public realm, but the case of goodness is much more extreme in this respect and therefore of greater relevance in our context. Only goodness must go into absolute hiding and flee all appearance if it is not to be destroyed. The philosopher, even if he decides with Plato to leave the "cave" of human affairs, does not have to hide from himself; on the contrary, under the sky of ideas he not only finds the true essences of everything that is,

---

84. Luke 18 : 19. The same thought occurs in Matt. 6 : 1–18, where Jesus warns against hypocrisy, against the open display of piety. Piety cannot "appear unto men" but only unto God, who "seeth in secret." God, it is true, "shall reward" man, but not, as the standard translation claims, "openly." The German word *Scheinheiligkeit* expresses this religious phenomenon, where mere appearance is already hypocrisy, quite adequately.

but also himself, in the dialogue between "me and myself" (*eme emautō*) in which Plato apparently saw the essence of thought.[85] To be in solitude means to be with one's self, and thinking, therefore, though it may be the most solitary of all activities, is never altogether without a partner and without company.

The man, however, who is in love with goodness can never afford to lead a solitary life, and yet his living with others and for others must remain essentially without testimony and lacks first of all the company of himself. He is not solitary, but lonely; when living with others he must hide from them and cannot even trust himself to witness what he is doing. The philosopher can always rely upon his thoughts to keep him company, whereas good deeds can never keep anybody company; they must be forgotten the moment they are done, because even memory will destroy their quality of being "good." Moreover, thinking, because it can be remembered, can crystallize into thought, and thoughts, like all things that owe their existence to remembrance, can be transformed into tangible objects which, like the written page or the printed book, become part of the human artifice. Good works, because they must be forgotten instantly, can never become part of the world; they come and go, leaving no trace. They truly are not of this world.

It is this worldlessness inherent in good works that makes the lover of goodness an essentially religious figure and that makes goodness, like wisdom in antiquity, an essentially non-human, superhuman quality. And yet love of goodness, unlike love of wisdom, is not restricted to the experience of the few, just as loneliness, unlike solitude, is within the range of every man's experience. In a sense, therefore, goodness and loneliness are of much greater relevance to politics than wisdom and solitude; yet only solitude can become an authentic way of life in the figure of the philosopher, whereas the much more general experience of loneliness is so contradictory to the human condition of plurality that it is simply unbearable for any length of time and needs the company of God, the only imaginable witness of good works, if it is not to annihilate human existence altogether. The otherworldiness of religious experience, in so far as it is truly the experience of love in the sense

85. One finds this idiom *passim* in Plato (see esp. *Gorgias* 482).

of an activity, and not the much more frequent one of beholding passively a revealed truth, manifests itself within the world itself; this, like all other activities, does not leave the world, but must be performed within it. But this manifestation, though it appears in the space where other activities are performed and depends upon it, is of an actively negative nature; fleeing the world and hiding from its inhabitants, it negates the space the world offers to men, and most of all that public part of it where everything and everybody are seen and heard by others.

Goodness, therefore, as a consistent way of life, is not only impossible within the confines of the public realm, it is even destructive of it. Nobody perhaps has been more sharply aware of this ruinous quality of doing good than Machiavelli, who, in a famous passage, dared to teach men "how not to be good."[86] Needless to add, he did not say and did not mean that men must be taught how to be bad; the criminal act, though for other reasons, must also flee being seen and heard by others. Machiavelli's criterion for political action was glory, the same as in classical antiquity, and badness can no more shine in glory than goodness. Therefore all methods by which "one may indeed gain power, but not glory" are bad.[87] Badness that comes out of hiding is impudent and directly destroys the common world; goodness that comes out of hiding and assumes a public role is no longer good, but corrupt in its own terms and will carry its own corruption wherever it goes. Thus, for Machiavelli, the reason for the Church's becoming a corrupting influence in Italian politics was her participation in secular affairs as such and not the individual corruptness of bishops and prelates. To him, the alternative posed by the problem of religious rule over the secular realm was inescapably this: either the public realm corrupted the religious body and thereby became itself corrupt, or the religious body remained uncorrupt and destroyed the public realm altogether. A reformed Church therefore was even more dangerous in Machiavelli's eyes, and he looked with great respect but greater apprehension upon the religious revival of his time, the "new orders" which, by "saving religion from being destroyed by the licentious-

86. *Prince*, ch. 15.
87. *Ibid.*, ch. 8.

ness of the prelates and heads of the Church," teach people to be good and not "to resist evil"—with the result that "wicked rulers do as much evil as they please."[88]

We chose the admittedly extreme example of doing good works, extreme because this activity is not even at home in the realm of privacy, in order to indicate that the historical judgments of political communities, by which each determined which of the activities of the *vita activa* should be shown in public and which be hidden in privacy, may have their correspondence in the nature of these activities themselves. By raising this question, I do not intend to attempt an exhaustive analysis of the activities of the *vita activa*, whose articulations have been curiously neglected by a tradition which considered it chiefly from the standpoint of the *vita contemplativa*, but to try to determine with some measure of assurance their political significance.

88. *Discourses*, Book III, ch. 1.

CHAPTER III

Labor

In the following chapter, Karl Marx will be criticized. This is unfortunate at a time when so many writers who once made their living by explicit or tacit borrowing from the great wealth of Marxian ideas and insights have decided to become professional anti-Marxists, in the process of which one of them even discovered that Karl Marx himself was unable to make a living, forgetting for the moment the generations of authors whom he has "supported." In this difficulty, I may recall a statement Benjamin Constant made when he felt compelled to attack Rousseau: "J'éviterai certes de me joindre aux détracteurs d'un grand homme. Quand le hasard fait qu'en apparence je me rencontre avec eux sur un seul point, je suis en défiance de moi-même; et pour me consoler de paraître un instant de leur avis . . . j'ai besoin de désavouer et de flétrir, autant qu'il est en moi, ces prétendus auxiliaires." ("Certainly, I shall avoid the company of detractors of a great man. If I happen to agree with them on a single point I grow suspicious of myself; and in order to console myself for having seemed to be of their opinion . . . I feel I must disavow and keep these false friends away from me as much as I can.")[1]

II

"THE LABOUR OF OUR BODY AND THE WORK OF OUR HANDS"[2]

The distinction between labor and work which I propose is unusual. The phenomenal evidence in its favor is too striking to be

1. See "De la liberté des anciens comparée a celle des modernes" (1819), reprinted in *Cours de politique constitutionnelle* (1872), II, 549.

2. Locke, *Second Treatise of Civil Government*, sec. 26.

ignored, and yet historically it is a fact that apart from a few scattered remarks, which moreover were never developed even in the theories of their authors, there is hardly anything in either the premodern tradition of political thought or in the large body of modern labor theories to support it. Against this scarcity of historical evidence, however, stands one very articulate and obstinate testimony, namely, the simple fact that every European language, ancient and modern, contains two etymologically unrelated words for what we have to come to think of as the same activity, and retains them in the face of their persistent synonymous usage.[3]

Thus, Locke's distinction between working hands and a laboring body is somewhat reminiscent of the ancient Greek distinction between the *cheirotechnēs*, the craftsman, to whom the German *Handwerker* corresponds, and those who, like "slaves and tame animals with their bodies minister to the necessities of life,"[4] or in the Greek idiom, *tō sōmati ergazesthai*, work with their bodies (yet even here, labor and work are already treated as identical, since the word used is not *ponein* [labor] but *ergazesthai* [work]). Only in one respect, which, however, is linguistically the most important one, did ancient and modern usage of the two words as synonyms fail altogether, namely in the formation of a corresponding noun. Here again we find complete unanimity; the word "labor," understood as a noun, never designates the finished product, the result of laboring, but remains a verbal noun to be classed with the gerund, whereas the product itself is invariably derived from the word for work, even when current usage has followed the

3. Thus, the Greek language distinguishes between *ponein* and *ergazesthai*, the Latin between *laborare* and *facere* or *fabricari*, which have the same etymological root, the French between *travailler* and *ouvrer*, the German between *arbeiten* and *werken*. In all these cases, only the equivalents for "labor" have an unequivocal connotation of pain and trouble. The German *Arbeit* applied originally only to farm labor executed by serfs and not to the work of the craftsman, which was called *Werk*. The French *travailler* replaced the older *labourer* and is derived from *tripalium*, a kind of torture. See Grimm, *Wörterbuch*, pp. 1854 ff., and Lucien Fèbre, "Travail: évolution d'un mot et d'une idée," *Journal de psychologie normale et pathologique*, Vol. XLI, No. 1 (1948).

4. Aristotle *Politics* |1254b25.

actual modern development so closely that the verb form of the word "work" has become rather obsolete.[5]

The reason why this distinction should have been overlooked in ancient times and its significance remained unexplored seems obvious enough. Contempt for laboring, originally arising out of a passionate striving for freedom from necessity and a no less passionate impatience with every effort that left no trace, no monument, no great work worthy of remembrance, spread with the increasing demands of *polis* life upon the time of the citizens and its insistence on their abstention (*skholē*) from all but political activities, until it covered everything that demanded an effort. Earlier political custom, prior to the full development of the city-state, merely distinguished between slaves, vanquished enemies (*dmōes* or *douloi*), who were carried off to the victor's household with other loot where as household inmates (*oiketai* or *familiares*) they slaved for their own and their master's life, and the *dēmiourgoi*, the workmen of the people at large, who moved freely outside the private realm and within the public.[6] A later time even changed the name for these artisans, whom Solon had still described as sons of Athena and Hephaestus, and called them *banausoi*, that is, men whose chief interest is their craft and not the market place. It is only from the late fifth century onward that the *polis* began to classify occupations according to the amount of effort required, so that Aristotle called those occupations the meanest "in which the

---

5. This is the case for the French *ouvrer* and the German *werken*. In both languages, as distinguished from the current English usage of the word "labor," the words *travailler* and *arbeiten* have almost lost the original significance of pain and trouble; Grimm (*op. cit.*) had already noted this development in the middle of the last century: "Während in älterer Sprache die Bedeutung von *molestia* und schwerer Arbeit vorherrschte, die von *opus, opera*, zurücktrat, tritt umgekehrt in der heutigen diese vor und jene erscheint seltener." It is also interesting that the nouns "work," *œuvre, Werk*, show an increasing tendency to be used for works of art in all three languages.

6. See J.-P. Vernant, "Travail et nature dans la Grèce ancienne" (*Journal de psychologie normale et pathologique*, LII, No. 1 [January-March, 1955]): "Le terme [*dēmiourgoi*], chez Homère et Hésiode, ne qualifie pas à l'origine l'artisan en tant que tel, comme 'ouvrier': il définit toutes les activités qui s'exercent en dehors du cadre de l'*oikos*, en faveur d'un public, *dēmos*: les artisans—charpentiers et forgerons—mais non moins qu'eux les devins, les hérauts, les aèdes."

body is most deteriorated." Although he refused to admit *banausoi* to citizenship, he would have accepted shepherds and painters (but neither peasants nor sculptors).[7]

We shall see later that, quite apart from their contempt for labor, the Greeks had reasons of their own to mistrust the craftsman, or rather, the *homo faber* mentality. This mistrust, however, is true only of certain periods, whereas all ancient estimates of human activities, including those which, like Hesiod, supposedly praise

7. *Politics* 1258b35 ff. For Aristotle's discussion about admission of *banausoi* to citizenship see *Politics* iii. 5. His theory corresponds closely to reality: it is estimated that up to 80 per cent of free labor, work, and commerce consisted of non-citizens, either "strangers" (*katoikountes* and *metoikoi*) or emancipated slaves who advanced into these classes (see Fritz Heichelheim, *Wirtschafts-geschichte des Altertums* [1938], I, 398 ff.). Jacob Burckhardt, who in his *Griechische Kulturgeschichte* (Vol. II, secs. 6 and 8) relates Greek current opinion of who does and who does not belong to the class of *banausoi*, also notices that we do not know of any treatise about sculpture. In view of the many essays on music and poetry, this probably is no more an accident of tradition than the fact that we know so many stories about the great feeling of superiority and even arrogance among the famous painters which are not matched by anecdotes about sculptors. This estimate of painters and sculptors survived many centuries. It is still found in the Renaissance, where sculpturing is counted among the servile arts whereas painting takes up a middle position between liberal and servile arts (see Otto Neurath, "Beiträge zur Geschichte der Opera Servilia," *Archiv für Sozialwissenschaft und Sozialpolitik*, Vol. XLI, No. 2 [1915]).

That Greek public opinion in the city-states judged occupations according to the effort required and the time consumed is supported by a remark of Aristotle about the life of shepherds: "There are great differences in human ways of life. The laziest are shepherds; for they get their food without labor [*ponos*] from tame animals and have leisure [*skholazousin*]" (*Politics* 1256a30 ff.). It is interesting that Aristotle, probably following current opinion, here mentions laziness (*aergia*) together with, and somehow as a condition for, *skholē*, abstention from certain activities which is the condition for a political life. Generally, the modern reader must be aware that *aergia* and *skholē* are not the same. Laziness had the same connotations it has for us, and a life of *skholē* was not considered to be a lazy life. The equation, however, of *skholē* and idleness is characteristic of a development within the *polis*. Thus Xenophon reports that Socrates was accused of having quoted Hesiod's line: "Work is no disgrace, but laziness [*aergia*] is a disgrace." The accusation meant that Socrates had instilled in his pupils a slavish spirit (*Memorabilia* i. 2. 56). Historically, it is important to keep in mind the distinction between the contempt of the Greek city-states for all non-political occupations which arose out of the enormous demands upon the time and energy of the citizens, and the earlier, more original, and more general contempt for labor,[8] rest on the conviction that the labor of our body which is necessitated by its needs is slavish. Hence, occupations which did not consist in laboring, yet were undertaken not for their own sake but in order to provide the necessities of life, were assimilated to the status of labor, and this explains changes and variations in their estimation and classification at different periods and in different places. The opinion that labor and work were despised in antiquity because only slaves were engaged in them is a prejudice of modern historians. The ancients reasoned the other way around and felt it necessary to possess slaves because of the slavish nature of all occupations that served the needs for the maintenance of life.[9] It was precisely on these grounds that the institution of slavery was defended and justified. To labor meant to be enslaved by necessity,

---

activities which serve only to sustain life—*ad vitae sustentationem* as the *opera servilia* are still defined in the eighteenth century. In the world of Homer, Paris and Odysseus help in the building of their houses, Nausicaä herself washes the linen of her brothers, etc. All this belongs to the self-sufficiency of the Homeric hero, to his independence and the autonomic supremacy of his person. No work is sordid if it means greater independence; the selfsame activity might well be a sign of slavishness if not personal independence but sheer survival is at stake, if it is not an expression of sovereignty but of subjection to necessity. The different estimate of craftsmanship in Homer is of course well known. But its actual meaning is beautifully presented in a recent essay by Richard Harder, *Eigenart der Griechen* (1949).

8. Labor and work (*ponos* and *ergon*) are distinguished in Hesiod; only work is due to Eris, the goddess of good strife (*Works and Days* 20–26), but labor, like all other evils, came out of Pandora's box (90 ff.) and is a punishment of Zeus because Prometheus "the crafty deceived him." Since then, "the gods have hidden life from men" (42 ff.) and their curse hits "the bread-eating men" (82). Hesiod, moreover, assumes as a matter of course that the actual farm labor is done by slaves and tame animals. He praises everyday life—which for a Greek is already extraordinary enough—but his ideal is a gentleman-farmer, rather than a laborer, who stays at home, keeps away from adventures of the sea as well as public business on the *agora* (29 ff.), and minds his own business.

9. Aristotle begins his famous discussion of slavery (*Politics* 1253b25) with the statement that "without the necessaries life as well as good life is impossible." To be a master of slaves is the human way to master necessity and therefore not *para physin*, against nature; life itself demands it. Peasants, therefore, who provided the necessities of life, are classed by Plato as well as Aristotle with the slaves (see Robert Schlaifer, "Greek Theories of Slavery from Homer to Aristotle," *Harvard Studies in Classical Philology*, Vol. XLVII [1936]).

and this enslavement was inherent in the conditions of human life. Because men were dominated by the necessities of life, they could win their freedom only through the domination of those whom they subjected to necessity by force. The slave's degradation was a blow of fate and a fate worse than death, because it carried with it a metamorphosis of man into something akin to a tame animal.[10] A change in a slave's status, therefore, such as manumission by his master or a change in general political circumstance that elevated certain occupations to public relevance, automatically entailed a change in the slave's "nature."[11]

The institution of slavery in antiquity, though not in later times, was not a device for cheap labor or an instrument of exploitation for profit but rather the attempt to exclude labor from the conditions of man's life. What men share with all other forms of animal life was not considered to be human. (This, incidentally, was also the reason for the much misunderstood Greek theory of the non-human nature of the slave. Aristotle, who argued this theory so explicitly, and then, on his deathbed, freed his slaves, may not have been so inconsistent as moderns are inclined to think. He denied not the slave's capacity to be human, but only the use of the word "men" for members of the species man-kind as long as they are totally subject to necessity.)[12] And it is true that the use of the word "animal" in the concept of *animal laborans*, as distinguished from the very questionable use of the same word in the term *animal rationale*, is fully justified. The *animal laborans* is indeed only one, at best the highest, of the animal species which populate the earth.

10. It is in this sense that Euripides calls all slaves "bad": they see everything from the viewpoint of the stomach (*Supplementum Euripideum*, ed. Arnim, frag. 49, no. 2).

11. Thus Aristotle recommended that slaves who were intrusted with "free occupations" (*ta eleuthera tōn ergōn*) be treated with more dignity and not like slaves. When, on the other hand, in the first centuries of the Roman Empire certain public functions which always had been performed by public slaves rose in esteem and relevance, these *servi publici*—who actually performed the tasks of civil servants—were permitted to wear the toga and to marry free women.

12. The two qualities that the slave, according to Aristotle, lacks—and it is because of these defects that he is not human—are the faculty to deliberate and decide (*to bouleutikon*) and to foresee and to choose (*proairesis*). This, of course, is but a more explicit way of saying that the slave is subject to necessity.

Labor

It is not surprising that the distinction between labor and work was ignored in classical antiquity. The differentiation between the private household and the public political realm, between the household inmate who was a slave and the household head who was a citizen, between activities which should be hidden in privacy and those which were worth being seen, heard, and remembered, overshadowed and predetermined all other distinctions until only one criterion was left: is the greater amount of time and effort spent in private or in public? is the occupation motivated by *cura privati negotii* or *cura rei publicae*, care for private or for public business?[13] With the rise of political theory, the philosophers overruled even these distinctions, which had at least distinguished between activities, by opposing contemplation to all kinds of activity alike. With them, even political activity was leveled to the rank of necessity, which henceforth became the common denominator of all articulations within the *vita activa*. Nor can we reasonably expect any help from Christian political thought, which accepted the philosophers' distinction, refined it, and, religion being for the many and philosophy only for the few, gave it general validity, binding for all men.

It is surprising at first glance, however, that the modern age— with its reversal of all traditions, the traditional rank of action and contemplation no less than the traditional hierarchy within the *vita activa* itself, with its glorification of labor as the source of all values and its elevation of the *animal laborans* to the position traditionally held by the *animal rationale*—should not have brought forth a single theory in which *animal laborans* and *homo faber*, "the labour of our body and the work of our hands," are clearly distinguished. Instead, we find first the distinction between productive and unproductive labor, then somewhat later the differentiation between skilled and unskilled work, and, finally, outranking both because seemingly of more elementary significance, the division of all activities into manual and intellectual labor. Of the three, however, only the distinction between productive and unproductive labor goes to the heart of the matter, and it is no accident that the two greatest theorists in the field, Adam Smith and Karl Marx, based the whole structure of their argument upon it. The very

13. Cicero *De re publica* v. 2.

reason for the elevation of labor in the modern age was its "productivity," and the seemingly blasphemous notion of Marx that labor (and not God) created man or that labor (and not reason) distinguished man from the other animals was only the most radical and consistent formulation of something upon which the whole modern age was agreed.[14]

Moreover, both Smith and Marx were in agreement with modern public opinion when they despised unproductive labor as parasitical, actually a kind of perversion of labor, as though nothing were worthy of this name which did not enrich the world. Marx certainly shared Smith's contempt for the "menial servants" who like "idle guests . . . leave nothing behind them in return for their consumption."[15] Yet it was precisely these menial servants, these household inmates, *oiketai* or *familiares*, laboring for sheer subsistence and needed for effortless consumption rather than for pro-

14. "The creation of man through human labor" was one of the most persistent ideas of Marx since his youth. It can be found in many variations in the *Jugendschriften* (where in the "Kritik der Hegelschen Dialektik" he credits Hegel with it). (See *Marx-Engels Gesamtausgabe*, Part I, Vol. 5 [Berlin, 1932], pp. 156 and 167.) That Marx actually meant to replace the traditional definition of man as an *animal rationale* by defining him as an *animal laborans* is manifest in the context. The theory is strengthened by a sentence from the *Deutsche Ideologie* which was later deleted: "Der erste geschichtliche Akt dieser Individuen, wodurch sie sich von den Tieren unterscheiden, ist nicht, dass sie denken, sondern, dass sie anfangen ihre Lebensmittel zu produzieren" (*ibid.*, p. 568). Similar formulations occur in the "Ökonomisch-philosophische Manuskripte" (*ibid.*, p. 125), and in "Die heilige Familie" (*ibid.*, p. 189). Engels used similar formulations many times, for instance in the Preface of 1884 to *Ursprung der Familie* or in the newspaper article of 1876, "Labour in the Transition from Ape to Man" (see Marx and Engels, *Selected Works* [London, 1950], Vol. II).

It seems that Hume, and not Marx, was the first to insist that labor distinguishes man from animal (Adriano Tilgher, *Homo faber* [1929]; English ed.: *Work: What It Has Meant to Men through the Ages* [1930]). As labor does not play any significant role in Hume's philosophy, this is of historical interest only; to him, this characteristic did not make human life more productive, but only harsher and more painful than animal life. It is, however, interesting in this context to note with what care Hume repeatedly insisted that neither thinking nor reasoning distinguishes man from animal and that the behavior of beasts demonstrates that they are capable of both.

15. *Wealth of Nations* (Everyman's ed.), II, 302.

duction, whom all ages prior to the modern had in mind when they identified the laboring condition with slavery. What they left behind them in return for their consumption was nothing more or less than their masters' freedom or, in modern language, their masters' potential productivity.

In other words, the distinction between productive and unproductive labor contains, albeit in a prejudicial manner, the more fundamental distinction between work and labor.[16] It is indeed the mark of all laboring that it leaves nothing behind, that the result of its effort is almost as quickly consumed as the effort is spent. And yet this effort, despite its futility, is born of a great urgency and motivated by a more powerful drive than anything else, because life itself depends upon it. The modern age in general and Karl Marx in particular, overwhelmed, as it were, by the unprecedented actual productivity of Western mankind, had an almost irresistible tendency to look upon all labor as work and to speak of the *animal laborans* in terms much more fitting for *homo faber*, hoping all the time that only one more step was needed to eliminate labor and necessity altogether.[17]

No doubt the actual historical development that brought labor out of hiding and into the public realm, where it could be organized

16. The distinction between productive and unproductive labor is due to the physiocrats, who distinguished between producing, property-owning, and sterile classes. Since they held that the original source of all productivity lies in the natural forces of the earth, their standard for productivity was related to the creation of new objects and not to the needs and wants of men. Thus, the Marquis de Mirabeau, father of the famous orator, calls sterile "la classe d'ouvriers dont les travaux, quoique nécessaires aux besoins des hommes et utiles à la société, ne sont pas néanmoins productifs" and illustrates his distinction between sterile and productive work by comparing it to the difference between cutting a stone and producing it (see Jean Dautry, "La notion de travail chez Saint-Simon et Fourier," *Journal de psychologie normale et pathologique*, Vol. LII, No. 1 [January–March, 1955]).

17. This hope accompanied Marx from beginning to end. We find it already in the *Deutsche Ideologie:* "Es handelt sich nicht darum die Arbeit zu befreien, sondern sie aufzuheben" (*Gesamtausgabe*, Part I, Vol. 3, p. 185) and many decades later in the third volume of *Das Kapital*, ch. 48: "Das Reich der Freiheit beginnt in der Tat erst da, wo das Arbeiten . . . aufhört" (*Marx-Engels Gesamtausgabe*, Part II [Zürich, 1933], p. 873).

and "divided,"[18] constituted a powerful argument in the development of these theories. Yet an even more significant fact in this respect, already sensed by the classical economists and clearly discovered and articulated by Karl Marx, is that the laboring activity itself, regardless of historical circumstances and independent of its location in the private or the public realm, possesses indeed a "productivity" of its own, no matter how futile and non-durable its products may be. This productivity does not lie in any of labor's products but in the human "power," whose strength is not exhausted when it has produced the means of its own subsistence and survival but is capable of producing a "surplus," that is, more than is necessary for its own "reproduction." It is because not labor itself but the surplus of human "labor *power*" (*Arbeits*kraft) explains labor's productivity that Marx's introduction of this term, as Engels rightly remarked, constituted the most original and revolutionary element of his whole system.[19] Unlike the productivity of work, which adds new objects to the human artifice, the productivity of labor power produces objects only incidentally and is primarily concerned with the means of its own reproduction; since its power is not exhausted when its own reproduction has been secured, it can be used for the reproduction of more than one life process, but it never "produces" anything but life.[20] Through violent oppression in a slave society or exploitation in the capitalist society of Marx's own time, it can be channeled in such a way that the labor of some suffices for the life of all.

From this purely social viewpoint, which is the viewpoint of the whole modern age but which received its most coherent and great-

18. In his Introduction to the second book of the *Wealth of Nations* (Everyman's ed., I, 241 ff.), Adam Smith emphasizes that productivity is due to the division of labor rather than to labor itself.

19. See Engels' Introduction to Marx's "Wage, Labour and Capital" (in Marx and Engels, *Selected Works* [London, 1950], I, 384), where Marx had introduced the new term with a certain emphasis.

20. Marx stressed always, and especially in his youth, that the chief function of labor was the "production of life" and therefore saw labor together with procreation (see *Deutsche Ideologie*, p. 19; also "Wage, Labour and Capital," p. 77).

est expression in Marx's work, all laboring is "productive," and the earlier distinction between the performance of "menial tasks" that leave no trace and the production of things durable enough to be accumulated loses its validity. The social viewpoint is identical, as we saw before, with an interpretation that takes nothing into account but the life process of mankind, and within its frame of reference all things become objects of consumption. Within a completely "socialized mankind," whose sole purpose would be the entertaining of the life process—and this is the unfortunately quite unutopian ideal that guides Marx's theories[21]—the distinction between labor and work would have completely disappeared; all work would have become labor because all things would be understood, not in their worldly, objective quality, but as results of living labor power and functions of the life process.[22]

It is interesting to note that the distinctions between skilled and unskilled and between intellectual and manual work play no role in either classical political economy or in Marx's work. Compared

21. The terms *vergesellschafteter Mensch* or *gesellschaftliche Menschheit* were frequently used by Marx to indicate the goal of socialism (see, for instance, the third volume of *Das Kapital*, p. 873, and the tenth of the "Theses on Feuerbach": "The standpoint of the old materialism is 'civil' society; the standpoint of the new is *human* society, or socialized humanity" [*Selected Works*, II, 367]). It consisted in the elimination of the gap between the individual and social existence of man, so that man "in his most individual being would be at the same time a social being [a *Gemeinwesen*]" (*Jugendschriften*, p. 113). Marx frequently calls this social nature of man his *Gattungswesen*, his being a member of the species, and the famous Marxian "self-alienation" is first of all man's alienation from being a *Gattungswesen* (*ibid.*, p. 89: "Eine unmittelbare Konsequenz davon, dass der Mensch dem Produkt seiner Arbeit, seiner Lebenstätigkeit, seinem Gattungswesen entfremdet, ist die Entfremdung des Menschen von *dem* Menschen"). The ideal society is a state of affairs where all human activities derive as naturally from human "nature" as the secretion of wax by bees for making the honeycomb; to live and to labor for life will have become one and the same, and life will no longer "begin for [the laborer] where [the activity of laboring] ceases" ("Wage, Labour and Capital," p. 77).

22. Marx's original charge against capitalist society was not merely its transformation of all objects into commodities, but that "the laborer behaves toward the product of his labor as to an alien object" ("dass der Arbeiter zum Produkt seiner Arbeit als einem fremden Gegenstand sich verhält" [*Jugendschriften*, p. 83])—in other words, that the things of the world, once they have been produced by men, are to an extent independent of, "alien" to, human life.

[*89*]

with the productivity of labor, they are indeed of secondary importance. Every activity requires a certain amount of skill, the activity of cleaning and cooking no less than the writing of a book or the building of a house. The distinction does not apply to different activities but notes only certain stages and qualities within each of them. It could acquire a certain importance through the modern division of labor, where tasks formerly assigned to the young and inexperienced were frozen into lifelong occupations. But this consequence of the division of labor, where one activity is divided into so many minute parts that each specialized performer needs but a minimum of skill, tends to abolish skilled labor altogether, as Marx rightly predicted. Its result is that what is bought and sold in the labor market is not individual skill but "labor power," of which each living human being should possess approximately the same amount. Moreover, since unskilled work is a contradiction in terms, the distinction itself is valid only for the laboring activity, and the attempt to use it as a major frame of reference already indicates that the distinction between labor and work has been abandoned in favor of labor.

Quite different is the case of the more popular category of manual and intellectual work. Here, the underlying tie between the laborer of the hand and the laborer of the head is again the laboring process, in one case performed by the head, in the other by some other part of the body. Thinking, however, which is presumably the activity of the head, though it is in some way like laboring—also a process which probably comes to an end only with life itself—is even less "productive" than labor; if labor leaves no permanent trace, thinking leaves nothing tangible at all. By itself, thinking never materializes into any objects. Whenever the intellectual worker wishes to manifest his thoughts, he must use his hands and acquire manual skills just like any other worker. In other words, thinking and working are two different activities which never quite coincide; the thinker who wants the world to know the "content" of his thoughts must first of all stop thinking and remember his thoughts. Remembrance in this, as in all other cases, prepares the intangible and the futile for their eventual materialization; it is the beginning of the work process, and like the craftsman's consideration of the model which will guide his work, its most immaterial stage. The work itself then always requires some material upon which it will be performed and which through fabrication, the activity of *homo faber*, will be transformed into a worldly object. The specific work quality of intellectual work is no less due to the "work of our hands" than any other kind of work.

It seems plausible and is indeed quite common to connect and justify the modern distinction between intellectual and manual labor with the ancient distinction between "liberal" and "servile arts." Yet the distinguishing mark between liberal and servile arts is not at all "a higher degree of intelligence," or that the "liberal artist" works with his brain and the "sordid tradesman" with his hands. The ancient criterion is primarily political. Occupations involving *prudentia*, the capacity for prudent judgment which is the virtue of statesmen, and professions of public relevance *(ad hominum utilitatem)*[23] such as architecture, medicine, and agriculture,[24] are liberal. All trades, the trade of a scribe no less than that of a carpenter, are "sordid," unbecoming for a full-fledged citizen, and the worst are those we would deem most useful, such as "fishmongers, butchers, cooks, poulterers and fishermen."[25] But not even these are necessarily sheer laboring. There is still a third category where the toil and effort itself (the *operae* as distinguished from the *opus*, the mere activity as distinguished from the

23. For convenience' sake, I shall follow Cicero's discussion of liberal and servile occupations in *De officiis* i. 50–54. The criteria of *prudentia* and *utilitas* or *utilitas hominum* are stated in pars. 151 and 155. (The translation of *prudentia* as "a higher degree of intelligence" by Walter Miller in the Loeb Classical Library edition seems to me to be misleading.)

24. The classification of agriculture among the liberal arts is, of course, specifically Roman. It is not due to any special "usefulness" of farming as we would understand it, but much rather related to the Roman idea of *patria*, according to which the *ager Romanus* and not only the city of Rome is the place occupied by the public realm.

25. It is this usefulness for sheer living which Cicero calls *mediocris utilitas* (par. 151) and eliminates from liberal arts. The translation again seems to me to miss the point; these are not "professions . . . from which no small benefit to society is derived," but occupations which, in clear opposition to those mentioned before, transcend the vulgar usefulness of consumer goods.

work) is paid, and in these cases "the very wage is a pledge of slavery."[26]

The distinction between manual and intellectual work, though its origin can be traced back to the Middle Ages,[27] is modern and has two quite different causes, both of which, however, are equally characteristic of the general climate of the modern age. Since under modern conditions every occupation had to prove its "usefulness" for society at large, and since the usefulness of the intellectual occupations had become more than doubtful because of the modern glorification of labor, it was only natural that intellectuals, too, should desire to be counted among the working population. At the same time, however, and only in seeming contradiction to this development, the need and esteem of this society for certain "intellectual" performances rose to a degree unprecedented in our history except in the centuries of the decline of the Roman Empire. It may be well to remember in this context that throughout ancient history the "intellectual" services of the scribes, whether they served the needs of the public or the private realm, were performed by slaves and rated accordingly. Only the bureaucratization of the Roman Empire and the concomitant social and political rise of the Emperors brought a re-evaluation of "intellectual" services.[28] In so

26. The Romans deemed the difference between *opus* and *operae* to be so decisive that they had two different forms of contract, the *locatio operis* and the *locatio operarum*, of which the latter played an insignificant role because most laboring was done by slaves (see Edgar Loening, in *Handwörterbuch der Staatswissenschaften* [1890], I, 742 ff.).

27. The *opera liberalia* were identified with intellectual or rather spiritual work in the Middle Ages (see Otto Neurath, "Beiträge zur Geschichte der Opera Servilia," *Archiv für Sozialwissenschaft und Sozialpolitik*, Vol. XLI [1915], No. 2).

28. H. Wallon describes this process under the rule of Diocletian: "... les fonctions jadis serviles se trouvèrent anoblies, élevées au premier rang de l'État. Cette haute considération qui de l'empereur se répandait sur les premiers serviteurs du palais, sur les plus hauts dignitaires de l'empire, descendait à tous les degrés des fonctions publiques ... ; le service public devint un office public." "Les charges les plus serviles, ... les noms que nous avons cités aux fonctions de l'esclavage, sont revêtus de l'éclat qui rejaillit de la personne du prince" (*Histoire de l'esclavage dans l'antiquité* [1847], III, 126 and 131). Before this elevation of the services, the scribes had been classified with the watchmen of public buildings or even with the men who led the prize fighters down to the arena

far as the intellectual is indeed not a "worker"—who like all other workers, from the humblest craftsman to the greatest artist, is engaged in adding one more, if possible durable, thing to the human artifice—he resembles perhaps nobody so much as Adam Smith's "menial servant," although his function is less to keep the life process intact and provide for its regeneration than to care for the upkeep of the various gigantic bureaucratic machines whose processes consume their services and devour their products as quickly and mercilessly as the biological life process itself.[29]

12

THE THING-CHARACTER OF THE WORLD

The contempt for labor in ancient theory and its glorification in modern theory both take their bearing from the subjective attitude or activity of the laborer, mistrusting his painful effort or praising his productivity. The subjectivity of the approach may be more obvious in the distinction between easy and hard work, but we saw that at least in the case of Marx—who, as the greatest of modern labor theorists, necessarily provides a kind of touchstone in these discussions—labor's productivity is measured and gauged against the requirements of the life process for its own reproduction; it resides in the potential surplus inherent in human labor power, not in the quality or character of the things it produces. Similarly, Greek opinion, which ranked painters higher than sculptors, certainly did not rest upon a higher regard for paintings.[30] It seems

---

(*ibid.*, p. 171). It seems noteworthy that the elevation of the "intellectuals" coincided with the establishment of a bureaucracy.

29. "The labour of some of the most respectable orders in the society is, like that of menial servants, unproductive of any value," says Adam Smith and ranks among them "the whole army and navy," the "servants of the public," and the liberal professions, such as "churchmen, lawyers, physicians, men of letters of all kinds." Their work, "like the declamation of the actors, the harangue of the orator, or the tune of the musician . . . perishes in the very instant of its production" (*op. cit.*, I, 295–96). Obviously, Smith would not have had any difficulty classifying our "white-collar jobs."

30. On the contrary, it is doubtful whether any painting was ever as much admired as Phidias' statue of Zeus at Olympia, whose magical power was cred-

that the distinction between labor and work, which our theorists have so obstinately neglected and our languages so stubbornly preserved, indeed becomes merely a difference in degree if the worldly character of the produced thing—its location, function, and length of stay in the world—is not taken into account. The distinction between a bread, whose "life expectancy" in the world is hardly more than a day, and a table, which may easily survive generations of men, is certainly much more obvious and decisive than the difference between a baker and a carpenter.

The curious discrepancy between language and theory which we noted at the outset therefore turns out to be a discrepancy between the world-oriented, "objective" language we speak and the man-oriented, subjective theories we use in our attempts at understanding. It is language, and the fundamental human experiences underlying it, rather than theory, that teaches us that the things of the world, among which the *vita activa* spends itself, are of a very different nature and produced by quite different kinds of activities. Viewed as part of the world, the products of work—and not the products of labor—guarantee the permanence and durability without which a world would not be possible at all. It is within this world of durable things that we find the consumer goods through which life assures the means of its own survival. Needed by our bodies and produced by its laboring, but without stability of their own, these things for incessant consumption appear and disappear in an environment of things that are not consumed but used, and to which, as we use them, we become used and accustomed. As such, they give rise to the familiarity of the world, its customs and habits of intercourse between men and things as well as between men and men. What consumer goods are for the life of man, use objects are for his world. From them, consumer goods derive their thing-character; and language, which does not permit the laboring activity to form anything so solid and non-verbal as a noun, hints at the strong probability that we would not even know what a thing is without having before us "the work of our hands."

Distinguished from both, consumer goods and use objects, there

---

ited to make one forget all trouble and sorrow; whoever had not seen it had lived in vain, etc.

are finally the "products" of action and speech, which together constitute the fabric of human relationships and affairs. Left to themselves, they lack not only the tangibility of other things, but are even less durable and more futile than what we produce for consumption. Their reality depends entirely upon human plurality, upon the constant presence of others who can see and hear and therefore testify to their existence. Acting and speaking are still outward manifestations of human life, which knows only one activity that, though related to the exterior world in many ways, is not necessarily manifest in it and needs neither to be seen nor heard nor used nor consumed in order to be real: the activity of thought.

Viewed, however, in their worldliness, action, speech, and thought have much more in common than any one of them has with work or labor. They themselves do not "produce," bring forth anything, they are as futile as life itself. In order to become worldly things, that is, deeds and facts and events and patterns of thoughts or ideas, they must first be seen, heard, and remembered and then transformed, reified as it were, into things—into sayings of poetry, the written page or the printed book, into paintings or sculpture, into all sorts of records, documents, and monuments. The whole factual world of human affairs depends for its reality and its continued existence, first, upon the presence of others who have seen and heard and will remember, and, second, on the transformation of the intangible into the tangibility of things. Without remembrance and without the reification which remembrance needs for its own fulfilment and which makes it, indeed, as the Greeks held, the mother of all arts, the living activities of action, speech, and thought would lose their reality at the end of each process and disappear as though they never had been. The materialization they have to undergo in order to remain in the world at all is paid for in that always the "dead letter" replaces something which grew out of and for a fleeting moment indeed existed as the "living spirit." They must pay this price because they themselves are of an entirely unworldly nature and therefore need the help of an activity of an altogether different nature; they depend for their reality and materialization upon the same workmanship that builds the other things in the human artifice.

The reality and reliability of the human world rest primarily on the fact that we are surrounded by things more permanent than the activity by which they were produced, and potentially even more permanent than the lives of their authors. Human life, in so far as it is world-building, is engaged in a constant process of reification, and the degree of worldliness of produced things, which all together form the human artifice, depends upon their greater or lesser permanence in the world itself.

13

LABOR AND LIFE

The least durable of tangible things are those needed for the life process itself. Their consumption barely survives the act of their production; in the words of Locke, all those "good things" which are "really useful to the life of man," to the "necessity of subsisting," are "generally of short duration, such as—if they are not consumed by use—will decay and perish by themselves."[31] After a brief stay in the world, they return into the natural process which yielded them either through absorption into the life process of the human animal or through decay; in their man-made shape, through which they acquired their ephemeral place in the world of man-made things, they disappear more quickly than any other part of the world. Considered in their worldliness, they are the least worldly and at the same time the most natural of all things. Although they are man-made, they come and go, are produced and consumed, in accordance with the ever-recurrent cyclical movement of nature. Cyclical, too, is the movement of the living organism, the human body not excluded, as long as it can withstand the process that permeates its being and makes it alive. Life is a process that everywhere uses up durability, wears it down, makes it disappear, until eventually dead matter, the result of small, single, cyclical, life processes, returns into the over-all gigantic circle of nature herself, where no beginning and no end exist and where all natural things swing in changeless, deathless repetition.

Nature and the cyclical movement into which she forces all living things know neither birth nor death as we understand them. The birth and death of human beings are not simple natural oc-

31. Locke, *op. cit.*, sec. 46.

currences, but are related to a world into which single individuals, unique, unexchangeable, and unrepeatable entities, appear and from which they depart. Birth and death presuppose a world which is not in constant movement, but whose durability and relative permanence makes appearance and disappearance possible, which existed before any one individual appeared into it and will survive his eventual departure. Without a world into which men are born and from which they die, there would be nothing but changeless eternal recurrence, the deathless everlastingness of the human as of all other animal species. A philosophy of life that does not arrive, as did Nietzsche, at the affirmation of "eternal recurrence" (*ewige Wiederkehr*) as the highest principle of all being, simply does not know what it is talking about.

The word "life," however, has an altogether different meaning if it is related to the world and meant to designate the time interval between birth and death. Limited by a beginning and an end, that is, by the two supreme events of appearance and disappearance within the world, it follows a strictly linear movement whose very motion nevertheless is driven by the motor of biological life which man shares with other living things and which forever retains the cyclical movement of nature. The chief characteristic of this specifically human life, whose appearance and disappearance constitute worldly events, is that it is itself always full of events which ultimately can be told as a story, establish a biography; it is of this life, *bios* as distinguished from mere *zōē*, that Aristotle said that it "somehow is a kind of *praxis*."[32] For action and speech, which, as we saw before, belonged close together in the Greek understanding of politics, are indeed the two activities whose end result will always be a story with enough coherence to be told, no matter how accidental or haphazard the single events and their causation may appear to be.

It is only within the human world that nature's cyclical movement manifests itself as growth and decay. Like birth and death, they, too, are not natural occurrences, properly speaking; they have no place in the unceasing, indefatigable cycle in which the whole household of nature swings perpetually. Only when they enter the man-made world can nature's processes be characterized

32. *Politics* 1254a7.

by growth and decay; only if we consider nature's products, this tree or this dog, as individual things, thereby already removing them from their "natural" surroundings and putting them into our world, do they begin to grow and to decay. While nature manifests itself in human existence through the circular movement of our bodily functions, she makes her presence felt in the man-made world through the constant threat of overgrowing or decaying it. The common characteristic of both, the biological process in man and the process of growth and decay in the world, is that they are part of the cyclical movement of nature and therefore endlessly repetitive; all human activities which arise out of the necessity to cope with them are bound to the recurring cycles of nature and have in themselves no beginning and no end, properly speaking; unlike *working*, whose end has come when the object is finished, ready to be added to the common world of things, *laboring* always moves in the same circle, which is prescribed by the biological process of the living organism and the end of its "toil and trouble" comes only with the death of this organism.[33]

When Marx defined labor as "man's metabolism with nature,"

33. In the earlier literature on labor up to the last third of the nineteenth century, it was not uncommon to insist on the connection between labor and the cyclical movement of the life process. Thus, Schulze-Delitzsch, in a lecture *Die Arbeit* (Leipzig, 1863), begins with a description of the cycle of desire-effort-satisfaction—"Beim letzten Bissen fängt schon die Verdauung an." However, in the huge post-Marxian literature on the labor problem, the only author who emphasizes and theorizes about this most elementary aspect of the laboring activity is Pierre Naville, whose *La vie de travail et ses problèmes* (1954) is one of the most interesting and perhaps the most original recent contribution. Discussing the particular traits of the workday as distinguished from other measurement of labor time, he says as follows: "Le trait principal est son caractère cyclique ou rythmique. Ce caractère est lié à la fois à l'esprit naturel et cosmologique de la journée ... et au caractère des fonctions physiologiques de l'être humain, qu'il a en commun avec les espèces animales supérieures. ... Il est évident que le travail devait être de prime abord lié à des rythmes et fonctions naturels." From this follows the cyclical character in the expenditure and reproduction of labor power that determines the time unit of the workday. Naville's most important insight is that the time character of human life, inasmuch as it is not merely part of the life of the species, stands in stark contrast to the cyclical time character of the workday. "Les limites naturelles supérieures de la vie ... ne sont pas dictées, comme celle de la journée, par la nécessité et la possibilité de se reproduire, mais au contraire, par l'impossibilité de se renouveler,

in whose process "nature's material [is] adapted by a change of form to the wants of man," so that "labour has incorporated itself with its subject," he indicated clearly that he was "speaking physiologically" and that labor and consumption are but two stages of the ever-recurring cycle of biological life.[34] This cycle needs to be sustained through consumption, and the activity which provides the means of consumption is laboring.[35] Whatever labor produces is meant to be fed into the human life process almost immediately, and this consumption, regenerating the life process, produces—or rather, reproduces—new "labor power," needed for the further sustenance of the body.[36] From the viewpoint of the exigencies of

---

sinon à l'échelle de l'espèce. Le cycle s'accomplit en une fois, et ne se renouvelle pas" (pp. 19–24).

34. *Capital* (Modern Library ed.), p. 201. This formula is frequent in Marx's work and always repeated almost *verbatim:* Labor is the eternal natural necessity to effect the metabolism between man and nature. (See, for instance, *Das Kapital*, Vol. I, Part 1, ch. 1, sec. 2, and Part 3, ch. 5. The standard English translation, Modern Library ed., pp. 50, 205, falls short of Marx's precision.) We find almost the same formulation in Vol. III of *Das Kapital*, p. 872. Obviously, when Marx speaks as he frequently does of the "life process of society," he is not thinking in metaphors.

35. Marx called labor "productive consumption" (*Capital* [Modern Library ed.], p. 204) and never lost sight of its being a physiological condition.

36. Marx's whole theory hinges on the early insight that the laborer first of all reproduces his own life by producing his means of subsistence. In his early writings he thought "that men begin to distinguish themselves from animals when they begin to produce their means of subsistence" (*Deutsche Ideologie*, p. 10). This indeed is the very content of the definition of man as *animal laborans*. It is all the more noteworthy that in other passages Marx is not satisfied with this definition because it does not distinguish man sharply enough from animals. "A spider conducts operations that resemble those of a weaver, and a bee puts to shame many an architect in the construction of her cells. But what distinguishes the worst architect from the best of bees is this, that the architect raises his structure in imagination before he erects it in reality. At the end of every labour-process, we get a result that already existed in the imagination of the labourer at its commencement" (*Capital* [Modern Library ed.], p. 198). Obviously, Marx no longer speaks of labor, but of work—with which he is not concerned; and the best proof of this is that the apparently all-important element of "imagination" plays no role whatsoever in his labor theory. In the third volume of *Das Kapital* he repeats that surplus labor beyond immediate needs serves the "progressive extension of the reproduction process" (pp. 872, 278). Despite occasional hesi-

the life process itself, the "necessity of subsisting," as Locke put it, laboring and consuming follow each other so closely that they almost constitute one and the same movement, which is hardly ended when it must be started all over again. The "necessity of subsisting" rules over both labor and consumption, and labor, when it incorporates, "gathers," and bodily "mixes with" the things provided by nature,[37] does actively what the body does even more intimately when it consumes its nourishment. Both are devouring processes that seize and destroy matter, and the "work" done by labor upon its material is only the preparation for its eventual destruction.

This destructive, devouring aspect of the laboring activity, to be sure, is visible only from the standpoint of the world and in distinction from work, which does not prepare matter for incorporation but changes it into material in order to work upon it and use the finished product. From the viewpoint of nature, it is work rather than labor that is destructive, since the work process takes matter out of nature's hands without giving it back to her in the swift course of the natural metabolism of the living body.

Equally bound up with the recurring cycles of natural movements, but not quite so urgently imposed upon man by "the condition of human life" itself,[38] is the second task of laboring—its constant, unending fight against the processes of growth and decay through which nature forever invades the human artifice, threatening the durability of the world and its fitness for human use. The protection and preservation of the world against natural processes are among the toils which need the monotonous performance of daily repeated chores. This laboring fight, as distinguished from the essentially peaceful fulfilment in which labor obeys the orders of immediate bodily needs, although it may be even less "productive" than man's direct metabolism with nature, has a much closer connection with the world, which it defends against

---

tations, Marx remained convinced that "Milton produced *Paradise Lost* for the same reason a silk worm produces silk" (*Theories of Surplus Value* [London, 1951], p. 186).

37. Locke, *op. cit.*, secs. 46, 26, and 27, respectively.

38. *Ibid.*, sec. 34.

nature. In old tales and mythological stories it has often assumed the grandeur of heroic fights against overwhelming odds, as in the account of Hercules, whose cleaning of the Augean stables is among the twelve heroic "labors." A similar connotation of heroic deeds requiring great strength and courage and performed in a fighting spirit is manifest in the medieval use of the word: labor, *travail, arebeit.* However, the daily fight in which the human body is engaged to keep the world clean and prevent its decay bears little resemblance to heroic deeds; the endurance it needs to repair every day anew the waste of yesterday is not courage, and what makes the effort painful is not danger but its relentless repetition. The Herculean "labors" share with all great deeds that they are unique; but unfortunately it is only the mythological Augean stable that will remain clean once the effort is made and the task achieved.

14

LABOR AND FERTILITY

The sudden, spectacular rise of labor from the lowest, most despised position to the highest rank, as the most esteemed of all human activities, began when Locke discovered that labor is the source of all property. It followed its course when Adam Smith asserted that labor was the source of all wealth and found its climax in Marx's "system of labor,"[39] where labor became the source of all productivity and the expression of the very humanity of man. Of the three, however, only Marx was interested in labor as such; Locke was concerned with the institution of private property as the root of society and Smith wished to explain and to secure the unhampered progress of a limitless accumulation of wealth. But all three, though Marx with greatest force and consistency, held that labor was considered to be the supreme world-building capacity of man, and since labor actually is the most natural and least worldly of man's activities, each of them, and again none more than Marx, found himself in the grip of certain genuine contradictions. It seems to lie in the very nature of this matter that

39. The expression is Karl Dunkmann's (*Soziologie der Arbeit* [1933], p. 71), who rightly remarks that the title of Marx's great work is a misnomer and should better have been called *System der Arbeit.*

the most obvious solution of these contradictions, or rather the most obvious reason why these great authors should have remained unaware of them is their equation of work with labor, so that labor is endowed by them with certain faculties which only work possesses. This equation always leads into patent absurdities, though they usually are not so neatly manifest as in the following sentence of Veblen: "The lasting evidence of productive labor is its material product—commonly some article of consumption,"[40] where the "lasting evidence" with which he begins, because he needs it for the alleged productivity of labor, is immediately destroyed by the "consumption" of the product with which he ends, forced, as it were, by the factual evidence of the phenomenon itself.

Thus Locke, in order to save labor from its manifest disgrace of producing only "things of short duration," had to introduce money —a "lasting thing which men may keep without spoiling"—a kind of *deus ex machina* without which the laboring body, in its obedience to the life process, could never have become the origin of anything so permanent and lasting as property, because there are no "durable things" to be kept to survive the activity of the laboring process. And even Marx, who actually defined man as an *animal laborans*, had to admit that productivity of labor, properly speaking, begins only with reification (*Vergegenständlichung*), with "the erection of an objective world of things" (*Erzeugung einer gegenständlichen Welt*).[41] But the effort of labor never frees the labor-

40. The curious formulation occurs in Thorstein Veblen, *The Theory of the Leisure Class* (1917), p. 44.

41. The term *vergegenständlichen* occurs not very frequently in Marx, but always in a crucial context. Cf. *Jugendschriften*, p. 88: "Das praktische Erzeugen einer gegenständlichen Welt, die Bearbeitung der unorganischen Natur ist die Bewährung des Menschen als eines bewussten Gattungswesens. ... [Das Tier] produziert unter der Herrschaft des unmittelbaren Bedürfnisses, während der Mensch selbst frei vom physischen Bedürfnis produziert und erst wahrhaft produziert in der Freiheit von demselben." Here, as in the passage from *Capital* quoted in note 36, Marx obviously introduces an altogether different concept of labor, that is, speaks about work and fabrication. The same reification is mentioned in *Das Kapital* (Vol. I, Part 3, ch. 5), though somewhat equivocally: "[Die Arbeit] ist vergegenständlicht und der Gegenstand ist verarbeitet." The play on words with the term *Gegenstand* obscures what actually happens in the process: through reification, a new thing has been produced, but the "object" that this process transformed into a thing is, from the viewpoint of the process,

ing animal from repeating it all over again and remains therefore an "eternal necessity imposed by nature."[42] When Marx insists that the labor "process comes to its end in the product,"[43] he forgets his own definition of this process as the "metabolism between man and nature" into which the product is immediately "incorporated," consumed, and annihilated by the body's life process.

Since neither Locke nor Smith is concerned with labor as such, they can afford to admit certain distinctions which actually would amount to a distinction in principle between labor and work, if it were not for an interpretation that treats of the genuine traits of laboring as merely irrelevant. Thus, Smith calls "unproductive labor" all activities connected with consumption, as though this were a negligible and accidental trait of something whose true nature was to be productive. The very contempt with which he describes how "menial tasks and services generally perish in the instant of their performance and seldom leave any trace or value behind them"[44] is much more closely related to premodern opinion on this matter than to its modern glorification. Smith and Locke were still quite aware of the fact that not every kind of labor "puts the difference of value on everything"[45] and that there exists a kind of activity which adds nothing "to the value of the materials which [it] works upon."[46] To be sure, labor, too, joins to nature something of man's own, but the proportion between what nature gives— the "good things"—and what man adds is the very opposite in the products of labor and the products of work. The "good things" for consumption never lose their naturalness altogether, and the grain never quite disappears in the bread as the tree has disappeared in the table. Thus, Locke, although he paid little attention to his own distinction between "the labour of our body and the work of our

---

only material and not a thing. (The English translation, Modern Library ed., p. 201, misses the meaning of the German text and therefore escapes the equivocality.)

42. This is a recurrent formulation in Marx's works. See, for instance, *Das Kapital*, Vol. I (Modern Library ed., p. 50) and Vol. III, pp. 873–74.

43. "Des Prozess erlischt im Produkt" (*Das Kapital*, Vol. I, Part 3, ch. 5).

44. Adam Smith, *op. cit.*, I, 295.

45. Locke, *op. cit.*, sec. 40.   46. Adam Smith, *op. cit.*, I, 294.

hands," had to acknowledge the distinction between things "of short duration" and those "lasting" long enough "that men might keep them without spoiling."[47] The difficulty for Smith and Locke was the same; their "products" had to stay long enough in the world of tangible things to become "valuable," whereby it is immaterial whether value is defined by Locke as something which can be kept and becomes property or by Smith as something which lasts long enough to be exchangeable for something else.

These certainly are minor points if compared with the fundamental contradiction which runs like a red thread through the whole of Marx's thought, and is present no less in the third volume of *Capital* than in the writings of the young Marx. Marx's attitude toward labor, and that is toward the very center of his thought, has never ceased to be equivocal.[48] While it was an "eternal necessity imposed by nature" and the most human and productive of man's activities, the revolution, according to Marx, has not the task of emancipating the laboring classes but of emancipating man from labor; only when labor is abolished can the "realm of freedom" supplant the "realm of necessity." For "the realm of freedom begins only where labor determined through want and external utility ceases," where "the rule of immediate physical needs" ends.[49] Such fundamental and flagrant contradictions rarely occur

47. *Op. cit.*, secs. 46 and 47.

48. Jules Vuillemin's *L'être et le travail* (1949) is a good example of what happens if one tries to resolve the central contradictions and equivocalities of Marx's thoughts. This is possible only if one abandons the phenomenal evidence altogether and begins to treat Marx's concepts as though they constituted in themselves a complicated jigsaw puzzle of abstractions. Thus, labor "springs apparently from necessity" but "actually realizes the work of liberty and affirms our power"; in labor "necessity expresses [for man] a hidden freedom" (pp. 15, 16). Against these attempts at a sophisticated vulgarization, one may remember Marx's own sovereign attitude toward his work as Kautsky reports it in the following anecdote: Kautsky asked Marx in 1881 if he did not contemplate an edition of his complete works, whereupon Marx replied: "These works must first be written" (Kautsky, *Aus der Frühzeit des Marxmismus* [1935], p. 53).

49. *Das Kapital*, III, 873. In the *Deutsche Ideologie* Marx states that "die kommunistische Revolution . . . die Arbeit beseitigt" (p. 59), after having stated some pages earlier (p. 10) that only through labor does man distinguish himself from animals.

in second-rate writers; in the work of the great authors they lead into the very center of their work. In the case of Marx, whose loyalty and integrity in describing phenomena as they presented themselves to his view cannot be doubted, the important discrepancies in his work, noted by all Marx scholars, can neither be blamed upon the difference "between the scientific point of view of the historian and the moral point of view of the prophet"[50] nor on a dialectical movement which needs the negative, or evil, to produce the positive, or good. The fact remains that in all stages of his work he defines man as an *animal laborans* and then leads him into a society in which this greatest and most human power is no longer necessary. We are left with the rather distressing alternative between productive slavery and unproductive freedom.

Thus, the question arises why Locke and all his successors, their own insights notwithstanding, clung so obstinately to labor as the origin of property, of wealth, of all values and, finally, of the very humanity of man. Or, to put it another way, what were the experiences inherent in the laboring activity that proved of such great importance to the modern age?

Historically, political theorists from the seventeenth century onward were confronted with a hitherto unheard-of process of growing wealth, growing property, growing acquisition. In the attempt to account for this steady growth, their attention was naturally drawn to the phenomenon of a progressing process itself, so that, for reasons we shall have to discuss later,[51] the concept of process became the very key term of the new age as well as the sciences, historical and natural, developed by it. From its beginning, this process, because of its apparent endlessness, was understood as a natural process and more specifically in the image of the life process itself. The crudest superstition of the modern age— that "money begets money"—as well as its sharpest political insight—that power generates power—owes its plausibility to the underlying metaphor of the natural fertility of life. Of all human activities, only labor, and neither action nor work, is unending,

50. The formulation is Edmund Wilson's in *To the Finland Station* (Anchor ed., 1953), but this criticism is familiar in Marxian literature.

51. See ch. vi, § 42, below.

progressing automatically in accordance with life itself and outside the range of wilful decisions or humanly meaningful purposes.

Perhaps nothing indicates more clearly the level of Marx's thought and the faithfulness of his descriptions to phenomenal reality than that he based his whole theory on the understanding of laboring and begetting as two modes of the same fertile life process. Labor was to him the "reproduction of one's own life" which assured the survival of the individual, and begetting was the production "of foreign life" which assured the survival of the species.[52] This insight is chronologically the never-forgotten origin of his theory, which he then elaborated by substituting for "abstract labor" the labor power of a living organism and by understanding labor's surplus as that amount of labor power still extant after the means for the laborer's own reproduction have been produced. With it, he sounded a depth of experience reached by none of his predecessors—to whom he otherwise owed almost all his decisive inspirations—and none of his successors. He squared his theory, the theory of the modern age, with the oldest and most persistent insights into the nature of labor, which, according to the Hebrew as well as the classical tradition, was as intimately bound up with life as giving birth. By the same token, the true meaning of labor's newly discovered productivity becomes manifest only in Marx's work, where it rests on the equation of productivity with fertility, so that the famous development of mankind's "productive forces" into a society of an abundance of "good things" actually obeys no other law and is subject to no other necessity than the aboriginal command, "Be ye fruitful and multiply," in which it is as though the voice of nature herself speaks to us.

The fertility of the human metabolism with nature, growing out of the natural redundancy of labor power, still partakes of the superabundance we see everywhere in nature's household. The "blessing or the joy" of labor is the human way to experience the sheer bliss of being alive which we share with all living creatures, and it is even the only way men, too, can remain and swing contentedly in nature's prescribed cycle, toiling and resting, laboring and consuming, with the same happy and purposeless regularity with which day and night and life and death follow each other.

52. *Deutsche Ideologie*, p. 17.

The reward of toil and trouble lies in nature's fertility, in the quiet confidence that he who in "toil and trouble" has done his part, remains a part of nature in the future of his children and his children's children. The Old Testament, which, unlike classical antiquity, held life to be sacred and therefore neither death nor labor to be an evil (and least of all an argument against life),[53] shows in the stories of the patriarchs how unconcerned about death their lives were, how they needed neither an individual, earthly immortality nor an assurance of the eternity of their souls, how death came to them in the familiar shape of night and quiet and eternal rest "in a good old age and full of years."

The blessing of life as a whole, inherent in labor, can never be found in work and should not be mistaken for the inevitably brief spell of relief and joy which follows accomplishment and attends achievement. The blessing of labor is that effort and gratification follow each other as closely as producing and consuming the means

53. Nowhere in the Old Testament is death "the wage of sin." Nor did the curse by which man was expelled from paradise punish him with labor and birth; it only made labor harsh and birth full of sorrow. According to Genesis, man (*adam*) had been created to take care and watch over the soil (*adamah*), as even his name, the masculine form of "soil," indicates (see Gen. 2:5, 15). "And *Adam* was not to till *adamah* . . . and He, God, created Adam of the dust of *adamah*. . . . He, God, took Adam and put him into the garden of Eden to till and to watch it" (I follow the translation of Martin Buber and Franz Rosenzweig, *Die Schrift* [Berlin, n.d.]). The word for "tilling" which later became the word for laboring in Hebrew, *leawod*, has the connotation of "to serve." The curse (3:17–19) does not mention this word, but the meaning is clear: the service for which man was created now became servitude. The current popular misunderstanding of the curse is due to an unconscious interpretation of the Old Testament in the light of Greek thinking. The misunderstanding is usually avoided by Catholic writers. See, for instance, Jacques Leclercq, *Leçons de droit naturel*, Vol. IV, Part 2, "Travail, Propriété," (1946), p. 31: "La peine du travail est le résultat du péché original. . . . L'homme non déchu eût travaillé dans la joie, mais il eût travaillé"; or J. Chr. Nattermann, *Die moderne Arbeit, soziologisch und theologisch betrachtet* (1953), p. 9. It is interesting in this context to compare the curse of the Old Testament with the seemingly similar explanation of the harshness of labor in Hesiod. Hesiod reports that the gods, in order to punish man, hid life from him (see n. 8) so that he had to search for it, while before, he apparently did not have to do anything but pluck the fruits of the earth from fields and trees. Here the curse consists not only in the harshness of labor but in labor itself.

of subsistence, so that happiness is a concomitant of the process it-self, just as pleasure is a concomitant of the functioning of a healthy body. The "happiness of the greatest number," into which we have generalized and vulgarized the felicity with which earthly life has always been blessed, conceptualized into an "ideal" the fundamental reality of a laboring humanity. The right to the pur-suit of this happiness is indeed as undeniable as the right to life; it is even identical with it. But it has nothing in common with good fortune, which is rare and never lasts and cannot be pursued, be-cause fortune depends on luck and what chance gives and takes, although most people in their "pursuit of happiness" run after good fortune and make themselves unhappy even when it befalls them, because they want to keep and enjoy luck as though it were an inexhaustible abundance of "good things." There is no lasting hap-piness outside the prescribed cycle of painful exhaustion and pleas-urable regeneration, and whatever throws this cycle out of balance —poverty and misery where exhaustion is followed by wretched-ness instead of regeneration, or great riches and an entirely effort-less life where boredom takes the place of exhaustion and where the mills of necessity, of consumption and digestion, grind an im-potent human body mercilessly and barrenly to death—ruins the elemental happiness that comes from being alive.

The force of life is fertility. The living organism is not ex-hausted when it has provided for its own reproduction, and its "surplus" lies in its potential multiplication. Marx's consistent naturalism discovered "labor power" as the specifically human mode of the life force which is as capable of creating a "surplus" as nature herself. Since he was almost exclusively interested in this process itself, the process of the "productive forces of society," in whose life, as in the life of every animal species, production and consumption always strike a balance, the question of a separate existence of worldly things, whose durability will survive and withstand the devouring processes of life, does not occur to him at all. From the viewpoint of the life of the species, all activities in-deed find their common denominator in laboring, and the only dis-tinguishing criterion left is the abundance or scarcity of the goods to be fed into the life process. When every thing has become an object for consumption, the fact that labor's surplus does not change the nature, the "short duration," of the products themselves loses all importance, and this loss is manifest in Marx's work in the contempt with which he treats the belabored distinctions of his predecessors between productive and unproductive, or skilled and unskilled labor.

The reason why Marx's predecessors were not able to rid themselves of these distinctions, which essentially are equivalent to the more fundamental distinction between work and labor, was not that they were less "scientific" but that they were still writing on the assumption of private property, or at least individual appropriation of national wealth. For the establishment of property, mere abundance can never be enough; labor's products do not become more durable by their abundance and cannot be "heaped up" and stored away to become part of a man's property; on the contrary, they are only too likely to disappear in the process of appropriation or to "perish uselessly" if they are not consumed "before they spoil."

15

THE PRIVACY OF PROPERTY
AND WEALTH

At first glance it must seem strange indeed that a theory which so conclusively ended in the abolition of all property should have taken its departure from the theoretical establishment of private property. This strangeness, however, is somewhat mitigated if we remember the sharply polemical aspect of the modern age's concern with property, whose rights were asserted explicitly against the common realm and against the state. Since no political theory prior to socialism and communism had proposed to establish an entirely propertyless society, and no government prior to the twentieth century had shown serious inclinations to expropriate its citizens, the content of the new theory could not possibly be prompted by the need to protect property rights against possible intrusion of government administration. The point is that then, unlike now when all property theories are obviously on the defensive, the economists were not on the defensive at all but on the contrary openly hostile to the whole sphere of government, which at best was considered a "necessary evil" and a "reflection on human nature,"[54] at worst a parasite on the otherwise healthy life of society.[55] What the modern age so heatedly defended was never property as such but the unhampered pursuit of more property or of appropriation; as against all organs that stood for the "dead" permanence of a common world, it fought its battles in the name of life, the life of society.

There is no doubt that, as the natural process of life is located in the body, there is no more immediately life-bound activity than laboring. Locke could neither remain satisfied with the traditional explanation of labor, according to which it is the natural and inevitable consequence of poverty and never a means of its abolition, nor with the traditional explanation of the origin of property through acquisition, conquest, or an original division of the common world.[56] What he actually was concerned with was appropria-

54. The writers of the modern age are all agreed that the "good" and "productive" side of human nature is reflected in society, while its wickedness makes government necessary. As Thomas Paine stated it: "Society is produced by our wants, and government by our wickedness; the former promotes our happiness positively by uniting our affections, the latter negatively by restraining our vices. . . . Society in every state is a blessing, but government, even in the best state, a necessary evil" (*Common Sense*, 1776). Or Madison: "But what is government itself but the greatest of all reflections on human nature? If men were angels, no government would be necessary. If angels were to govern men, neither external or internal controls would be necessary" (*The Federalist* [Modern Library ed.], p. 337).

55. This was the opinion of Adam Smith, for instance, who was very indignant about "the public extravagance of government": "The whole, or almost the whole public revenue, is in most countries employed in maintaining unproductive hands" (*op. cit.*, I, 306).

56. No doubt, "before 1690 no one understood that a man had a natural right to property created by his labour; after 1690 the idea came to be an axiom of social science" (Richard Schlatter, *Private Property: The History of an Idea* [1951], p. 156). The concept of labor and property was even mutually exclusive, whereas labor and poverty (*ponos* and *penia*, *Arbeit* and *Armut*) belonged together in the sense that the activity corresponding to the status of poverty was laboring. Plato, therefore, who held that laboring slaves were "bad" because they were not masters of the animal part within them, said almost the same about the status of poverty. The poor man is "not master of himself" (*penēs ōn kai heautou mē kratōn* [*Seventh Letter* 351A]). None of the classical writers ever thought of labor as a possible source of wealth. According to Cicero—and he probably only sums

tion and what he had to find was a world-appropriating activity whose privacy at the same time must be beyond doubt and dispute.

Nothing, to be sure, is more private than the bodily functions of the life process, its fertility not excluded, and it is quite noteworthy that the few instances where even a "socialized mankind" respects and imposes strict privacy concern precisely such "activities" as are imposed by the life process itself. Of these, labor, because it is an activity and not merely a function, is the least private, so to speak, the only one we feel need not be hidden; yet it is still close enough to the life process to make plausible the argument for the privacy of appropriation as distinguished from the very different argument for the privacy of property.[57] Locke founded private property on the most privately owned thing there is, "the property [of man] in his own person," that is, in his own body.[58] "The labour of our body and the work of our hands" become one and the same, because both are the "means" to "appropriate" what "God . . . hath given . . . to men in common." And these means, body and hands and mouth, are the natural appropriators because they do not "belong to mankind in common" but are given to each man for his private use.[59]

Just as Marx had to introduce a natural force, the "labor power" of the body, to account for labor's productivity and a progressing process of growing wealth, Locke, albeit less explicitly, had to trace property to a natural origin of appropriation in order to force open those stable, worldly boundaries that "enclose" each person's privately owned share of the world "from the common."[60] What Marx still had in common with Locke was that he wished to see the process of growing wealth as a natural process, automatically following its own laws and beyond wilful decisions and purposes. If any human activity was to be involved in the process at all, it could only be a bodily "activity" whose natural functioning could not be checked even if one wanted to do so. To check these "activi-

up contemporary opinion—property comes about either through ancient conquest or victory or legal division (*aut vetere occupatione aut victoria aut lege* [*De officiis* i. 21]).

57. See § 8 above.

58. *Op. cit.*, sec. 26.

59. *Ibid.*, sec. 25.

60. *Ibid.*, sec. 31.

ties" is indeed to destroy nature, and for the whole modern age, whether it holds fast to the institution of private property or considers it to be an impediment to the growth of wealth, a check or control of the process of wealth was equivalent to an attempt to destroy the very life of society.

The development of the modern age and the rise of society, where the most private of all human activities, laboring, has become public and been permitted to establish its own common realm, may make it doubtful whether the very existence of property as a privately held place within the world can withstand the relentless process of growing wealth. But it is true, nevertheless, that the very privacy of one's holdings, that is, their complete independence "from the common," could not be better guaranteed than by the transformation of property into appropriation or by an interpretation of the "enclosure from the common" which sees it as the result, the "product," of bodily activity. In this aspect, the body becomes indeed the quintessence of all property because it is the only thing one could not share even if one wanted to. Nothing, in fact, is less common and less communicable, and therefore more securely shielded against the visibility and audibility of the public realm, than what goes on within the confines of the body, its pleasures and its pains, its laboring and consuming. Nothing, by the same token, ejects one more radically from the world than exclusive concentration upon the body's life, a concentration forced upon man in slavery or in the extremity of unbearable pain. Whoever wishes, for whatever reason, to make human existence entirely "private," independent of the world and aware only of its own being alive, must rest his arguments on these experiences; and since the relentless drudgery of slave labor is not "natural" but man-made and in contradiction to the natural fertility of the *animal laborans*, whose strength is not exhausted and whose time is not consumed when it has reproduced his own life, the "natural" experience underlying the Stoic as well as the Epicurean independence of the world is not labor or slavery but pain. The happiness achieved in isolation from the world and enjoyed within the confines of one's own private existence can never be anything but the famous "absence of pain," a definition on which all variations of consistent sensualism must agree. Hedonism, the doctrine that only bodily sensations are real, is but the most radical form of a non-political, totally private way of life, the true fulfilment of Epicurus' *lathe biōsas kai mē politeuesthai* ("live in hiding and do not care about the world").

Normally, absence of pain is no more than the bodily condition for experiencing the world; only if the body is not irritated and, through irritation, thrown back upon itself, can our bodily senses function normally, receive what is given to them. Absence of pain is usually "felt" only in the short intermediate stage between pain and non-pain, and the sensation which corresponds to the sensualists' concept of happiness is release from pain rather than its absence. The intensity of this sensation is beyond doubt; it is, indeed, matched only by the sensation of pain itself.[61] The mental effort required by philosophies which for various reasons wish to "liberate" man from the world is always an act of imagination in which the mere absence of pain is experienced and actualized into a feeling of being released from it.[62]

61. It seems to me that certain types of mild and rather frequent drug addictions, which usually are blamed upon the habit-forming properties of drugs, might perhaps be due to the desire to repeat the once experienced pleasure of relief from pain with its intense feeling of euphoria. The phenomenon itself was well known in antiquity, whereas in modern literature I found the only support for my assumption in Isak Dinesen's "Converse at Night in Copenhagen" (*Last Tales* [1957], pp. 338 ff.), where she counts "cessation from pain" among the "three kinds of perfect happiness." Plato already argues against those who "when drawn away from pain firmly believe that they have reached the goal of . . . pleasure" (*Republic* 585A), but concedes that these "mixed pleasures" which follow pain or privation are more intense than the pure pleasures, such as smelling an exquisite aroma or contemplating geometrical figures. Curiously enough, it was the hedonists who confused the issue and did not want to admit that the pleasure of release from pain is greater in intensity than "pure pleasure," let alone mere absence of pain. Thus Cicero accused Epicurus of having confused mere absence of pain with the pleasure of release from it (see V. Brochard, *Études de philosophie ancienne et de philosophie moderne* [1912], pp. 252 ff.). And Lucretius exclaimed: "Do you not see that nature is clamouring for two things only, a body free from pain, a mind released from worry . . . ?" (*The Nature of the Universe* [Penguin ed.], p. 60).

62. Brochard (*op. cit.*) gives an excellent summary of the philosophers of late antiquity, especially of Epicurus. The way to unshaken sensual happiness lies in the soul's capacity "to escape into a happier world which it creates, so that

In any event, pain and the concomitant experience of release from pain are the only sense experiences that are so independent from the world that they do not contain the experience of any worldly object. The pain caused by a sword or the tickling caused by a feather indeed tells me nothing whatsoever of the quality or even the worldly existence of a sword or a feather.[63] Only an irresistible distrust in the capacity of human senses for an adequate experience of the world—and this distrust is the origin of all specifically modern philosophy—can explain the strange and even

---

with the help of imagination it can always persuade the body to experience the same pleasure which it once has known" (pp. 278 and 294 ff.).

63. It is characteristic of all theories that argue against the world-giving capacity of the senses that they remove vision from its position as the highest and most noble of the senses and substitute touch or taste, which are indeed the most private senses, that is, those in which the body primarily senses itself while perceiving an object. All thinkers who deny the reality of the outer world would have agreed with Lucretius, who said: "For touch and nothing but touch (by all that men call holy) is the essence of all our bodily sensations" (op. cit., p. 72). This, however, is not enough; touch or taste in a non-irritated body still give too much of the reality of the world: when I eat a dish of strawberries, I taste strawberries and not the taste itself, or, to take an example from Galileo, when "I pass a hand, first over a marble statue, then over a living man," I am aware of marble and a living body, and not primarily of my own hand that touches them. Galileo, therefore, when he wishes to demonstrate that the secondary qualities, such as colors, tastes, odors, are "nothing else than mere names [having] their residence solely in the sensitive body," has to give up his own example and introduce the sensation of being tickled by a feather, whereupon he concludes: "Of precisely a similar and not greater existence do I believe these various qualities to be possessed, which are attributed to natural bodies, such as tastes, odours, colours and others" (Il Saggiatore, in Opere, IV, 333 ff.; translation quoted from E. A. Burtt, Metaphysical Foundations of Modern Science [1932]).

This argument can base itself only upon sense experiences in which the body is clearly thrown back upon itself and therefore, as it were, ejected from the world in which it normally moves. The stronger the inner bodily sensation, the more plausible becomes the argument. Descartes in the same line of argument says as follows: "The motion merely of a sword cutting a part of our skin causes pain but does not on that account make us aware of the motion or the figure of the sword. And it is certain that this sensation of pain is not less different from the motion that causes it . . . than are the sensation we have of colour, sound, odour, or taste" (Principles, Part 4; translated by Haldane and Ross, Philosophical Works [1911]).

absurd choice that uses phenomena which, like pain or tickling, obviously prevent our senses' functioning normally, as examples of all sense experience, and can derive from them the subjectivity of "secondary" and even "primary" qualities. If we had no other sense perceptions than these in which the body senses itself, the reality of the outer world would not only be open to doubt, we would not even possess any notion of a world at all.

The only activity which corresponds strictly to the experience of worldlessness, or rather to the loss of world that occurs in pain, is laboring, where the human body, its activity notwithstanding, is also thrown back upon itself, concentrates upon nothing but its own being alive, and remains imprisoned in its metabolism with nature without ever transcending or freeing itself from the recurring cycle of its own functioning. We mentioned before the twofold pain connected with the life process for which language has but one word and which according to the Bible was imposed upon the life of man together, the painful effort involved in the reproduction of one's own life and the life of the species. If this painful effort of living and fertility were the true origin of property, then the privacy of this property would be indeed as worldless as the unequaled privacy of having a body and of experiencing pain.

This privacy, however, while it is essentially the privacy of appropriation, is by no means what Locke, whose concepts were still essentially those of the premodern tradition, understood by private property. No matter what its origin, this property was to him still an "enclosure from the common," that is, primarily a place in the world where that which is private can be hidden and protected against the public realm. As such, it remained in contact with the common world even at a time when growing wealth and appropriation began to threaten the common world with extinction. Property does not strengthen but rather mitigates the unrelatedness to the world of the laboring process, because of its own worldly security. By the same token, the process character of laboring, the relentlessness with which labor is urged and driven by the life process itself, is checked by the acquisition of property. In a society of property-owners, as distinguished from a society of laborers or jobholders, it is still the world, and neither natural abundance nor the sheer necessity of life, which stands at the center of human care and worry.

The matter becomes altogether different if the leading interest is no longer property but the growth of wealth and the process of accumulation as such. This process can be as infinite as the life process of the species, and its very infinity is constantly challenged and interrupted by the inconvenient fact that private individuals do not live forever and have no infinite time before them. Only if the life of society as a whole, instead of the limited lives of individual men, is considered to be the gigantic subject of the accumulation process can this process go on in full freedom and at full speed, unhampered by limitations imposed by the individual life-span and individually held property. Only when man no longer acts as an individual, concerned only with his own survival, but as a "member of the species," a *Gattungswesen* as Marx used to say, only when the reproduction of individual life is absorbed into the life process of man-kind, can the collective life process of a "socialized man-kind" follow its own "necessity," that is, its automatic course of fertility in the twofold sense of multiplication of lives and the increasing abundance of goods needed by them.

The coincidence of Marx's labor philosophy with the evolution and development theories of the nineteenth century—the natural evolution of a single life process from the lowest forms of organic life to the emergence of the human animal and the historical development of a life process of mankind as a whole—is striking and was early observed by Engels, who called Marx "the Darwin of history." What all these theories in the various sciences—economics, history, biology, geology—have in common is the concept of process, which was virtually unknown prior to the modern age. Since the discovery of processes by the natural sciences had coincided with the discovery of introspection in philosophy, it is only natural that the biological process within ourselves should eventually become the very model of the new concept; within the framework of experiences given to introspection, we know of no other process but the life process within our bodies, and the only activity into which we can translate it and which corresponds to it is labor.

Hence, it may seem almost inevitable that the equation of productivity with fertility in the labor philosophy of the modern age should have been succeeded by the different varieties of life philosophy which rest on the same equation.[64] The difference between the earlier labor theories and the later life philosophies is chiefly that the latter have lost sight of the only activity necessary to sustain the life process. Yet even this loss seems to correspond to the factual historical development which made labor more effortless than ever before and therefore even more similar to the automatically functioning life process. If at the turn of the century (with Nietzsche and Bergson) life and not labor was proclaimed to be "the creator of all values," this glorification of the sheer dynamism of the life process excluded that minimum of initiative present even in those activities which, like laboring and begetting, are urged upon man by necessity.

However, neither the enormous increase in fertility nor the socialization of the process, that is, the substitution of society or collective man-kind for individual men as its subject, can eliminate the character of strict and even cruel privacy from the experience of bodily processes in which life manifests itself, or from the activity of laboring itself. Neither abundance of goods nor the shortening of the time actually spent in laboring are likely to result in the establishment of a common world, and the expropriated *animal laborans* becomes no less private because he has been deprived of a private place of his own to hide and be protected from the common realm. Marx predicted correctly, though with an unjustified glee, "the withering away" of the public realm under conditions of unhampered development of the "productive forces of society," and he was equally right, that is, consistent with his conception of man as an *animal laborans*, when he foresaw that "socialized men" would

64. This connection was dimly perceived by Bergson's pupils in France (see esp. Édouard Berth, *Les méfaits des intellectuels* [1914], ch. 1, and Georges Sorel, *D'Aristote à Marx* [1935]). In the same school belongs the work of the Italian scholar Adriano Tilgher (*op. cit.*) who emphasizes that the idea of labor is central and constitutes the key to the new concept and image of life (English ed., p. 55). The school of Bergson, like its master, idealizes labor by equating it with work and fabrication. Yet the similarity between the motor of biological life and Bergson's *élan vital* is striking.

spend their freedom from laboring in those strictly private and essentially worldless activities that we now call "hobbies."[65]

16

THE INSTRUMENTS OF WORK AND
THE DIVISION OF LABOR

Unfortunately, it seems to be in the nature of the conditions of life as it has been given to man that the only possible advantage of the fertility of human labor power lies in its ability to procure the necessities of life for more than one man or one family. Labor's products, the products of man's metabolism with nature, do not stay in the world long enough to become a part of it, and the laboring activity itself, concentrated exclusively on life and its maintenance, is oblivious of the world to the point of worldlessness. The *animal laborans*, driven by the needs of its body, does not use this body freely as *homo faber* uses his hands, his primordial tools, which is why Plato suggested that laborers and slaves were not only subject to necessity and incapable of freedom but also unable to rule the "animal" part within them.[66] A mass society of laborers, such as Marx had in mind when he spoke of "socialized mankind," consists of worldless specimens of the species mankind, whether they are household slaves, driven into their predicament by the violence of others, or free, performing their functions willingly.

This worldlessness of the *animal laborans*, to be sure, is entirely different from the active flight from the publicity of the world which we found inherent in the activity of "good works." The *animal laborans* does not flee the world but is ejected from it in so far as he is imprisoned in the privacy of his own body, caught in the

65. In communist or socialist society, all professions would, as it were, become hobbies: there would be no painters but only people who among other things spend their time also on painting; people, that is, who "do this today and that tomorrow, who hunt in the morning, go fishing in the afternoon, raise cattle in the evening, are critics after dinner, as they see fit, without for that matter ever becoming hunters, fisherman, shepherds or critics" (*Deutsche Ideologie*, pp. 22 and 373).

66. *Republic* 590C.

fulfilment of needs in which nobody can share and which nobody can fully communicate. The fact that slavery and banishment into the household was, by and large, the social condition of all laborers prior to the modern age is primarily due to the human condition itself; life, which for all other animal species is the very essence of their being, becomes a burden to man because of his innate "repugnance to futility."[67] This burden is all the heavier since none of the so-called "loftier desires" has the same urgency, is actually forced upon man by necessity, as the elementary needs of life. Slavery became the social condition of the laboring classes because it was felt that it was the natural condition of life itself. *Omnis vita servitium est.*[68]

The burden of biological life, weighing down and consuming the specifically human life-span between birth and death, can be eliminated only by the use of servants, and the chief function of ancient slaves was rather to carry the burden of consumption in the household than to produce for society at large.[69] The reason why slave labor could play such an enormous role in ancient societies and why its wastefulness and unproductivity were not discovered is that the ancient city-state was primarily a "consumption center," unlike medieval cities which were chiefly production centers.[70] The price for the elimination of life's burden from the shoulders of all citizens was enormous and by no means consisted only in the violent injustice of forcing one part of humanity into the darkness of pain and necessity. Since this darkness is natural, inherent in the human condition—only the act of violence, when one group of men tries to rid itself of the shackles binding all of us to pain and necessity, is man-made—the price for absolute freedom from necessity

67. Veblen, *op. cit.*, p. 33.

68. Seneca *De tranquillitate animae* ii. 3.

69. See the excellent analysis in Winston Ashley, *The Theory of Natural Slavery, according to Aristotle and St. Thomas* (Dissertation, University of Notre Dame [1941], ch. 5), who rightly emphasizes: "It would be wholly to miss Aristotle's argument, therefore, to believe that he considered slaves as universally necessary merely as productive tools. He emphasizes rather their necessity for consumption."

70. Max Weber, "Agrarverhältnisse im Altertum," in *Gesammelte Aufsätze zur Sozial- und Wirtschaftsgeschichte* (1924), p. 13.

is, in a sense, life itself, or rather the substitution of vicarious life for real life. Under the conditions of slavery, the great of the earth could even use their senses vicariously, could "see and hear through their slaves," as the Greek idiom used by Herodotus expressed it.[71]

On its most elementary level the "toil and trouble" of obtaining and the pleasures of "incorporating" the necessities of life are so closely bound together in the biological life cycle, whose recurrent rhythm conditions human life in its unique and unilinear movement, that the perfect elimination of the pain and effort of labor would not only rob biological life of its most natural pleasures but deprive the specifically human life of its very liveliness and vitality. The human condition is such that pain and effort are not just symptoms which can be removed without changing life itself; they are rather the modes in which life itself, together with the necessity to which it is bound, makes itself felt. For mortals, the "easy life of the gods" would be a lifeless life.

For our trust in the reality of life and in the reality of the world is not the same. The latter derives primarily from the permanence and durability of the world, which is far superior to that of mortal life. If one knew that the world would come to an end with or soon after his own death, it would lose all its reality, as it did for the early Christians as long as they were convinced of the immediate fulfilment of their eschatological expectations. Trust in the reality of life, on the contrary, depends almost exclusively on the intensity with which life is felt, on the impact with which it makes itself felt. This intensity is so great and its force so elementary that wherever it prevails, in bliss or sorrow, it blacks out all other worldly reality. That the life of the rich loses in vitality, in closeness to the "good things" of nature, what it gains in refinement, in sensitivity to the beautiful things in the world, has often been noted. The fact is that the human capacity for life in the world always implies an

71. Herodotus i. 113 for instance: *eide te dia toutōn*, and *passim*. A similar expression occurs in Plinius, *Naturalis historia* xxix. 19: *alienis pedibus ambulamus; alienis oculis agnoscimus; aliena memoria salutamus; aliena vivimus opera* (quoted from R. H. Barrow, *Slavery in the Roman Empire* [1928], p. 26). "We walk with alien feet; we see with alien eyes; we recognize and greet people with an alien memory; we live from alien labor."

ability to transcend and to be alienated from the processes of life itself, while vitality and liveliness can be conserved only to the extent that men are willing to take the burden, the toil and trouble of life, upon themselves.

It is true that the enormous improvement in our labor tools—the mute robots with which *homo faber* has come to the help of the *animal laborans*, as distinguished from the human, speaking instruments (the *instrumentum vocale*, as the slaves in ancient households were called) whom the man of action had to rule and oppress when he wanted to liberate the *animal laborans* from its bondage—has made the twofold labor of life, the effort of its sustenance and the pain of giving birth, easier and less painful than it has ever been. This, of course, has not eliminated compulsion from the laboring activity or the condition of being subject to need and necessity from human life. But, in distinction from slave society, where the "curse" of necessity remained a vivid reality because the life of a slave testified daily to the fact that "life is slavery," this condition is no longer fully manifest and its lack of appearance has made it much more difficult to notice and remember. The danger here is obvious. Man cannot be free if he does not know that he is subject to necessity, because his freedom is always won in his never wholly successful attempts to liberate himself from necessity. And while it may be true that his strongest impulse toward this liberation comes from his "repugnance to futility," it is also likely that the impulse may grow weaker as this "futility" appears easier, as it requires less effort. For it is still probable that the enormous changes of the industrial revolution behind us and the even greater changes of the atomic revolution before us will remain changes of the world, and not changes in the basic condition of human life on earth.

Tools and instruments which can ease the effort of labor considerably are themselves not a product of labor but of work; they do not belong in the process of consumption but are part and parcel of the world of use objects. Their role, no matter how great it may be in the labor of any given civilization, can never attain the fundamental importance of tools for all kinds of work. No work can be produced without tools, and the birth of *homo faber* and the coming into being of a man-made world of things are actually coeval with the discovery of tools and instruments. From the standpoint of labor, tools strengthen and multiply human strength to the point of almost replacing it, as in all cases where natural forces, such as tame animals or water power or electricity, and not mere material things, are brought under a human master. By the same token, they increase the natural fertility of the *animal laborans* and provide an abundance of consumer goods. But all these changes are of a quantitative order, whereas the very quality of fabricated things, from the simplest use object to the masterwork of art, depends intimately on the existence of adequate instruments.

Moreover, the limitations of instruments in the easing of life's labor—the simple fact that the services of one servant can never be fully replaced by a hundred gadgets in the kitchen and half a dozen robots in the cellar—are of a fundamental nature. A curious and unexpected testimony to this is that it could be predicted thousands of years before the fabulous modern development of tools and machines had taken place. In a half-fanciful, half-ironical mood, Aristotle once imagined what has long since become a reality, namely that "every tool could perform its own work when ordered . . . like the statues of Daedalus or the tripods of Hephaestus, which, says the poet, 'of their own accord entered the assembly of the gods.' " Then, "the shuttle would weave and the plectrum touch the lyre without a hand to guide them." This, he goes on to say, would indeed mean that the craftsman would no longer need human assistants, but it would not mean that household slaves could be dispensed with. For slaves are not instruments of making things or of production, but of living, which constantly consumes their services.[72] The process of making a thing is limited and the function of the instrument comes to a predictable, controllable end with the finished product; the process of life that requires laboring is an endless activity and the only "instrument" equal to it would have to be a *perpetuum mobile*, that is, the *instrumentum vocale* which is as alive and "active" as the living organism which it serves. It is precisely because from "the instruments of the household nothing else results except the use of the possession itself" that they cannot be replaced by tools and instruments of workmanship "from which results something more than the mere use of the instrument."[73]

72. Aristotle *Politics* 1253b30–1254a18.
73. Winston Ashley, *op. cit.*, ch. 5.

Labor

While tools and instruments, designed to produce more and something altogether different from their mere use, are of secondary importance for laboring, the same is not true for the other great principle in the human labor process, the division of labor. Division of labor indeed grows directly out of the laboring process and should not be mistaken for the apparently similar principle of specialization which prevails in working processes and with which it is usually equated. Specialization of work and division of labor have in common only the general principle of organization, which itself has nothing to do with either work or labor but owes its origin to the strictly political sphere of life, to the fact of man's capacity to act and to act together and in concert. Only within the framework of political organization, where men not merely live, but act, together, can specialization of work and division of labor take place.

Yet, while specialization of work is essentially guided by the finished product itself, whose nature it is to require different skills which then are pooled and organized together, division of labor, on the contrary, presupposes the qualitative equivalence of all single activities for which no special skill is required, and these activities have no end in themselves, but actually represent only certain amounts of labor power which are added together in a purely quantitative way. Division of labor is based on the fact that two men can put their labor power together and "behave toward each other as though they were one."[74] This one-ness is the exact opposite of co-operation, it indicates the unity of the species with regard to which every single member is the same and exchangeable. (The formation of a labor collective where the laborers are socially organized in accordance with this principle of common and divisible labor power is the very opposite of the various workmen's organizations, from the old guilds and corporations to certain types of modern trade unions, whose members are bound together by the skills and specializations that distinguish them from others.) Since

74. See Viktor von Weizsäcker, "Zum Begriff der Arbeit," in *Festschrift für Alfred Weber* (1948), p. 739. The essay is noteworthy for certain scattered observations, but on the whole unfortunately useless, since Weizsäcker further obscures the concept of labor by the rather gratuitous assumption that the sick human being has to "perform labor" in order to get well.

none of the activities into which the process is divided has an end in itself, their "natural" end is exactly the same as in the case of "undivided" labor: either the simple reproduction of the means of subsistence, that is, the capacity for consumption of the laborers, or the exhaustion of human labor power. Neither of these two limitations, however, is final; exhaustion is part of the individual's, not of the collective's, life process, and the subject of the laboring process under the conditions of division of labor is a collective labor force, not individual labor power. The inexhaustibility of this labor force corresponds exactly to the deathlessness of the species, whose life process as a whole is also not interrupted by the individual births and deaths of its members.

More serious, it seems, is the limitation imposed by the capacity to consume, which remains bound to the individual even when a collective labor force has replaced individual labor power. The progress of accumulation of wealth may be limitless in a "socialized mankind" which has rid itself of the limitations of individual property and overcome the limitation of individual appropriation by dissolving all stable wealth, the possession of "heaped up" and "stored away" things, into money to spend and consume. We already live in a society where wealth is reckoned in terms of earning and spending power, which are only modifications of the two-fold metabolism of the human body. The problem therefore is how to attune individual consumption to an unlimited accumulation of wealth.

Since mankind as a whole is still very far from having reached the limit of abundance, the mode in which society may overcome this natural limitation of its own fertility can be perceived only tentatively and on a national scale. There, the solution seems to be simple enough. It consists in treating all use objects as though they were consumer goods, so that a chair or a table is now consumed as rapidly as a dress and a dress used up almost as quickly as food. This mode of intercourse with the things of the world, moreover, is perfectly adequate to the way they are produced. The industrial revolution has replaced all workmanship with labor, and the result has been that the things of the modern world have become labor products whose natural fate is to be consumed, instead of work products which are there to be used. Just as tools and instruments, though originating from work, were always employed in labor processes as well, so the division of labor, entirely appropriate and attuned to the laboring process, has become one of the chief characteristics of modern work processes, that is, of the fabrication and production of use objects. Division of labor rather than increased mechanization has replaced the rigorous specialization formerly required for all workmanship. Workmanship is required only for the design and fabrication of models before they go into mass production, which also depends on tools and machinery. But mass production would, in addition, be altogether impossible without the replacement of workmen and specialization with laborers and the division of labor.

Tools and instruments ease pain and effort and thereby change the modes in which the urgent necessity inherent in labor once was manifest to all. They do not change the necessity itself; they only serve to hide it from our senses. Something similar is true of labor's products, which do not become more durable through abundance. The case is altogether different in the corresponding modern transformation of the work process by the introduction of the principle of division of labor. Here the very nature of work is changed and the production process, although it by no means produces objects for consumption, assumes the character of labor. Although machines have forced us into an infinitely quicker rhythm of repetition than the cycle of natural processes prescribed—and this specifically modern acceleration is only too apt to make us disregard the repetitive character of all laboring—the repetition and the endlessness of the process itself put the unmistakable mark of laboring upon it. This is even more evident in the use objects produced by these techniques of laboring. Their very abundance transforms them into consumer goods. The endlessness of the laboring process is guaranteed by the ever-recurrent needs of consumption; the endlessness of production can be assured only if its products lose their use character and become more and more objects of consumption, or if, to put it in another way, the rate of use is so tremendously accelerated that the objective difference between use and consumption, between the relative durability of use objects and the swift coming and going of consumer goods, dwindles to insignificance.

In our need for more and more rapid replacement of the worldly things around us, we can no longer afford to use them, to respect and preserve their inherent durability; we must consume, devour, as it were, our houses and furniture and cars as though they were the "good things" of nature which spoil uselessly if they are not drawn swiftly into the never-ending cycle of man's metabolism with nature. It is as though we had forced open the distinguishing boundaries which protected the world, the human artifice, from nature, the biological process which goes on in its very midst as well as the natural cyclical processes which surround it, delivering and abandoning to them the always threatened stability of a human world.

The ideals of *homo faber*, the fabricator of the world, which are permanence, stability, and durability, have been sacrificed to abundance, the ideal of the *animal laborans*. We live in a laborers' society because only laboring, with its inherent fertility, is likely to bring about abundance; and we have changed work into laboring, broken it up into its minute particles until it has lent itself to division where the common denominator of the simplest performance is reached in order to eliminate from the path of human labor power —which is part of nature and perhaps even the most powerful of all natural forces—the obstacle of the "unnatural" and purely worldly stability of the human artifice.

17

A CONSUMERS' SOCIETY

It is frequently said that we live in a consumers' society, and since, as we saw, labor and consumption are but two stages of the same process, imposed upon man by the necessity of life, this is only another way of saying that we live in a society of laborers. This society did not come about through the emancipation of the laboring classes but by the emancipation of the laboring activity itself, which preceded by centuries the political emancipation of laborers. The point is not that for the first time in history laborers were admitted and given equal rights in the public realm, but that we have almost succeeded in leveling all human activities to the common denominator of securing the necessities of life and providing for their abundance. Whatever we do, we are supposed to do for the sake of "making a living"; such is the verdict of society, and the number of people, especially in the professions who might challenge it, has decreased rapidly. The only exception society is willing to grant is the artist, who, strictly speaking, is the only "worker" left in a laboring society. The same trend to level down all serious activities to the status of making a living is manifest in present-day labor theories, which almost unanimously define labor as the opposite of play. As a result, all serious activities, irrespective of their fruits, are called labor, and every activity which is not necessary either for the life of the individual or for the life process of society is subsumed under playfulness.[75] In these theories,

75. Although this labor-play category appears at first glance to be so general as to be meaningless, it is characteristic in another respect: the real opposite underlying it is the opposition of necessity and freedom, and it is indeed remarkable to see how plausible it is for modern thinking to consider playfulness to be the source of freedom. Aside from this generalization, the modern idealizations of labor may be said to fall roughly into the following categories: (1) Labor is a means to attain a higher end. This is generally the Catholic position, which has the great merit of not being able to escape from reality altogether, so that the intimate connections between labor and life and between labor and pain are usually at least mentioned. One outstanding representative is Jacques Leclercq of Louvain, especially his discussion of labor and property in *Leçons de droit naturel* (1946), Vol. IV, Part 2. (2) Labor is an act of shaping in which "a given structure is transformed into another, higher structure." This is the central thesis of the famous work by Otto Lipmann, *Grundriss der Arbeitswissenschaft* (1926). (3) Labor in a laboring society is pure pleasure or "can be made fully as satisfying as leisure-time activities" (see Glen W. Cleeton, *Making Work Human* [1949]). This position is taken today by Corrado Gini in his *Ecconomica Lavorista* (1954), who considers the United States to be a "laboring society" (*società lavorista*) where "labor is a pleasure and where all men want to labor." (For a summary of his position in German see *Zeitschrift für die gesamte Staatswissenschaft*, CIX [1953] and CX [1954].) This theory, incidentally, is less new than it seems. It was first formulated by F. Nitti ("Le travail humain et ses lois," *Revue internationale de sociologie* [1895]), who even then maintained that the "idea that labor is painful is a psychological rather than a physiological fact," so that pain will disappear in a society where everybody works. (4) Labor, finally, is man's confirmation of himself against nature, which is brought under his domination through labor. This is the assumption which underlies—explicitly or implicitly—the new, especially French trend of a humanism of labor. Its best-known representative is Georges Friedmann.

After all these theories and academic discussions, it is rather refreshing to learn that a large majority of workers, if asked "why does man work?" answer

which by echoing the current estimate of a laboring society on the theoretical level sharpen it and drive it into its inherent extreme, not even the "work" of the artist is left; it is dissolved into play and has lost its worldly meaning. The playfulness of the artist is felt to fulfil the same function in the laboring life process of society as the playing of tennis or the pursuit of a hobby fulfils in the life of the individual. The emancipation of labor has not resulted in an equality of this activity with the other activities of the *vita activa*, but in its almost undisputed predominance. From the standpoint of "making a living," every activity unconnected with labor becomes a "hobby."[76]

In order to dispel the plausibility of this self-interpretation of modern man, it may be well to remember that all civilizations prior to our own would rather have agreed with Plato that the "art of earning money" (*technē mistharnētikē*) is entirely unconnected with the actual content even of such arts as medicine, navigation, or architecture, which were attended by monetary rewards. It was in order to explain this monetary reward, which obviously is of an altogether different nature from health, the object of medicine, or the erection of buildings, the object of architecture, that Plato introduced one more art to accompany them all. This additional art is by no means understood as the element of labor in the otherwise free arts, but, on the contrary, the one art through which the "artist," the professional worker, as we would say, keeps himself free from the necessity to labor.[77] This art is in the same category

simply "in order to be able to live" or "to make money" (see Helmut Schelsky, *Arbeiterjugend Gestern und Heute* [1955], whose publications are remarkably free of prejudices and idealizations).

76. The role of the hobby in modern labor society is quite striking and may be the root of experience in the labor-play theories. What is especially noteworthy in this context is that Marx, who had no inkling of this development, expected that in his utopian, laborless society all activities would be performed in a manner which very closely resembles the manner of hobby activities.

77. *Republic* 346. Therefore, "the art of acquisition wards off poverty as medicine wards off disease" (*Gorgias* 478). Since payment for their services was voluntary (Loening, *op. cit.*), the liberal professions must indeed have attained a remarkable perfection in the "art of making money."

with the art required of the master of a household who must know how to exert authority and use violence in his rule over slaves. Its aim is to remain free from having "to make a living," and the aims of the other arts are even farther removed from this elementary necessity.

The emancipation of labor and the concomitant emancipation of the laboring classes from oppression and exploitation certainly meant progress in the direction of non-violence. It is much less certain that it was also progress in the direction of freedom. No man-exerted violence, except the violence used in torture, can match the natural force with which necessity itself compels. It is for this reason that the Greeks derived their word for torture from *necessity*, calling it *anagkai*, and not from *bia*, used for violence as exerted by man over man, just as this is the reason for the historical fact that throughout occidental antiquity torture, the "necessity no man can withstand," could be applied only to slaves, who were subject to necessity anyhow.[78] It was the arts of violence, the arts of war, piracy, and ultimately absolute rule, which brought the defeated into the services of the victors and thereby held necessity in abeyance for the longer period of recorded history.[79] The modern age, much more markedly than Christianity, has brought about —together with its glorification of labor—a tremendous degradation in the estimation of these arts and a less great but not less important actual decrease in the use of the instruments of violence in

78. The current modern explanation of this custom which was characteristic of the whole of Greek and Latin antiquity—that its origin is to be found in "the belief that the slave is unable to tell the truth except on the rack" (Barrow, *op. cit.*, p. 31)—is quite erroneous. The belief, on the contrary, is that nobody can invent a lie under torture: "On croyait recueillir la voix même de la nature dans les cris de la douleur. Plus la douleur pénétrait avant, plus intime et plus vrai sembla être ce témoignage de la chair et du sang" (Wallon, *op. cit.*, I, 325). Ancient psychology was much more aware than we are of the element of freedom, of free invention, in telling lies. The "necessities" of torture were supposed to destroy this freedom and therefore could not be applied to free citizens.

79. The older of the Greek words for slaves, *douloi* and *dmōes*, still signify the defeated enemy. About wars and the sale of prisoners of war as the chief source of slavery in antiquity, see W. L. Westermann, "Sklaverei," in Pauly-Wissowa.

human affairs generally.[80] The elevation of labor and the necessity inherent in the laboring metabolism with nature appear to be intimately connected with the downgrading of all activities which either spring directly from violence, as the use of force in human relations, or harbor an element of violence within themselves, which, as we shall see, is the case for all workmanship. It is as though the growing elimination of violence throughout the modern age almost automatically opened the doors for the re-entry of necessity on its most elementary level. What already happened once in our history, in the centuries of the declining Roman Empire, may be happening again. Even then, labor became an occupation of the free classes, "only to bring to them the obligations of the servile classes."[81]

The danger that the modern age's emancipation of labor will not only fail to usher in an age of freedom for all but will result, on the contrary, in forcing all mankind for the first time under the yoke of necessity, was already clearly perceived by Marx when he insisted that the aim of a revolution could not possibly be the already-accomplished emancipation of the laboring classes, but must consist in the emancipation of man from labor. At first glance, this aim seems utopian, and the only strictly utopian element in Marx's

80. Today, because of the new developments of instruments of war and destruction, we are likely to overlook this rather important trend in the modern age. As a matter of fact, the nineteenth century was one of the most peaceful centuries in history.

81. Wallon, *op. cit.*, III, 265. Wallon shows brilliantly how the late Stoic generalization that all men are slaves rested on the development of the Roman Empire, where the old freedom was gradually abolished by the imperial government, so that eventually nobody was free and everybody had his master. The turning point is when first Caligula and then Trajan consented to being called *dominus*, a word formerly used only for the master of the household. The so-called slave morality of late antiquity and its assumption that no real difference existed between the life of a slave and that of a free man had a very realistic background. Now the slave could indeed tell his master: Nobody is free, everybody has a master. In the words of Wallon: "Les condamnés aux mines ont pour confrères, à un moindre degré de peine, les condamnés aux moulins, aux boulangeries, aux relais publics, à tout autre travail faisant l'objet d'une corporation particulière" (p. 216). "C'est le droit de l'esclavage qui gouverne maintenant le citoyen; et nous avons retrouvé toute la législation propre aux esclaves dans les règlements qui concernent sa personne, sa famille ou ses biens" (pp. 219–20).

teachings.[82] Emancipation from labor, in Marx's own terms, is emancipation from necessity, and this would ultimately mean emancipation from consumption as well, that is, from the metabolism with nature which is the very condition of human life.[83] Yet the developments of the last decade, and especially the possibilities opened up through the further development of automation, give us reason to wonder whether the utopia of yesterday will not turn into the reality of tomorrow, so that eventually only the effort of consumption will be left of "the toil and trouble" inherent in the biological cycle to whose motor human life is bound.

However, not even this utopia could change the essential worldly futility of the life process. The two stages through which the ever-recurrent cycle of biological life must pass, the stages of labor and consumption, may change their proportion even to the point where nearly all human "labor power" is spent in consuming, with the concomitant serious social problem of leisure, that is, essentially the problem of how to provide enough opportunity for daily exhaustion to keep the capacity for consumption intact.[84]

82. The classless and stateless society of Marx is not utopian. Quite apart from the fact that modern developments have an unmistakable tendency to do away with class distinctions in society and to replace government by that "administration of things" which according to Engels was to be the hallmark of socialist society, these ideals in Marx himself were obviously conceived in accordance with Athenian democracy, except that in communist society the privileges of the free citizens were to be extended to all.

83. It is perhaps no exaggeration to say that Simone Weil's *La condition ouvrière* (1951) is the only book in the huge literature on the labor question which deals with the problem without prejudice and sentimentality. She chose as the motto for her diary, relating from day to day her experiences in a factory, the line from Homer: *poll' aekadzomenē, kraterē d'epikeiset' anagkē* ("much against your own will, since necessity lies more mightily upon you"), and concludes that the hope for an eventual liberation from labor and necessity is the only utopian element of Marxism and at the same time the actual motor of all Marx-inspired revolutionary labor movements. It is the "opium of the people" which Marx had believed religion to be.

84. This leisure, needless to say, is not at all the same, as current opinion has it, as the *skholē* of antiquity, which was not a phenomenon of consumption, "conspicuous" or not, and did not come about through the emergence of "spare time" saved from laboring, but was on the contrary a conscious "abstention from" all activities connected with mere being alive, the consuming activity no less than the laboring. The touchstone of this *skholē*, as distinguished from the

Painless and effortless consumption would not change but would only increase the devouring character of biological life until a mankind altogether "liberated" from the shackles of pain and effort would be free to "consume" the whole world and to reproduce daily all things it wished to consume. How many things would appear and disappear daily and hourly in the life process of such a society would at best be immaterial for the world, if the world and its thing-character could withstand the reckless dynamism of a wholly motorized life process at all. The danger of future automation is less the much deplored mechanization and artificialization of natural life than that, its artificiality notwithstanding, all human productivity would be sucked into an enormously intensified life process and would follow automatically, without pain or effort, its ever-recurrent natural cycle. The rhythm of machines would magnify and intensify the natural rhythm of life enormously, but it would not change, only make more deadly, life's chief character with respect to the world, which is to wear down durability.

It is a long way from the gradual decrease of working hours, which has progressed steadily for nearly a century, to this utopia. The progress, moreover, has been rather overrated, because it was measured against the quite exceptionally inhuman conditions of exploitation prevailing during the early stages of capitalism. If we think in somewhat longer periods, the total yearly amount of individual free time enjoyed at present appears less an achievement of modernity than a belated approximation to normality.[85] In this as

---

modern ideal of leisure, is the well-known and frequently described frugality of Greek life in the classical period. Thus, it is characteristic that the maritime trade, which more than anything else was responsible for wealth in Athens, was felt to be suspect, so that Plato, following Hesiod, recommended the foundation of new city-states far away from the sea.

85. During the Middle Ages, it is estimated that one hardly worked more than half of the days of the year. Official holidays numbered 141 days (see Levasseur, *op. cit.*, p. 329; see also Liesse, *Le Travail* [1899], p. 253, for the number of working days in France before the Revolution). The monstrous extension of the working day is characteristic of the beginning of the industrial revolution, when the laborers had to compete with newly introduced machines. Before that, the length of the working day amounted to eleven or twelve hours in fifteenth-century England and to ten hours in the seventeenth (see H. Herkner, "Arbeitszeit," in *Handwörterbuch für die Staatswissenschaft* [1923], I, 889 ff.). In

in other respects, the specter of a true consumers' society is more alarming as an ideal of present-day society than as an already existing reality. The ideal is not new; it was clearly indicated in the unquestioned assumption of classical political economy that the ultimate goal of the *vita activa* is growing wealth, abundance, and the "happiness of the greatest number." And what else, finally, is this ideal of modern society but the age-old dream of the poor and destitute, which can have a charm of its own so long as it is a dream, but turns into a fool's paradise as soon as it is realized.

The hope that inspired Marx and the best men of the various workers' movements—that free time eventually will emancipate men from necessity and make the *animal laborans* productive—rests on the illusion of a mechanistic philosophy which assumes that labor power, like any other energy, can never be lost, so that if it is not spent and exhausted in the drudgery of life it will automatically nourish other, "higher," activities. The guiding model of this hope in Marx was doubtless the Athens of Pericles which, in the future, with the help of the vastly increased productivity of human labor, would need no slaves to sustain itself but would become a reality for all. A hundred years after Marx we know the fallacy of this reasoning; the spare time of the *animal laborans* is never spent in anything but consumption, and the more time left to him, the greedier and more craving his appetites. That these appetites become more sophisticated, so that consumption is no longer restricted to the necessities but, on the contrary, mainly concentrates on the superfluities of life, does not change the character of this society, but harbors the grave danger that eventually no object of the world will be safe from consumption and annihilation through consumption.

The rather uncomfortable truth of the matter is that the triumph

brief, "les travailleurs ont connu, pendant la première moitié du 19e siècle, des conditions d'existences pires que celles subies auparavant par les plus infortunés" (Édouard Dolléans, *Histoire du travail en France* [1953]). The extent of progress achieved in our time is generally overrated, since we measure it against a very "dark age" indeed. It may, for instance, be that the life expectancy of the most highly civilized countries today corresponds only to the life expectancy in certain centuries of antiquity. We do not know, of course, but a reflection upon the age of death in the biographies of famous people invites this suspicion.

the modern world has achieved over necessity is due to the emancipation of labor, that is, to the fact that the *animal laborans* was permitted to occupy the public realm; and yet, as long as the *animal laborans* remains in possession of it, there can be no true public realm, but only private activities displayed in the open. The outcome is what is euphemistically called mass culture, and its deep-rooted trouble is a universal unhappiness, due on one side to the troubled balance between laboring and consumption and, on the other, to the persistent demands of the *animal laborans* to obtain a happiness which can be achieved only where life's processes of exhaustion and regeneration, of pain and release from pain, strike a perfect balance. The universal demand for happiness and the widespread unhappiness in our society (and these are but two sides of the same coin) are among the most persuasive signs that we have begun to live in a labor society which lacks enough laboring to keep it contented. For only the *animal laborans*, and neither the craftsman nor the man of action, has ever demanded to be "happy" or thought that mortal men could be happy.

One of the obvious danger signs that we may be on our way to bring into existence the ideal of the *animal laborans* is the extent to which our whole economy has become a waste economy, in which things must be almost as quickly devoured and discarded as they have appeared in the world, if the process itself is not to come to a sudden catastrophic end. But if the ideal were already in existence and we were truly nothing but members of a consumers' society, we would no longer live in a world at all but simply be driven by a process in whose ever-recurring cycles things appear and disappear, manifest themselves and vanish, never to last long enough to surround the life process in their midst.

The world, the man-made home erected on earth and made of the material which earthly nature delivers into human hands, consists not of things that are consumed but of things that are used. If nature and the earth generally constitute the condition of human *life*, then the world and the things of the world constitute the condition under which this specifically human life can be at home on earth. Nature seen through the eyes of the *animal laborans* is the great provider of all "good things," which belong equally to all her children, who "take [them] out of [her] hands" and "mix with" them in labor and consumption.[86] The same nature seen through the eyes of *homo faber*, the builder of the world, "furnishes only the almost worthless materials as in themselves," whose whole value lies in the work performed upon them.[87] Without taking things out of nature's hands and consuming them, and without defending himself against the natural processes of growth and decay, the *animal laborans* could never survive. But without being at home in the midst of things whose durability makes them fit for use and for erecting a world whose very permanence stands in direct contrast to life, this life would never be human.

The easier that life has become in a consumers' or laborers' society, the more difficult it will be to remain aware of the urges of necessity by which it is driven, even when pain and effort, the outward manifestations of necessity, are hardly noticeable at all. The danger is that such a society, dazzled by the abundance of its growing fertility and caught in the smooth functioning of a never-ending process, would no longer be able to recognize its own futility—the futility of a life which "does not fix or realize itself in any permanent subject which endures after [its] labour is past."[88]

86. Locke, *op. cit.*, sec. 28.
87. *Ibid.*, sec. 43.
88. Adam Smith, *op. cit.*, I, 295.

CHAPTER IV

Work

18

THE DURABILITY OF THE WORLD

The work of our hands, as distinguished from the labor of our bodies—*homo faber* who makes and literally "works upon"[1] as distinguished from the *animal laborans* which labors and "mixes with" —fabricates the sheer unending variety of things whose sum total constitutes the human artifice. They are mostly, but not exclusively, objects for use and they possess the durability Locke needed for the establishment of property, the "value" Adam Smith needed for the exchange market, and they bear testimony to productivity, which Marx believed to be the test of human nature. Their proper use does not cause them to disappear and they give the human artifice the stability and solidity without which it could not be relied upon to house the unstable and mortal creature which is man.

The durability of the human artifice is not absolute; the use we make of it, even though we do not consume it, uses it up. The life process which permeates our whole being invades it, too, and if we do not use the things of the world, they also will eventually decay, return into the over-all natural process from which they were

1. The Latin word *faber*, probably related to *facere* ("to make something" in the sense of production), originally designated the fabricator and artist who works upon hard material, such as stone or wood; it also was used as translation for the Greek *tektōn*, which has the same connotation. The word *fabri*, often followed by *tignarii*, especially designates construction workers and carpenters. I have been unable to ascertain when and where the expression *homo faber*, certainly of modern, postmedieval origin, first appeared. Jean Leclercq ("Vers la société basée sur le travail," *Revue du travail*, Vol. LI, No. 3 [March, 1950]) suggests that only Bergson "threw the concept of *homo faber* into the circulation of ideas."

drawn and against which they were erected. If left to itself or discarded from the human world, the chair will again become wood, and the wood will decay and return to the soil from which the tree sprang before it was cut off to become the material upon which to work and with which to build. But though this may be the unavoidable end of all single things in the world, the sign of their being products of a mortal maker, it is not so certainly the eventual fate of the human artifice itself, where all single things can be constantly replaced with the change of generations which come and inhabit the man-made world and go away. Moreover, while usage is bound to use up these objects, this end is not their destiny in the same way as destruction is the inherent end of all things for consumption. What usage wears out is durability.

It is this durability which gives the things of the world their relative independence from men who produced and use them, their "objectivity" which makes them withstand, "stand against"[2] and endure, at least for a time, the voracious needs and wants of their living makers and users. From this viewpoint, the things of the world have the function of stabilizing human life, and their objectivity lies in the fact that—in contradiction to the Heraclitean saying that the same man can never enter the same stream—men, their ever-changing nature notwithstanding, can retrieve their sameness, that is, their identity, by being related to the same chair and the same table. In other words, against the subjectivity of men stands the objectivity of the man-made world rather than the sublime indifference of an untouched nature, whose overwhelming elementary force, on the contrary, will compel them to swing relentlessly in the circle of their own biological movement, which fits so closely into the over-all cyclical movement of nature's household. Only we who have erected the objectivity of a world of our own from what nature gives us, who have built it into the environment of nature so that we are protected from her, can look upon nature as something "objective." Without a world between men and nature, there is eternal movement, but no objectivity.

Although use and consumption, like work and labor, are not the

2. This is implied in the Latin verb *obicere*, from which our "object" is a late derivation, and in the German word for object, *Gegenstand*. "Object" means, literally, "something thrown" or "put against."

same, they seem to overlap in certain important areas to such an extent that the unanimous agreement with which both public and learned opinion have identified these two different matters seems well justified. Use, indeed, does contain an element of consumption, in so far as the wearing-out process comes about through the contact of the use object with the living consuming organism, and the closer the contact between the body and the used thing, the more plausible will an equation of the two appear. If one construes, for instance, the nature of use objects in terms of wearing apparel, he will be tempted to conclude that use is nothing but consumption at a slower pace. Against this stands what we mentioned before, that destruction, though unavoidable, is incidental to use but inherent in consumption. What distinguishes the most flimsy pair of shoes from mere consumer goods is that they do not spoil if I do not wear them, that they have an independence of their own, however modest, which enables them to survive even for a considerable time the changing moods of their owner. Used or unused, they will remain in the world for a certain while unless they are wantonly destroyed.

A similar, much more famous and much more plausible, argument can be raised in favor of an identification of work and labor. The most necessary and elementary labor of man, the tilling of the soil, seems to be a perfect example of labor transforming itself into work in the process, as it were. This seems so because tilling the soil, its close relation to the biological cycle and its utter dependence upon the larger cycle of nature notwithstanding, leaves some product behind which outlasts its own activity and forms a durable addition to the human artifice: the same task, performed year in and year out, will eventually transform the wilderness into cultivated land. The example figures prominently in all ancient and modern theories of laboring precisely for this reason. Yet, despite an undeniable similarity and although doubtless the time-honored dignity of agriculture arises from the fact that tilling the soil not only procures means of subsistence but in this process prepares the earth for the building of the world, even in this case the distinction remains quite clear: the cultivated land is not, properly speaking, a use object, which is there in its own durability and requires for its permanence no more than ordinary care in preservation; the tilled soil, if it is to remain cultivated, needs to be labored upon time and again. A true reification, in other words, in which the produced thing in its existence is secured once and for all, has never come to pass; it needs to be reproduced again and again in order to remain within the human world at all.

19

REIFICATION

Fabrication, the work of *homo faber*, consists in reification. Solidity, inherent in all, even the most fragile, things, comes from the material worked upon, but this material itself is not simply given and there, like the fruits of field and trees which we may gather or leave alone without changing the household of nature. Material is already a product of human hands which have removed it from its natural location, either killing a life process, as in the case of the tree which must be destroyed in order to provide wood, or interrupting one of nature's slower processes, as in the case of iron, stone, or marble torn out of the womb of the earth. This element of violation and violence is present in all fabrication, and *homo faber*, the creator of the human artifice, has always been a destroyer of nature. The *animal laborans*, which with its body and the help of tame animals nourishes life, may be the lord and master of all living creatures, but he still remains the servant of nature and the earth; only *homo faber* conducts himself as lord and master of the whole earth. Since his productivity was seen in the image of a Creator-God, so that where God creates *ex nihilo*, man creates out of given substance, human productivity was by definition bound to result in a Promethean revolt because it could erect a man-made world only after destroying part of God-created nature.[3]

3. This interpretation of human creativity is medieval, whereas the notion of man as lord of the earth is characteristic of the modern age. Both are in contradiction to the spirit of the Bible. According to the Old Testament, man is the master of all living creatures (Gen. 1), which were created to help him (2:19). But nowhere is he made the lord and master of the earth; on the contrary, he was put into the garden of Eden to serve and preserve it (2:15). It is interesting to note that Luther, consciously rejecting the scholastic compromise with Greek and Latin antiquity, tries to eliminate from human work and labor all elements of production and making. Human labor according to him is only "finding" the

The experience of this violence is the most elemental experience of human strength and, therefore, the very opposite of the painful, exhausting effort experienced in sheer labor. It can provide self-assurance and satisfaction, and can even become a source of self-confidence throughout life, all of which are quite different from the bliss which can attend a life spent in labor and toil or from the fleeting, though intense pleasure of laboring itself which comes about if the effort is co-ordinated and rhythmically ordered, and which essentially is the same as the pleasure felt in other rhythmic body movements. Most descriptions of the "joys of labor," in so far as they are not late reflections of the biblical contented bliss of life and death and do not simply mistake the pride in having done a job with the "joy" of accomplishing it, are related to the elation felt by the violent exertion of a strength with which man measures himself against the overwhelming forces of the elements and which through the cunning invention of tools he knows how to multiply far beyond its natural measure.[4] Solidity is not the result of pleasure or exhaustion in earning one's bread "in the sweat of his brow," but of this strength, and it is not simply borrowed or plucked as a free gift from nature's own eternal presence, although it would be impossible without the material torn out of nature; it is already a product of man's hands.

The actual work of fabrication is performed under the guidance of a model in accordance with which the object is constructed. This model can be an image beheld by the eye of the mind or a blueprint in which the image has already found a tentative materialization through work. In either case, what guides the work of fabrication is outside the fabricator and precedes the actual work

---

treasures God has put into the earth. Following the Old Testament, he stresses the utter dependence of man upon the earth, not his mastery: "Sage an, wer legt das Silber und Gold in die Berge, dass man es findet? Wer legt in die Äcker solch grosses Gut als heraus wächst . . . ? Tut das Menschen Arbeit? Ja wohl, Arbeit findet es wohl; aber Gott muss es dahin legen, soll es die Arbeit finden. . . . So finden wir denn, dass alle unsere Arbeit nichts ist denn Gottes Güter finden und aufheben, nichts aber möge machen und erhalten" (*Werke*, ed. Walch, V, 1873).

4. Hendrik de Man, for instance, describes almost exclusively the satisfactions of making and workmanship under the misleading title: *Der Kampf um die Arbeitsfreude* (1927).

process in much the same way as the urgencies of the life process within the laborer precede the actual labor process. (This description is in flagrant contradiction to the findings of modern psychology, which tell us almost unanimously that the images of the mind are as safely located in our heads as the pangs of hunger are located in our stomachs. This subjectivization of modern science, which is only a reflection of an even more radical subjectivization of the modern world, has its justification in this case in the fact that, indeed, most work in the modern world is performed in the mode of labor, so that the worker, even if he wanted to, could not "labor for his work rather than for himself,"[5] and frequently is instrumental in the production of objects of whose ultimate shape he has not the slightest notion.[6] These circumstances, though of great historical importance, are irrelevant in a description of the fundamental articulations of the *vita activa*.) What claims our attention is the veritable gulf that separates all bodily sensations, pleasure or pain, desires and satisfactions—which are so "private" that they cannot even be adequately voiced, much less represented in the outside world, and therefore are altogether incapable of being reified—from mental images which lend themselves so easily and naturally to reification that we neither conceive of making a bed without first having some image, some "idea" of a bed before our inner eye, nor can imagine a bed without having recourse to some visual experience of a real thing.

It is of great importance to the role fabrication came to play within the hierarchy of the *vita activa* that the image or model whose shape guides the fabrication process not only precedes it, but does not disappear with the finished product, which it survives intact, present, as it were, to lend itself to an infinite continuation of fabrication. This potential multiplication, inherent in work, is

5. Yves Simon, *Trois leçons sur le travail* (Paris, n.d.). This type of idealization is frequent in liberal or left-wing Catholic thought in France (see especially Jean Lacroix, "La notion du travail," *La vie intellectuelle* [June, 1952], and the Dominican M. D. Chenu, "Pour une théologie du travail," *Esprit* [1952 and 1955]: "Le travailleur travaille pour son œuvre plutôt que pour lui-même: loi de générosité métaphysique, qui définit l'activité laborieuse").

6. Georges Friedmann (*Problèmes humains du machinisme industriel* [1946], p. 211) relates how frequently the workers in the great factories do not even know the name or the exact function of the piece produced by their machine.

different in principle from the repetition which is the mark of labor. This repetition is urged upon and remains subject to the biological cycle; the needs and wants of the human body come and go, and though they reappear again and again at regular intervals, they never remain for any length of time. Multiplication, in distinction from mere repetition, multiplies something that already possesses a relatively stable, relatively permanent existence in the world. This quality of permanence in the model or image, of being there before fabrication starts and remaining after it has come to an end, surviving all the possible use objects it continues to help into existence, had a powerful influence on Plato's doctrine of eternal ideas. In so far as his teaching was inspired by the word *idea* or *eidos* ("shape" or "form"), which he used for the first time in a philosophical context, it rested on experiences in *poiēsis* or fabrication, and although Plato used his theory to express quite different and perhaps much more "philosophical" experiences, he never failed to draw his examples from the field of making when he wanted to demonstrate the plausibility of what he was saying.[7]

7. Aristotle's testimony that Plato introduced the term *idea* into philosophic terminology occurs in the first book of his *Metaphysics* (987b8). An excellent account of the earlier usage of the word and of Plato's teaching is Gerard F. Else, "The Terminology of Ideas," *Harvard Studies in Classical Philology*, Vol. XLVII (1936). Else rightly insists that "what the doctrine of Ideas was in its final and complete form is something we cannot learn from the dialogues." We are equally uncertain about the doctrine's origin, but there the safest guide may still be the word itself which Plato so strikingly introduced into philosophic terminology, even though the word was not current in Attic speech. The words *eidos* and *idea* doubtlessly relate to visible forms or shapes, especially of living creatures; this makes it unlikely that Plato conceived the doctrine of ideas under the influence of geometrical forms. Francis M. Cornford's thesis (*Plato and Parmenides* [Liberal Arts ed.], pp. 69–100) that the doctrine is probably Socratic in origin, in so far as Socrates sought to define justice in itself or goodness in itself, which cannot be perceived with the senses, as well as Pythagorean, in so far as the doctrine of the ideas' eternal and separate existence (*chōrismos*) from all perishable things involves "the separate existence of a conscious and knowing soul, apart from the body and the senses," sounds to me very convincing. But my own presentation leaves all such assumptions in abeyance. It relates simply to the tenth book of the *Republic*, where Plato himself explains his doctrine by taking "the common instance" of a craftsman who makes beds and tables "in accordance with [their] idea," and then adds, "that is our way of speaking in this and similar instances." Obviously, to Plato the very word *idea* was suggestive,

The one eternal idea presiding over a multitude of perishable things derives its plausibility in Plato's teachings from the permanence and oneness of the model according to which many and perishable objects can be made.

The process of making is itself entirely determined by the categories of means and end. The fabricated thing is an end product in the twofold sense that the production process comes to an end in it ("the process disappears in the product," as Marx said) and that it is only a means to produce this end. Labor, to be sure, also produces for the end of consumption, but since this end, the thing to be consumed, lacks the worldly permanence of a piece of work, the end of the process is not determined by the end product but rather by the exhaustion of labor power, while the products themselves, on the other hand, immediately become means again, means of subsistence and reproduction of labor power. In the process of making, on the contrary, the end is beyond doubt: it has come when an entirely new thing with enough durability to remain in the world as an independent entity has been added to the human artifice. As far as the thing, the end product of fabrication, is concerned, the process need not be repeated. The impulse toward repetition comes from the craftsman's need to earn his means of subsistence, in which case his working coincides with his laboring; or it comes from a demand for multiplication in the market, in which case the craftsman who wishes to meet this demand has added, as Plato would have said, the art of earning money to his craft. The point here is that in either case the process is repeated for reasons outside itself and is unlike the compulsory repetition inherent in laboring, where one must eat in order to labor and must labor in order to eat.

To have a definite beginning and a definite, predictable end is the mark of fabrication, which through this characteristic alone dis-

---

and he wanted it to suggest "the craftsman who makes a couch or a table not by looking . . . at another couch or another table, but by looking at the idea of the couch" (Kurt von Fritz, *The Constitution of Athens* [1950], pp. 34–35). Needless to say, none of these explanations touches the root of the matter, that is, the specifically philosophic experience underlying the concept of ideas on the one hand, and their most striking quality on the other—their illuminating power, their being *to phanotaton* or *ekphanestaton*.

tinguishes itself from all other human activities. Labor, caught in the cyclical movement of the body's life process, has neither a beginning nor an end. Action, though it may have a definite beginning, never, as we shall see, has a predictable end. This great reliability of work is reflected in that the fabrication process, unlike action, is not irreversible: every thing produced by human hands can be destroyed by them, and no use object is so urgently needed in the life process that its maker cannot survive and afford its destruction. *Homo faber* is indeed a lord and master, not only because he is the master or has set himself up as the master of all nature but because he is master of himself and his doings. This is true neither of the *animal laborans*, which is subject to the necessity of its own life, nor of the man of action, who remains in dependence upon his fellow men. Alone with his image of the future product, *homo faber* is free to produce, and again facing alone the work of his hands, he is free to destroy.

20

INSTRUMENTALITY AND
Animal Laborans

From the standpoint of *homo faber*, who relies entirely on the primordial tools of his hands, man is, as Benjamin Franklin said, a "tool-maker." The same instruments, which only lighten the burden and mechanize the labor of the *animal laborans*, are designed and invented by *homo faber* for the erection of a world of things, and their fitness and precision are dictated by such "objective" aims as he may wish to invent rather than by subjective needs and wants. Tools and instruments are so intensely worldly objects that we can classify whole civilizations using them as criteria. Nowhere, however, is their worldly character more manifest than when they are used in labor processes, where they are indeed the only tangible things that survive both the labor and the consumption process itself. For the *animal laborans*, therefore, as it is subject to and constantly occupied with the devouring processes of life, the durability and stability of the world are primarily represented in the tools and instruments it uses, and in a society of laborers, tools are very likely to assume a more than mere instrumental character or function.

The frequent complaints we hear about the perversion of ends and means in modern society, about men becoming the servants of the machines they themselves invented and of being "adapted" to their requirements instead of using them as instruments for human needs and wants, have their roots in the factual situation of laboring. In this situation, where production consists primarily in preparation for consumption, the very distinction between means and ends, so highly characteristic of the activities of *homo faber*, simply does not make sense, and the instruments which *homo faber* invented and with which he came to the help of the labor of the *animal laborans* therefore lose their instrumental character once they are used by it. Within the life process itself, of which laboring remains an integral part and which it never transcends, it is idle to ask questions that presuppose the category of means and end, such as whether men live and consume in order to have strength to labor or whether they labor in order to have the means of consumption.

If we consider this loss of the faculty to distinguish clearly between means and ends in terms of human behavior, we can say that the free disposition and use of tools for a specific end product is replaced by rhythmic unification of the laboring body with its implement, the movement of laboring itself acting as the unifying force. Labor but not work requires for best results a rhythmically ordered performance and, in so far as many laborers gang together, needs a rhythmic co-ordination of all individual movements.[8] In

8. Karl Bücher's well-known compilation of rhythmic labor songs in 1897 (*Arbeit und Rhythmus* [6th ed.; 1924]) has been followed by a voluminous literature of a more scientific nature. One of the best of these studies (Joseph Schopp, *Das deutsche Arbeitslied* [1935]) stresses that there exist only labor songs, but no work songs. The songs of the craftsmen are social; they are sung after work. The fact is, of course, that there exists no "natural" rhythm for work. The striking resemblance between the "natural" rhythm inherent in every laboring operation and the rhythm of the machines is sometimes noticed, apart from the repeated complaints about the "artificial" rhythm which the machines impose upon the laborer. Such complaints, characteristically, are relatively rare among the laborers themselves, who, on the contrary, seem to find the same amount of pleasure in repetitive machine work as in other repetitive labor (see, for instance, Georges Friedmann, *Où va le travail humain?* [2d ed.; 1953], p. 233, and Hendrik de Man, *op. cit.*, p. 213). This confirms observations which were already made in

this motion, the tools lose their instrumental character, and the clear distinction between man and his implements, as well as his ends, becomes blurred. What dominates the labor process and all work processes which are performed in the mode of laboring is neither man's purposeful effort nor the product he may desire, but the motion of the process itself and the rhythm it imposes upon the laborers. Labor implements are drawn into this rhythm until body and tool swing in the same repetitive movement, that is, until, in the use of machines, which of all implements are best suited to the performance of the *animal laborans*, it is no longer the body's movement that determines the implement's movement but the machine's movement which enforces the movements of the body. The point is that nothing can be mechanized more easily and less artificially than the rhythm of the labor process, which in its turn corresponds to the equally automatic repetitive rhythm of the life process and its metabolism with nature. Precisely because the *animal laborans*

the Ford factories at the beginning of our century. Karl Bücher, who believed that "rhythmic labor is highly spiritual labor" (*vergeistigt*), already stated: "Aufreibend werden nur solche einförmigen Arbeiten, die sich nicht rhythmisch gestalten lassen" (*op. cit.*, p. 443). For though the speed of machine work undoubtedly is much higher and more repetitive than that of "natural" spontaneous labor, the fact of a rhythmic performance as such makes that machine labor and pre-industrial labor have more in common with each other than either of them has with work. Hendrik de Man, for instance, is well aware that "diese von Bücher . . . gepriesene Welt weniger die des . . . handwerksmässig schöpferischen Gewerbes als der der einfachen, schieren . . . Arbeitsfron [ist]" (*op. cit.*, p. 244).

All these theories appear highly questionable in view of the fact that the workers themselves give an altogether different reason for their preference for repetitive labor. They prefer it because it is mechanical and does not demand attention, so that while performing it they can think of something else. (They can "geistig wegtreten," as Berlin workers formulated it. See Thielicke and Pentzlin, *Mensch und Arbeit im technischen Zeitalter: Zum Problem der Rationalisierung* [1954], pp. 35 ff., who also report that according to an investigation of the *Max Planck Institut für Arbeitspsychologie*, about 90 per cent of the workers prefer monotonous tasks.) This explanation is all the more noteworthy, as it coincides with very early Christian recommendations of the merits of manual labor, which, because it demands less attention, is less likely to interfere with contemplation than other occupations and professions (see Étienne Delaruelle, "Le travail dans les règles monastiques occidentales du 4e au 9e siècle," *Journal de psychologie normale et pathologique*, Vol. XLI, No. 1 [1948]).

does not use tools and instruments in order to build a world but in order to ease the labors of its own life process, it has lived literally in a world of machines ever since the industrial revolution and the emancipation of labor replaced almost all hand tools with machines which in one way or another supplanted human labor power with the superior power of natural forces.

The decisive difference between tools and machines is perhaps best illustrated by the apparently endless discussion of whether man should be "adjusted" to the machine or the machines should be adjusted to the "nature" of man. We mentioned in the first chapter the chief reason why such a discussion must be sterile: if the human condition consists in man's being a conditioned being for whom everything, given or man-made, immediately becomes a condition of his further existence, then man "adjusted" himself to an environment of machines the moment he designed them. They certainly have become as inalienable a condition of our existence as tools and implements were in all previous ages. The interest of the discussion, from our point of view, therefore, lies rather in the fact that this question of adjustment could arise at all. There never was any doubt about man's being adjusted or needing special adjustment to the tools he used; one might as well have adjusted him to his hands. The case of the machines is entirely different. Unlike the tools of workmanship, which at every given moment in the work process remain the servants of the hand, the machines demand that the laborer serve them, that he adjust the natural rhythm of his body to their mechanical movement. This, certainly, does not imply that men as such adjust to or become the servants of their machines; but it does mean that, as long as the work at the machines lasts, the mechanical process has replaced the rhythm of the human body. Even the most refined tool remains a servant, unable to guide or to replace the hand. Even the most primitive machine guides the body's labor and eventually replaces it altogether.

As is so frequently the case with historical developments, it seems as though the actual implications of technology, that is, of the replacement of tools and implements with machinery, have come to light only in its last stage, with the advent of automation. For our purposes it may be useful to recall, however briefly, the main stages of modern technology's development since the beginning of the modern age. The first stage, the invention of the steam engine, which led into the industrial revolution, was still characterized by an imitation of natural processes and the use of natural forces for human purposes, which did not differ in principle from the old use of water and wind power. Not the principle of the steam engine was new but rather the discovery and use of the coal mines to feed it.[9] The machine tools of this early stage reflect this imitation of naturally known processes; they, too, imitate and put to more powerful use the natural activities of the human hand. But today we are told that "the greatest pitfall to avoid is the assumption that the design aim is reproduction of the hand movements of the operator or laborer."[10]

The next stage is chiefly characterized by the use of electricity, and, indeed, electricity still determines the present stage of technical development. This stage can no longer be described in terms of a gigantic enlargement and continuation of the old arts and crafts, and it is only to this world that the categories of *homo faber*, to whom every instrument is a means to achieve a prescribed end, no longer apply. For here we no longer use material as nature yields it to us, killing natural processes or interrupting or imitating them. In all these instances, we changed and denaturalized nature for our own worldly ends, so that the human world or artifice on one hand and nature on the other remained two distinctly separate entities. Today we have begun to "create," as it were, that is, to unchain natural processes of our own which would never have happened without us, and instead of carefully surrounding the human artifice with defenses against nature's elementary forces, keeping them as

9. One of the important material conditions of the industrial revolution was the extinction of the forests and the discovery of coal as a substitute for wood. The solution which R. H. Barrow (in his *Slavery in the Roman Empire* [1928]) proposed to "the well-known puzzle in the study of the economic history of the ancient world that industry developed up to a certain point, but stopped short of making progress which might have been expected," is quite interesting and rather convincing in this connection. He maintains that the only factor that "hindered the application of machinery to industry [was] . . . the absence of cheap and good fuel, . . . no abundant supply of coal [being] close at hand" (p. 123).

10. John Diebold, *Automation: The Advent of the Automatic Factory* (1952), p. 67.

far as possible outside the man-made world, we have channeled these forces, along with their elementary power, into the world itself. The result has been a veritable revolution in the concept of fabrication; manufacturing, which always had been "a series of separate steps," has become "a continuous process," the process of the conveyor belt and the assembly line.[11]

Automation is the most recent stage in this development, which indeed "illuminates the whole history of machinism."[12] It certainly will remain the culminating point of the modern development, even if the atomic age and a technology based upon nuclear discoveries puts a rather rapid end to it. The first instruments of nuclear technology, the various types of atom bombs, which, if released in suf-

11. *Ibid.*, p. 69.

12. Friedmann, *Problèmes humains du machinisme industriel*, p. 168. This, in fact, is the most obvious conclusion to be drawn from Diebold's book: The assembly line is the result of "the concept of manufacturing as a continuous process," and automation, one may add, is the result of the machinization of the assembly line. To the release of human labor power in the earlier stage of industrialization, automation adds the release of human brain power, because "the monitoring and control tasks now humanly performed will be done by machines" (*op. cit.*, p. 140). The one as well as the other releases labor, and not work. The worker or the "self-respecting craftsman," whose "human and psychological values" (p. 164) almost every author in the field tries desperately to save—and sometimes with a grain of involuntary irony, as when Diebold and others earnestly believe that repair work, which perhaps will never be entirely automatic, can inspire the same contentment as fabrication and production of a new object—does not belong in this picture for the simple reason that he was eliminated from the factory long before anybody knew about automation. The workers in a factory have always been laborers, and though they may have excellent reasons for self-respect, it certainly cannot arise from the work they do. One can only hope that they themselves will not accept the social substitutes for contentment and self-respect offered them by labor theorists, who by now really believe that the interest in work and the satisfaction of craftsmanship can be replaced by "human relations" and by the respect workers "earn from their fellow workers" (p. 164). Automation, after all, should at least have the advantage of demonstrating the absurdities of all "humanisms of labor"; if the verbal and historical meaning of the word "humanism" is at all taken into account, the very term "humanism of labor" is clearly a contradiction in terms. (For an excellent criticism of the vogue of "human relations" see Daniel Bell, *Work and Its Discontents* [1956], ch. 5, and R. P. Genelli, "Facteur humain ou facteur social du travail," *Revue française du travail*, Vol. VII, Nos. 1–3 [January–March, 1952], where one also finds a very determined denunciation of the "terrible illusion" of the "joy of labor.")

ficient and not even very great quantities, could destroy all organic life on earth, present sufficient evidence for the enormous scale on which such a change might take place. Here it would no longer be a question of unchaining and letting loose elementary natural processes, but of handling on the earth and in everyday life energies and forces such as occur only outside the earth, in the universe; this is already done, but only in the research laboratories of nuclear physicists.[13] If present technology consists of channeling natural forces into the world of the human artifice, future technology may yet consist of channeling the universal forces of the cosmos around us into the nature of the earth. It remains to be seen whether these future techniques will transform the household of nature as we have known it since the beginning of our world to the same extent or even more than the present technology has changed the very worldliness of the human artifice.

The channeling of natural forces into the human world has shattered the very purposefulness of the world, the fact that objects are the ends for which tools and implements are designed. It is characteristic of all natural processes that they come into being without the help of man, and those things are natural which are not "made" but grow by themselves into whatever they become. (This is also the authentic meaning of our word "nature," whether we derive it from its latin root *nasci*, to be born, or trace it back to its Greek origin, *physis*, which comes from *phyein*, to grow out of, to appear by itself.) Unlike the products of human hands, which must be realized step by step and for which the fabrication process is entirely distinct from the existence of the fabricated thing itself, the natural thing's existence is not separate but is somehow identical with the process through which it comes into being: the seed contains and, in a certain sense, already *is* the tree, and the tree stops being if the process of growth through which it came into existence

13. Günther Anders, in an interesting essay on the atom bomb (*Die Anti-quiertheit des Menschen* [1956]), argues convincingly that the term "experiment" is no longer applicable to nuclear experiments involving explosions of the new bombs. For it was characteristic of experiments that the space where they took place was strictly limited and isolated against the surrounding world. The effects of the bombs are so enormous that "their laboratory becomes co-extensive with the globe" (p. 260).

stops. If we see these processes against the background of human purposes, which have a willed beginning and a definite end, they assume the character of automatism. We call automatic all courses of movement which are self-moving and therefore outside the range of wilful and purposeful interference. In the mode of production ushered in by automation, the distinction between operation and product, as well as the product's precedence over the operation (which is only the means to produce the end), no longer make sense and have become obsolete.[14] The categories of *homo faber* and his world apply here no more than they ever could apply to nature and the natural universe. This is, incidentally, why modern advocates of automation usually take a very determined stand against the mechanistic view of nature and against the practical utilitarianism of the eighteenth century, which were so eminently characteristic of the one-sided, single-minded work orientation of *homo faber*.

The discussion of the whole problem of technology, that is, of the transformation of life and world through the introduction of the machine, has been strangely led astray through an all-too-exclusive concentration upon the service or disservice the machines render to men. The assumption here is that every tool and implement is primarily designed to make human life easier and human labor less painful. Their instrumentality is understood exclusively in this anthropocentric sense. But the instrumentality of tools and implements is much more closely related to the object it is designed to produce, and their sheer "human value" is restricted to the use the *animal laborans* makes of them. In other words, *homo faber*, the toolmaker, invented tools and implements in order to erect a world, not—at least, not primarily—to help the human life process. The question therefore is not so much whether we are the masters or the slaves of our machines, but whether machines still serve the world and its things, or if, on the contrary, they and the automatic motion of their processes have begun to rule and even destroy world and things.

One thing is certain: the continuous automatic process of manufacturing has not only done away with the "unwarranted assumption" that "human hands guided by human brains represent the

14. Diebold, *op. cit.*, pp. 59–60.

optimum efficiency,"[15] but with the much more important assumption that the things of the world around us should depend upon human design and be built in accordance with human standards of either utility or beauty. In place of both utility and beauty, which are standards of the world, we have come to design products that still fulfil certain "basic functions" but whose shape will be primarily determined by the operation of the machine. The "basic functions" are of course the functions of the human animal's life process, since no other function is basically necessary, but the product itself—not only its variations but even the "total change to a new product"—will depend entirely upon the capacity of the machine.[16]

To design objects for the operational capacity of the machine instead of designing machines for the production of certain objects would indeed be the exact reversal of the means-end category, if this category still made any sense. But even the most general end, the release of manpower, that was usually assigned to machines, is now thought to be a secondary and obsolete aim, inadequate to and limiting potential "startling increases in efficiency."[17] As matters stand today, it has become as senseless to describe this world of machines in terms of means and ends as it has always been senseless to ask nature if she produced the seed to produce a tree or the tree to produce the seed. By the same token it is quite probable that the continuous process pursuant to the channeling of nature's never-ending processes into the human world, though it may very well destroy the world *qua* world as human artifice, will as reliably and limitlessly provide the species man-kind with the necessities of life as nature herself did before men erected their artificial home on earth and set up a barrier between nature and themselves.

For a society of laborers, the world of machines has become a substitute for the real world, even though this pseudo world cannot fulfil the most important task of the human artifice, which is to offer mortals a dwelling place more permanent and more stable than themselves. In the continuous process of operation, this world of machines is even losing that independent worldly character which the tools and implements and the early machinery of the

15. *Ibid.*, p. 67.　　16. *Ibid.*, pp. 38–45.　　17. *Ibid.*, pp. 110 and 157.

modern age so eminently possessed. The natural processes on which it feeds increasingly relate it to the biological process itself, so that the apparatuses we once handled freely begin to look as though they were "shells belonging to the human body as the shell belongs to the body of a turtle." Seen from the vantage point of this development, technology in fact no longer appears "as the product of a conscious human effort to enlarge material power, but rather like a biological development of mankind in which the innate structures of the human organism are transplanted in an ever-increasing measure into the environment of man."[18]

21

INSTRUMENTALITY AND *Homo Faber*

The implements and tools of *homo faber*, from which the most fundamental experience of instrumentality arises, determine all work and fabrication. Here it is indeed true that the end justifies the means; it does more, it produces and organizes them. The end justifies the violence done to nature to win the material, as the wood justifies killing the tree and the table justifies destroying the wood. Because of the end product, tools are designed and implements invented, and the same end product organizes the work process itself, decides about the needed specialists, the measure of co-operation, the number of assistants, etc. During the work process, everything is judged in terms of suitability and usefulness for the desired end, and for nothing else.

The same standards of means and end apply to the product itself. Though it is an end with respect to the means by which it was produced and is the end of the fabrication process, it never becomes, so to speak, an end in itself, at least not as long as it remains an object for use. The chair which is the end of carpentering can show its usefulness only by again becoming a means, either as a thing whose durability permits its use as a means for comfortable living or as a means of exchange. The trouble with the utility standard inherent in the very activity of fabrication is that the relationship between means and end on which it relies is very much like a chain whose every end can serve again as a means in some

18. Werner Heisenberg, *Das Naturbild der heutigen Physik* (1955), pp. 14–15.

other context. In other words, in a strictly utilitarian world, all ends are bound to be of short duration and to be transformed into means for some further ends.[19]

This perplexity, inherent in all consistent utilitarianism, the philosophy of *homo faber* par excellence, can be diagnosed theoretically as an innate incapacity to understand the distinction between utility and meaningfulness, which we express linguistically by distinguishing between "in order to" and "for the sake of." Thus the ideal of usefulness permeating a society of craftsmen— like the ideal of comfort in a society of laborers or the ideal of acquisition ruling commercial societies—is actually no longer a matter of utility but of meaning. It is "for the sake of" usefulness in general that *homo faber* judges and does everything in terms of "in order to." The ideal of usefulness itself, like the ideals of other societies, can no longer be conceived as something needed in order to have something else; it simply defies questioning about its own use. Obviously there is no answer to the question which Lessing once put to the utilitarian philosophers of his time: "And what is the use of use?" The perplexity of utilitarianism is that it gets caught in the unending chain of means and ends without ever arriving at some principle which could justify the category of means and end, that is, of utility itself. The "in order to" has become the content of the "for the sake of"; in other words, utility established as meaning generates meaninglessness.

Within the category of means and end, and among the experiences of instrumentality which rules over the whole world of use objects and utility, there is no way to end the chain of means and ends and prevent all ends from eventually being used again as means, except to declare that one thing or another is "an end in itself." In the world of *homo faber*, where everything must be of some use, that is, must lend itself as an instrument to achieve something else, meaning itself can appear only as an end, as an "end in itself" which actually is either a tautology applying to all ends or a contradiction in terms. For an end, once it is attained, ceases to be an end and loses its capacity to guide and justify the

19. About the endlessness of the means-end chain (the *"Zweckprogressus in infinitum"*) and its inherent destruction of meaning, compare Nietzsche, Aph. 666 in *Wille zur Macht.*

choice of means, to organize and produce them. It has now become an object among objects, that is, it has been added to the huge arsenal of the given from which *homo faber* selects freely his means to pursue his ends. Meaning, on the contrary, must be permanent and lose nothing of its character, whether it is achieved or, rather, found by man or fails man and is missed by him. *Homo faber*, in so far as he is nothing but a fabricator and thinks in no terms but those of means and ends which arise directly out of his work activity, is just as incapable of understanding meaning as the *animal laborans* is incapable of understanding instrumentality. And just as the implements and tools *homo faber* uses to erect the world become for the *animal laborans* the world itself, thus the meaningfulness of this world, which actually is beyond the reach of *homo faber*, becomes for him the paradoxical "end in itself."

The only way out of the dilemma of meaninglessness in all strictly utilitarian philosophy is to turn away from the objective world of use things and fall back upon the subjectivity of use itself. Only in a strictly anthropocentric world, where the user, that is, man himself, becomes the ultimate end which puts a stop to the unending chain of ends and means, can utility as such acquire the dignity of meaningfulness. Yet the tragedy is that in the moment *homo faber* seems to have found fulfilment in terms of his own activity, he begins to degrade the world of things, the end and end product of his own mind and hands; if man the user is the highest end, "the measure of all things," then not only nature, treated by *homo faber* as the almost "worthless material" upon which to work, but the "valuable" things themselves have become mere means, losing thereby their own intrinsic "value."

The anthropocentric utilitarianism of *homo faber* has found its greatest expression in the Kantian formula that no man must ever become a means to an end, that every human being is an end in himself. Although we find earlier (for instance, in Locke's insistence that no man can be permitted to possess another man's body or use his bodily strength) an awareness of the fateful consequences which an unhampered and unguided thinking in terms of means and ends must invariably entail in the political realm, it is only in Kant that the philosophy of the earlier stages of the modern age frees itself entirely of the common sense platitudes which we always find where *homo faber* rules the standards of society. The reason is, of course, that Kant did not mean to formulate or conceptualize the tenets of the utilitarianism of his time, but on the contrary wanted first of all to relegate the means-end category to its proper place and prevent its use in the field of political action. His formula, however, can no more deny its origin in utilitarian thinking than his other famous and also inherently paradoxical interpretation of man's attitude toward the only objects that are not "for use," namely works of art, in which he said we take "pleasure without any interest."[20] For the same operation which establishes man as the "supreme end" permits him "if he can [to] subject the whole of nature to it,"[21] that is, to degrade nature and the world into mere means, robbing both of their independent dignity. Not even Kant could solve the perplexity or enlighten the blindness of *homo faber* with respect to the problem of meaning without turning to the paradoxical "end in itself," and this perplexity lies in the fact that while only fabrication with its instrumentality is capable of building a world, this same world becomes as worthless as the employed material, a mere means for further ends, if the standards which governed its coming into being are permitted to rule it after its establishment.

Man, in so far as he is *homo faber*, instrumentalizes, and his instrumentalization implies a degradation of all things into means, their loss of intrinsic and independent value, so that eventually not only the objects of fabrication but also "the earth in general and all forces of nature," which clearly came into being without the help of man and have an existence independent of the human world, lose their "value because [they] do not present the reification which comes from work."[22] It was for no other reason than this attitude of *homo faber* to the world that the Greeks in their classical period declared the whole field of the arts and crafts, where men work with instruments and do something not for its own sake but

20. Kant's term is "ein Wohlgefallen ohne alles Interesse" (*Kritik der Urteilskraft* [Cassirer ed.], V, 272).

21. *Ibid.*, p. 515.

22. "Der Wasserfall, wie die Erde überhaupt, wie alle Naturkraft hat keinen Wert, weil er keine in ihm vergegenständlichte Arbeit darstellt" (*Das Kapital*, III [*Marx-Engels Gesamtausgabe*, Abt. II, Zürich, 1933], 698).

in order to produce something else, to be *banausic*, a term perhaps best translated by "philistine," implying vulgarity of thinking and acting in terms of expediency. The vehemence of this contempt will never cease to startle us if we realize that the great masters of Greek sculpture and architecture were by no means excepted from the verdict.

The issue at stake is, of course, not instrumentality, the use of means to achieve an end, as such, but rather the generalization of the fabrication experience in which usefulness and utility are established as the ultimate standards for life and the world of men. This generalization is inherent in the activity of *homo faber* because the experience of means and end, as it is present in fabrication, does not disappear with the finished product but is extended to its ultimate destination, which is to serve as a use object. The instrumentalization of the whole world and the earth, this limitless devaluation of everything given, this process of growing meaninglessness where every end is transformed into a means and which can be stopped only by making man himself the lord and master of all things, does not directly arise out of the fabrication process; for from the viewpoint of fabrication the finished product is as much an end in itself, an independent durable entity with an existence of its own, as man is an end in himself in Kant's political philosophy. Only in so far as fabrication chiefly fabricates use objects does the finished product again become a means, and only in so far as the life process takes hold of things and uses them for its purposes does the productive and limited instrumentality of fabrication change into the limitless instrumentalization of everything that exists.

It is quite obvious that the Greeks dreaded this devaluation of world and nature with its inherent anthropocentrism—the "absurd" opinion that man is the highest being and that everything else is subject to the exigencies of human life (Aristotle)—no less than they despised the sheer vulgarity of all consistent utilitarianism. To what extent they were aware of the consequences of seeing in *homo faber* the highest human possibility is perhaps best illustrated by Plato's famous argument against Protagoras and his apparently self-evident statement that "man is the measure of all use things (*chrēmata*), of the existence of those that are, and of the non-existence of those that are not."[23] (Protagoras evidently did not say: "Man is the measure of all things," as tradition and the standard translations have made him say.) The point of the matter is that Plato saw immediately that if one makes man the measure of all things for use, it is man the user and instrumentalizer, and not man the speaker and doer or man the thinker, to whom the world is being related. And since it is in the nature of man the user and instrumentalizer to look upon everything as means to an end—upon every tree as potential wood—this must eventually mean that man becomes the measure not only of things whose existence depends upon him but of literally everything there is.

In this Platonic interpretation, Protagoras in fact sounds like the earliest forerunner of Kant, for if man is the measure of all things, then man is the only thing outside the means-end relationship, the only end in himself who can use everything else as a means. Plato knew quite well that the possibilities of producing use objects and of treating all things of nature as potential use objects are as limitless as the wants and talents of human beings. If one permits the standards of *homo faber* to rule the finished world as they must necessarily rule the coming into being of this world, then *homo faber* will eventually help himself to everything and consider everything that is as a mere means for himself. He will judge every thing as though it belonged to the class of *chrēmata*, of use objects, so that, to follow Plato's own example, the wind will no longer be understood in its own right as a natural force but will be considered exclusively in accordance with human needs for warmth or refreshment—which, of course, means that the wind as something objectively given has been eliminated from human experience. It is because of these consequences that Plato, who at the end of his life recalls once more in the *Laws* the saying of Protagoras, replies with an almost paradoxical formula: not man—who because of his

23. *Theaetetus* 152, and *Cratylus* 385E. In these instances, as well as in other ancient quotations of the famous saying, Protagoras is always quoted as follows: *pantōn chrēmatōn metron estin anthrōpos* (see Diels, *Fragmente der Vorsokratiker* [4th ed.; 1922], frag. B1). The word *chrēmata* by no means signifies "all things," but specifically things used or needed or possessed by men. The supposed Protagorean saying, "Man is the measure of all things," would be rendered in Greek rather as *anthrōpos metron pantōn*, corresponding for instance to Heraclitus' *polemos patēr pantōn* ("strife is the father of all things").

wants and talents wishes to use everything and therefore ends by depriving all things of their intrinsic worth—but "the god is the measure [even] of mere use objects."[24]

22

THE EXCHANGE MARKET

Marx—in one of many asides which testify to his eminent historical sense—once remarked that Benjamin Franklin's definition of man as a toolmaker is as characteristic of "Yankeedom," that is, of the modern age, as the definition of man as a political animal was for antiquity.[25] The truth of this remark lies in the fact that the modern age was as intent on excluding political man, that is, man who acts and speaks, from its public realm as antiquity was on excluding *homo faber*. In both instances the exclusion was not a matter of course, as was the exclusion of laborers and the propertyless classes until their emancipation in the nineteenth century. The modern age was of course perfectly aware that the political realm was not always and need not necessarily be a mere function of "society," destined to protect the productive, social side of human nature through governmental administration; but it regarded everything beyond the enforcement of law and order as "idle talk" and "vain-glory." The human capacity on which it based its claim of the natural innate productivity of society was the unquestionable productivity of *homo faber*. Conversely, antiquity knew full well types of human communities in which not the citizen of the *polis* and not the *res publica* as such established and determined the content of the public realm, but where the public life of the ordinary man was restricted to "working for the people" at large, that is, to being a *dēmiourgos*, a worker for the people as distinguished from an *oiketēs*, a household laborer and therefore a slave.[26]

24. *Laws* 716D quotes the saying of Protagoras textually, except that for the word "man" (*anthrōpos*), "the god" (*ho theos*) appears.

25. *Capital* (Modern Library ed.), p. 358, n. 3.

26. Early medieval history, and particularly the history of the craft guilds, offers a good illustration of the inherent truth in the ancient understanding of laborers as household inmates, as against craftsmen, who were considered workers for the people at large. For the "appearance [of the guilds] marks the second

The hallmark of these non-political communities was that their public place, the *agora*, was not a meeting place of citizens, but a market place where craftsmen could show and exchange their products. In Greece, moreover, it was the ever-frustrated ambition of all tyrants to discourage the citizens from worrying about public affairs, from idling their time away in unproductive *agoreuein* and *politeuesthai*, and to transform the *agora* into an assemblage of shops like the bazaars of oriental despotism. What characterized these market places, and later characterized the medieval cities' trade and craft districts, was that the display of goods for sale was accompanied by a display of their production. "Conspicuous production" (if we may vary Veblen's term) is, in fact, no less a trait of a society of producers than "conspicuous consumption" is a characteristic of a laborers' society.

Unlike the *animal laborans*, whose social life is worldless and herdlike and who therefore is incapable of building or inhabiting a public, worldly realm, *homo faber* is fully capable of having a public realm of his own, even though it may not be a political realm, properly speaking. His public realm is the exchange market, where he can show the products of his hand and receive the esteem which is due him. This inclination to showmanship is closely connected with and probably no less deeply rooted than the "propensity to truck, barter and exchange one thing for another," which, according to Adam Smith, distinguishes man from animal.[27] The point is that *homo faber*, the builder of the world and the producer of things, can find his proper relationship to other people only by exchanging his products with theirs, because these products them-

stage in the history of industry, the transition from the family system to the artisan or guild system. In the former there was no class of artisans properly so called . . . because all the needs of a family or other domestic groups . . . were satisfied by the labours of the members of the group itself" (W. J. Ashley, *An Introduction to English Economic History and Theory* [1931], p. 76).

In medieval German, the word *Störer* is an exact equivalent to the Greek word *dēmiourgos*. "Der griechische *dēmiourgos* heisst 'Störer', er geht beim Volk arbeiten, er geht auf die Stör." *Stör* means *dēmos* ("people"). (See Jost Trier, "Arbeit und Gemeinschaft," *Studium Generale*, Vol. III, No. 11 [November, 1950].)

27. He adds rather emphatically: "Nobody ever saw a dog make a fair and deliberate exchange of one bone for another with another dog" (*Wealth of Nations* [Everyman's ed.], I, 12).

selves are always produced in isolation. The privacy which the early modern age demanded as the supreme right of each member of society was actually the guaranty of isolation, without which no work can be produced. Not the onlookers and spectators on the medieval market places, where the craftsman in his isolation was exposed to the light of the public, but only the rise of the social realm, where the others are not content with beholding, judging, and admiring but wish to be admitted to the company of the crafts-man and to participate as equals in the work process, threatened the "splendid isolation" of the worker and eventually undermined the very notions of competence and excellence. This isolation from others is the necessary life condition for every mastership which consists in being alone with the "idea," the mental image of the thing to be. This mastership, unlike political forms of domination, is primarily a mastery of things and material and not of people. The latter, in fact, is quite secondary to the activity of craftsman-ship, and the words "worker" and "master"—*ouvrier* and *maître*—were originally used synonymously.[28]

The only company that grows out of workmanship directly is in the need of the master for assistants or in his wish to educate others in his craft. But the distinction between his skill and the unskilled help is temporary, like the distinction between adults and children. There can be hardly anything more alien or even more destructive to workmanship than teamwork, which actually is only a variety of the division of labor and presupposes the "breakdown of opera-tions into their simple constituent motions."[29] The team, the multi-

28. E. Levasseur, *Histoire des classes ouvrières et de l'industrie en France avant 1789* (1900): "Les mots maître et ouvrier étaient encore pris comme synonymes au 14e siècle" (p. 564, n. 2), whereas "au 15e siècle . . . la maîtrise est devenue un titre auquel il n'est permis à tous d'aspirer" (p. 572). Originally, "le mot ouvrier s'appliquait d'ordinaire à quiconque ouvrait, faisait ouvrage, maître ou valet" (p. 309). In the workshops themselves and outside them in social life, there was no great distinction between the master or the owner of the shop and the workers (p. 313). (See also Pierre Brizon, *Histoire du travail et des travailleurs* [4th ed.; 1926], pp. 39 ff.)

29. Charles R. Walker and Robert H. Guest, *The Man on the Assembly Line* (1952), p. 10. Adam Smith's famous description of this principle in pin-making (*op. cit.*, I, 4 ff.) shows clearly how machine work was preceded by the division of labor and derives its principle from it.

headed subject of all production carried out according to the principle of division of labor, possesses the same togetherness as the parts which form the whole, and each attempt of isolation on the part of the members of the team would be fatal to the production itself. But it is not only this togetherness which the master and workman lacks while actively engaged in production; the specifically political forms of being together with others, acting in concert and speaking with each other, are completely outside the range of his productivity. Only when he stops working and his product is finished can he abandon his isolation.

Historically, the last public realm, the last meeting place which is at least connected with the activity of *homo faber*, is the exchange market on which his products are displayed. The commercial society, characteristic of the earlier stages of the modern age or the beginnings of manufacturing capitalism, sprang from this "conspicuous production" with its concomitant hunger for universal possibilities of truck and barter, and its end came with the rise of labor and the labor society which replaced conspicuous production and its pride with "conspicuous consumption" and its concomitant vanity.

The people who met on the exchange market, to be sure, were no longer the fabricators themselves, and they did not meet as persons but as owners of commodities and exchange values, as Marx abundantly pointed out. In a society where exchange of products has become the chief public activity, even the laborers, because they are confronted with "money or commodity owners," become proprietors, "owners of their labor power." It is only at this point that Marx's famous self-alienation, the degradation of men into commodities, sets in, and this degradation is characteristic of labor's situation in a manufacturing society which judges men not as persons but as producers, according to the quality of their products. A laboring society, on the contrary, judges men according to the functions they perform in the labor process; while labor power in the eyes of *homo faber* is only the means to produce the necessarily higher end, that is, either a use object or an object for exchange, laboring society bestows upon labor power the same higher value it reserves for the machine. In other words, this society is only seemingly more "humane," although it is true that under its conditions the price of human labor rises to such an extent that it may seem to be more valued and more valuable than any given material or matter; in fact, it only foreshadows something even more "valuable," namely, the smoother functioning of the machine whose tremendous power of processing first standardizes and then devaluates all things into consumer goods.

Commercial society, or capitalism in its earlier stages when it was still possessed by a fiercely competitive and acquisitive spirit, is still ruled by the standards of *homo faber*. When *homo faber* comes out of his isolation, he appears as a merchant and trader and establishes the exchange market in this capacity. This market must exist prior to the rise of a manufacturing class, which then produces exclusively for the market, that is, produces exchange objects rather than use things. In this process from isolated crafts-manship to manufacturing for the exchange market, the finished end product changes its quality somewhat but not altogether. Durability, which alone determines if a thing can exist as a thing and endure in the world as a distinct entity, remains the supreme criterion, although it no longer makes a thing fit for use but rather fit to "be stored up beforehand" for future exchange.[30]

This is the change in quality reflected in the current distinction between use and exchange value, whereby the latter is related to the former as the merchant and trader is related to the fabricator and manufacturer. In so far as *homo faber* fabricates use objects, he not only produces them in the privacy of isolation but also for the privacy of usage, from which they emerge and appear in the public realm when they become commodities in the exchange market. It has frequently been remarked and unfortunately as frequently been forgotten that value, being "an idea of proportion between the possession of one thing and the possession of another in the conception of man,"[31] "always means value in exchange."[32] For it is only in the exchange market, where everything can be exchanged for something else, that all things, whether they are products of labor

30. Adam Smith, *op. cit.*, II, 241.

31. This definition was given by the Italian economist Abbey Galiani. I quote from Hannah R. Sewall, *The Theory of Value before Adam Smith* (1901) ("Publications of the American Economic Association," 3d Ser., Vol. II, No. 3), p. 92.

32. Alfred Marshall, *Principles of Economics* (1920), I, 8.

or work, consumer goods or use objects, necessary for the life of the body or the convenience of living or the life of the mind, become "values." This value consists solely in the esteem of the public realm where the things appear as commodities, and it is neither labor, nor work, nor capital, nor profit, nor material, which bestows such value upon an object, but only and exclusively the public realm where it appears to be esteemed, demanded, or neglected. Value is the quality a thing can never possess in privacy but acquires automatically the moment it appears in public. This "marketable value," as Locke very clearly pointed out, has nothing to do with "the intrinsick natural worth of anything"[33] which is an objective quality of the thing itself, "outside the will of the individual purchaser or seller; something attached to the thing itself, existing whether he liked it or not, and that he ought to recognize."[34] This intrinsic worth of a thing can be changed only through the change of the thing itself—thus one ruins the worth of a table by depriving it of one of its legs—whereas "the marketable value" of a commodity is altered by "the alteration of some proportion which that commodity bears to something else."[35]

Values, in other words, in distinction from things or deeds or ideas, are never the products of a specific human activity, but come into being whenever any such products are drawn into the ever-changing relativity of exchange between the members of society.

33. "Considerations upon the Lowering of Interest and Raising the Value of Money," *Collected Works* (1801), II, 21.

34. W. J. Ashley (*op. cit.*, p. 140) remarks that "the fundamental difference between the medieval and modern point of view . . . is that, with us, value is something entirely subjective; it is what each individual cares to give for a thing. With Aquinas it was something objective." This is true only to an extent, for "the first thing upon which the medieval teachers insist is that value is not determined by the intrinsic excellence of the thing itself, because, if it were, a fly would be more valuable than a pearl as being intrinsically more excellent" (George O'Brien, *An Essay on Medieval Economic Teaching* [1920], p. 109). The discrepancy is resolved if one introduces Locke's distinction between "worth" and "value," calling the former *valor naturalis* and the latter *pretium* and also *valor*. This distinction exists, of course, in all but the most primitive societies, but in the modern age the former disappears more and more in favor of the latter. (For medieval teaching, see also Slater, "Value in Theology and Political Economy," *Irish Ecclesiastical Record* [September, 1901].)

35. Locke, *Second Treatise of Civil Government*, sec. 22.

Nobody, as Marx rightly insisted, seen "in his isolation produces values," and nobody, he could have added, in his isolation cares about them; things or ideas or moral ideals "become values only in their social relationship."[36]

The confusion in classical economics,[37] and the worse confusion arising from the use of the term "value" in philosophy, were originally caused by the fact that the older word "worth," which we still find in Locke, was supplanted by the seemingly more scientific term, "use value." Marx, too, accepted this terminology and, in line with his repugnance to the public realm, saw quite consistently in the change from use value to exchange value the original sin of capitalism. But against these sins of a commercial society, where indeed the exchange market is the most important public place and where therefore every thing becomes an exchangeable value, a commodity, Marx did not summon up the "intrinsick" objective worth of the thing in itself. In its stead he put the function things have in the consuming life process of men which knows neither objective and intrinsic worth nor subjective and socially determined value. In the socialist equal distribution of all goods to all who labor, every tangible thing dissolves into a mere function in the regeneration process of life and labor power.

However, this verbal confusion tells only one part of the story. The reason for Marx's stubborn retention of the term "use value," as well as for the numerous futile attempts to find some objective source—such as labor, or land, or profit—for the birth of values, was that nobody found it easy to accept the simple fact that no "absolute value" exists in the exchange market, which is the proper sphere for values, and that to look for it resembled nothing so much as the attempt to square the circle. The much deplored devaluation of all things, that is, the loss of all intrinsic worth, begins with their transformation into values or commodities, for from this moment on they exist only in relation to some other thing which can

36. *Das Kapital*, III, 689 (*Marx-Engels Gesamtausgabe*, Part II [Zürich, 1933]).

37. The clearest illustration of the confusion is Ricardo's theory of value especially his desperate belief in an absolute value. (The interpretations in Gunnar Myrdal, *The Political Element in the Development of Economic Theory* [1953], pp. 66 ff., and Walter A. Weisskopf, *The Psychology of Economics* [1955], ch. 3, are excellent.)

be acquired in their stead. Universal relativity, that a thing exists only in relation to other things, and loss of intrinsic worth, that nothing any longer possesses an "objective" value independent of the ever-changing estimations of supply and demand, are inherent in the very concept of value itself.[38] The reason why this development, which seems inevitable in a commercial society, became a deep source of uneasiness and eventually constituted the chief problem of the new science of economics was not even relativity as such, but rather the fact that *homo faber*, whose whole activity is determined by the constant use of yardsticks, measurements, rules, and standards, could not bear the loss of "absolute" standards or yardsticks. For money, which obviously serves as the common denominator for the variety of things so that they can be exchanged for each other, by no means possesses the independent and objective existence, transcending all uses and surviving all manipulation, that the yardstick or any other measurement possesses with regard to the things it is supposed to measure and to the men who handle them.

It is this loss of standards and universal rules, without which no world could ever be erected by man, that Plato already perceived in the Protagorean proposal to establish man, the fabricator of things, and the use he makes of them, as their supreme measure. This shows how closely the relativity of the exchange market is connected with the instrumentality arising out of the world of the craftsman and the experience of fabrication. The former, indeed, develops without break and consistently from the latter. Plato's reply, however—not man, a "god is the measure of all things"

38. The truth of Ashley's remark, which we quoted above (n. 34), lies in the fact that the Middle Ages did not know the exchange market, properly speaking. To the medieval teachers the value of a thing was either determined by its worth or by the objective needs of men—as for instance in Buridan: *valor rerum aestimatur secundum humanam indigentiam*—and the "just price" was normally the result of the common estimate, except that "on account of the varied and corrupt desires of man, it becomes expedient that the medium should be fixed according to the judgment of some wise men" (Gerson *De contractibus* i. 9, quoted from O'Brien, *op. cit.*, pp. 104 ff.). In the absence of an exchange market, it was inconceivable that the value of one thing should consist solely in its relationship or proportion to another thing. The question, therefore, is not so much whether value is objective or subjective, but whether it can be absolute or indicates only the relationship between things.

—would be an empty, moralizing gesture if it were really true, as the modern age assumed, that instrumentality under the disguise of usefulness rules the realm of the finished world as exclusively as it rules the activity through which the world and all things it contains came into being.

23

THE PERMANENCE OF THE WORLD AND THE WORK OF ART

Among the things that give the human artifice the stability without which it could never be a reliable home for men are a number of objects which are strictly without any utility whatsoever and which, moreover, because they are unique, are not exchangeable and therefore defy equalization through a common denominator such as money; if they enter the exchange market, they can only be arbitrarily priced. Moreover, the proper intercourse with a work of art is certainly not "using" it; on the contrary, it must be removed carefully from the whole context of ordinary use objects to attain its proper place in the world. By the same token, it must be removed from the exigencies and wants of daily life, with which it has less contact than any other thing. Whether this uselessness of art objects has always pertained or whether art formerly served the so-called religious needs of men as ordinary use objects serve more ordinary needs does not enter the argument. Even if the historical origin of art were of an exclusively religious or mythological character, the fact is that art has survived gloriously its severance from religion, magic, and myth.

Because of their outstanding permanence, works of art are the most intensely worldly of all tangible things; their durability is almost untouched by the corroding effect of natural processes, since they are not subject to the use of living creatures, a use which, indeed, far from actualizing their own inherent purpose—as the purpose of a chair is actualized when it is sat upon—can only destroy them. Thus, their durability is of a higher order than that which all things need in order to exist at all; it can attain permanence throughout the ages. In this permanence, the very stability of the human artifice, which, being inhabited and used by mortals, can never be absolute, achieves a representation of its own. Nowhere else does the sheer durability of the world of things appear in such purity and clarity, nowhere else therefore does this thing-world reveal itself so spectacularly as the non-mortal home for mortal beings. It is as though worldly stability had become transparent in the permanence of art, so that a premonition of immortality, not the immortality of the soul or of life but of something immortal achieved by mortal hands, has become tangibly present, to shine and to be seen, to sound and to be heard, to speak and to be read.

The immediate source of the art work is the human capacity for thought, as man's "propensity to truck and barter" is the source of exchange objects, and as his ability to use is the source of use things. These are capacities of man and not mere attributes of the human animal like feelings, wants, and needs, to which they are related and which often constitute their content. Such human properties are as unrelated to the world which man creates as his home on earth as the corresponding properties of other animal species, and if they were to constitute a man-made environment for the human animal, this would be a non-world, the product of emanation rather than of creation. Thought is related to feeling and transforms its mute and inarticulate despondency, as exchange transforms the naked greed of desire and usage transforms the desperate longing of needs—until they all are fit to enter the world and to be transformed into things, to become reified. In each instance, a human capacity which by its very nature is world-open and communicative transcends and releases into the world a passionate intensity from its imprisonment within the self.

In the case of art works, reification is more than mere transformation; it is transfiguration, a veritable metamorphosis in which it is as though the course of nature which wills that all fire burn to ashes is reverted and even dust can burst into flames.[39] Works of

39. The text refers to a poem by Rilke on art, which under the title "Magic," describes this transfiguration. It reads as follows: "Aus unbeschreiblicher Verwandlung stammen / solche Gebilde—: Fühl! und glaub! / Wir leidens oft: zu Asche werden Flammen, / doch, in der Kunst: zur Flamme wird der Staub. / Hier ist Magie. In das Bereich des Zaubers / scheint das gemeine Wort hinaufgestuft ... / und ist doch wirklich wie der Ruf des Taubers, / der nach der unsichtbaren Taube ruft" (in *Aus Taschen-Büchern und Merk-Blättern* [1950]).

art are thought things, but this does not prevent their being things. The thought process by itself no more produces and fabricates tangible things, such as books, paintings, sculptures, or compositions, than usage by itself produces and fabricates houses and furniture. The reification which occurs in writing something down, painting an image, modeling a figure, or composing a melody is of course related to the thought which preceded it, but what actually makes the thought a reality and fabricates things of thought is the same workmanship which, through the primordial instrument of human hands, builds the other durable things of the human artifice.

We mentioned before that this reification and materialization, without which no thought can become a tangible thing, is always paid for, and that the price is life itself: it is always the "dead letter" in which the "living spirit" must survive, a deadness from which it can be rescued only when the dead letter comes again into contact with a life willing to resurrect it, although this resurrection of the dead shares with all living things that it, too, will die again. This deadness, however, though somehow present in all art and indicating, as it were, the distance between thought's original home in the heart or head of man and its eventual destination in the world, varies in the different arts. In music and poetry, the least "materialistic" of the arts because their "material" consists of sounds and words, reification and the workmanship it demands are kept to a minimum. The young poet and the musical child prodigy can attain a perfection without much training and experience—a phenomenon hardly matched in painting, sculpture, or architecture.

Poetry, whose material is language, is perhaps the most human and least worldly of the arts, the one in which the end product remains closest to the thought that inspired it. The durability of a poem is produced through condensation, so that it is as though language spoken in utmost density and concentration were poetic in itself. Here, remembrance, *Mnēmosynē*, the mother of the muses, is directly transformed into memory, and the poet's means to achieve the transformation is rhythm, through which the poem becomes fixed in the recollection almost by itself. It is this closeness to living recollection that enables the poem to remain, to retain its durability, outside the printed or the written page, and though the "quality" of a poem may be subject to a variety of

standards, its "memorability" will inevitably determine its durability, that is, its chance to be permanently fixed in the recollection
of humanity. Of all things of thought, poetry is closest to thought,
and a poem is less a thing than any other work of art; yet even a
poem, no matter how long it existed as a living spoken word in the
recollection of the bard and those who listened to him, will eventually be "made," that is, written down and transformed into a
tangible thing among things, because remembrance and the gift of
recollection, from which all desire for imperishability springs, need
tangible things to remind them, lest they perish themselves.[40]

Thought and cognition are not the same. Thought, the source
of art works, is manifest without transformation or transfiguration
in all great philosophy, whereas the chief manifestation of the cognitive processes, by which we acquire and store up knowledge, is
the sciences. Cognition always pursues a definite aim, which can be
set by practical considerations as well as by "idle curiosity"; but
once this aim is reached, the cognitive process has come to an end.
Thought, on the contrary, has neither an end nor an aim outside
itself, and it does not even produce results; not only the utilitarian
philosophy of *homo faber* but also the men of action and the lovers
of results in the sciences have never tired of pointing out how entirely "useless" thought is—as useless, indeed, as the works of art
it inspires. And not even to these useless products can thought lay
claim, for they as well as the great philosophic systems can hardly
be called the results of pure thinking, strictly speaking, since it is
precisely the thought process which the artist or writing philosopher must interrupt and transform for the materializing reification

40. The idiomatic "make a poem" or *faire des vers* for the activity of the poet
already relates to this reification. The same is true for the German *dichten*, which
probably comes from the Latin *dictare:* "das ausgesonnene geistig Geschaffene
niederschreiben oder zum Niederschreiben vorsagen" (Grimm's *Wörterbuch*); the
same would be true if the word were derived, as is now suggested by the
Etymologisches Wörterbuch (1951) of Kluge/Götze, from *tichen*, an old word for
schaffen, which is perhaps related to the Latin *fingere*. In this case, the poetic
activity which produces the poem before it is written down is also understood
as "making." Thus Democritus praised the divine genius of Homer, who "framed
a cosmos out of all kinds of words"—*epeōn kosmon etektēnato pantoiōn* (Diels,
op. cit., B21). The same emphasis on the craftsmanship of poets is present in the
Greek idiom for the art of poetry: *tektōnes hymnōn.*

of his work. The activity of thinking is as relentless and repetitive as life itself, and the question whether thought has any meaning at all constitutes the same unanswerable riddle as the question for the meaning of life; its processes permeate the whole of human existence so intimately that its beginning and end coincide with the beginning and end of human life itself. Thought, therefore, although it inspires the highest worldly productivity of *homo faber*, is by no means his prerogative; it begins to assert itself as his source of inspiration only where he overreaches himself, as it were, and begins to produce useless things, objects which are unrelated to material or intellectual wants, to man's physical needs no less than to his thirst for knowledge. Cognition, on the other hand, belongs to all, and not only to intellectual or artistic work processes; like fabrication itself, it is a process with a beginning and end, whose usefulness can be tested, and which, if it produces no results, has failed, like a carpenter's workmanship has failed when he fabricates a two-legged table. The cognitive processes in the sciences are basically not different from the function of cognition in fabrication; scientific results produced through cognition are added to the human artifice like all other things.

Both thought and cognition, furthermore, must be distinguished from the power of logical reasoning which is manifest in such operations as deductions from axiomatic or self-evident statements, subsumption of particular occurrences under general rules, or the techniques of spinning out consistent chains of conclusions. In these human faculties we are actually confronted with a sort of brain power which in more than one respect resembles nothing so much as the labor power the human animal develops in its metabolism with nature. The mental processes which feed on brain power we usually call intelligence, and this intelligence can indeed be measured by intelligence tests as bodily strength can be measured by other devices. Their laws, the laws of logic, can be discovered like other laws of nature because they are ultimately rooted in the structure of the human brain, and they possess, for the normally healthy individual, the same force of compulsion as the driving necessity which regulates the other functions of our bodies. It is in the structure of the human brain to be compelled to admit that two and two equal four. If it were true that man is an *animal rationale* in the sense in which the modern age understood the term, namely, an animal species which differs from other animals in that it is endowed with superior brain power, then the newly invented electronic machines, which, sometimes to the dismay and sometimes to the confusion of their inventors, are so spectacularly more "intelligent" than human beings, would indeed be *homunculi*. As it is, they are, like all machines, mere substitutes and artificial improvers of human labor power, following the time-honored device of all division of labor to break down every operation into its simplest constituent motions, substituting, for instance, repeated addition for multiplication. The superior power of the machine is manifest in its speed, which is far greater than that of human brain power; because of this superior speed, the machine can dispense with multiplication, which is the pre-electronic technical device to speed up addition. All that the giant computers prove is that the modern age was wrong to believe with Hobbes that rationality, in the sense of "reckoning with consequences," is the highest and most human of man's capacities, and that the life and labor philosophers, Marx or Bergson or Nietzsche, were right to see in this type of intelligence, which they mistook for reason, a mere function of the life process itself, or, as Hume put it, a mere "slave of the passions." Obviously, this brain power and the compelling logical processes it generates are not capable of erecting a world, are as worldless as the compulsory processes of life, labor, and consumption.

One of the striking discrepancies in classical economics is that the same theorists who prided themselves on the consistency of their utilitarian outlook frequently took a very dim view of sheer utility. As a rule, they were well aware that the specific productivity of work lies less in its usefulness than in its capacity for producing durability. By this discrepancy, they tacitly admit the lack of realism in their own utilitarian philosophy. For although the durability of ordinary things is but a feeble reflection of the permanence of which the most worldly of all things, works of art, are capable, something of this quality—which to Plato was divine because it approaches immortality—is inherent in every thing as a thing, and it is precisely this quality or the lack of it that shines forth in its shape and makes it beautiful or ugly. To be sure, an ordinary use object is not and should not be intended to be beautiful; yet whatever has a shape at all and is seen cannot help being either beautiful, ugly, or something in-between. Everything that is, must appear, and nothing can appear without a shape of its own; hence there is in fact no thing that does not in some way transcend its functional use, and its transcendence, its beauty or ugliness, is identical with appearing publicly and being seen. By the same token, namely, in its sheer worldly existence, every thing also transcends the sphere of pure instrumentality once it is completed. The standard by which a thing's excellence is judged is never mere usefulness, as though an ugly table will fulfil the same function as a handsome one, but its adequacy or inadequacy to what it should *look* like, and this is, in Platonic language, nothing but its adequacy or inadequacy to the *eidos* or *idea*, the mental image, or rather the image seen by the inner eye, that preceded its coming into the world and survives its potential destruction. In other words, even use objects are judged not only according to the subjective needs of men but by the objective standards of the world where they will find their place, to last, to be seen, and to be used.

The man-made world of things, the human artifice erected by *homo faber*, becomes a home for mortal men, whose stability will endure and outlast the ever-changing movement of their lives and actions, only insomuch as it transcends both the sheer functionalism of things produced for consumption and the sheer utility of objects produced for use. Life in its non-biological sense, the span of time each man has between birth and death, manifests itself in action and speech, both of which share with life its essential futility. The "doing of great deeds and the speaking of great words" will leave no trace, no product that might endure after the moment of action and the spoken word has passed. If the *animal laborans* needs the help of *homo faber* to ease his labor and remove his pain, and if mortals need his help to erect a home on earth, acting and speaking men need the help of *homo faber* in his highest capacity, that is, the help of the artist, of poets and historiographers, of monument-builders or writers, because without them the only product of their activity, the story they enact and tell, would not survive at all. In order to be what the world is always meant to be, a home for men during their life on earth, the human artifice must be a place fit for action and speech, for activities not only entirely useless for the necessities of life but of an entirely different nature from the manifold activities of fabrication by which the world itself and all things in it are produced. We need not choose here between Plato and Protagoras, or decide whether man or a god should be the measure of all things; what is certain is that the measure can be neither the driving necessity of biological life and labor nor the utilitarian instrumentalism of fabrication and usage.

CHAPTER V

Action

All sorrows can be borne if you put them into a story or tell a story about them.

ISAK DINESEN

Nam in omni actione principaliter intenditur ab agente, sive necessitate naturae sive voluntarie agat, propriam similitudinem explicare; unde fit quod omne agens, in quantum huiusmodi, delectatur, quia, cum omne quod est appetat suum esse, ac in agendo agentis esse modammodo amplietur, sequitur de necessitate delectatio. . . . Nihil igitur agit nisi tale existens quale patiens fieri debet.

(For in every action what is primarily intended by the doer, whether he acts from natural necessity or out of free will, is the disclosure of his own image. Hence it comes about that every doer, in so far as he does, takes delight in doing; since everything that is desires its own being, and since in action the being of the doer is somehow intensified, delight necessarily follows. . . . Thus, nothing acts unless [by acting] it makes patent its latent self.)

DANTE

24

THE DISCLOSURE OF THE AGENT IN SPEECH AND ACTION

Human plurality, the basic condition of both action and speech, has the twofold character of equality and distinction. If men were not equal, they could neither understand each other and those who came before them nor plan for the future and foresee the needs of those who will come after them. If men were not distinct, each human being distinguished from any other who is, was, or will ever be, they would need neither speech nor action to make themselves understood. Signs and sounds to communicate immediate, identical needs and wants would be enough.

Human distinctness is not the same as otherness—the curious quality of *alteritas* possessed by everything that is and therefore, in medieval philosophy, one of the four basic, universal characteristics of Being, transcending every particular quality. Otherness, it is true, is an important aspect of plurality, the reason why all our definitions are distinctions, why we are unable to say what anything is without distinguishing it from something else. Otherness in its most abstract form is found only in the sheer multiplication of inorganic objects, whereas all organic life already shows variations and distinctions, even between specimens of the same species. But only man can express this distinction and distinguish himself, and only he can communicate himself and not merely something—thirst or hunger, affection or hostility or fear. In man, otherness, which he shares with everything that is, and distinctness, which he shares with everything alive, become uniqueness, and human plurality is the paradoxical plurality of unique beings.

Speech and action reveal this unique distinctness. Through them, men distinguish themselves instead of being merely distinct; they are the modes in which human beings appear to each other, not indeed as physical objects, but *qua* men. This appearance, as distinguished from mere bodily existence, rests on initiative, but it is an initiative from which no human being can refrain and still be human. This is true of no other activity in the *vita activa*. Men can very well live without laboring, they can force others to labor for them, and they can very well decide merely to use and enjoy the world of things without themselves adding a single useful object to it; the life of an exploiter or slaveholder and the life of a parasite may be unjust, but they certainly are human. A life without speech and without action, on the other hand—and this is the only way of life that in earnest has renounced all appearance and all vanity in the biblical sense of the word—is literally dead to the world; it has ceased to be a human life because it is no longer lived among men.

With word and deed we insert ourselves into the human world, and this insertion is like a second birth, in which we confirm and take upon ourselves the naked fact of our original physical appearance. This insertion is not forced upon us by necessity, like labor, and it is not prompted by utility, like work. It may be stimulated by the presence of others whose company we may wish to join, but it is never conditioned by them; its impulse springs from the beginning which came into the world when we were born and to which we respond by beginning something new on our own initiative.[1] To act, in its most general sense, means to take an initiative, to begin (as the Greek word *archein*, "to begin," "to lead," and eventually "to rule," indicates), to set something into motion (which is the original meaning of the Latin *agere*). Because they are *initium*, newcomers and beginners by virtue of birth, men take initiative, are prompted into action. [*Initium*] *ergo ut esset, creatus est homo, ante quem nullus fuit* ("that there be a beginning, man was created before whom there was nobody"), said Augustine in his political philosophy.[2] This beginning is not the same as the beginning of the world;[3] it is not the beginning of something but of somebody, who is a beginner himself. With the creation of man, the principle of beginning came into the world itself, which, of course, is only another way of saying that the principle of freedom was created when man was created but not before.

It is in the nature of beginning that something new is started

1. This description is supported by recent findings in psychology and biology which also stress the inner affinity between speech and action, their spontaneity and practical purposelessness. See especially Arnold Gehlen, *Der Mensch: Seine Natur und seine Stellung in der Welt* (1955), which gives an excellent summary of the results and interpretations of current scientific research and contains a wealth of valuable insights. That Gehlen, like the scientists upon whose results he bases his own theories, believes that these specifically human capabilities are also a "biological necessity," that is, necessary for a biologically weak and ill-fitted organism such as man, is another matter and need not concern us here.

2. *De civitate Dei* xii. 20.

3. According to Augustine, the two were so different that he used a different word to indicate the beginning which is man (*initium*), designating the beginning of the world by *principium*, which is the standard translation for the first Bible verse. As can be seen from *De civitate Dei* xi. 32, the word *principium* carried for Augustine a much less radical meaning; the beginning of the world "does not mean that nothing was made before (for the angels were)," whereas he adds explicitly in the phrase quoted above with reference to man that nobody was before him.

which cannot be expected from whatever may have happened before. This character of startling unexpectedness is inherent in all beginnings and in all origins. Thus, the origin of life from inorganic matter is an infinite improbability of inorganic processes, as is the coming into being of the earth viewed from the standpoint of processes in the universe, or the evolution of human out of animal life. The new always happens against the overwhelming odds of statistical laws and their probability, which for all practical, everyday purposes amounts to certainty; the new therefore always appears in the guise of a miracle. The fact that man is capable of action means that the unexpected can be expected from him, that he is able to perform what is infinitely improbable. And this again is possible only because each man is unique, so that with each birth something uniquely new comes into the world. With respect to this somebody who is unique it can be truly said that nobody was there before. If action as beginning corresponds to the fact of birth, if it is the actualization of the human condition of natality, then speech corresponds to the fact of distinctness and is the actualization of the human condition of plurality, that is, of living as a distinct and unique being among equals.

Action and speech are so closely related because the primordial and specifically human act must at the same time contain the answer to the question asked of every newcomer: "Who are you?" This disclosure of who somebody is, is implicit in both his words and his deeds; yet obviously the affinity between speech and revelation is much closer than that between action and revelation,[4] just as the affinity between action and beginning is closer than that between speech and beginning, although many, and even most acts, are performed in the manner of speech. Without the accompaniment of speech, at any rate, action would not only lose its revelatory character, but, and by the same token, it would lose its subject, as it were; not acting men but performing robots would achieve what, humanly speaking, would remain incomprehensible. Speechless action would no longer be action because there would no longer be an actor, and the actor, the doer of

4. This is the reason why Plato says that *lexis* ("speech") adheres more closely to truth than *praxis*.

deeds, is possible only if he is at the same time the speaker of words. The action he begins is humanly disclosed by the word, and though his deed can be perceived in its brute physical appearance without verbal accompaniment, it becomes relevant only through the spoken word in which he identifies himself as the actor, announcing what he does, has done, and intends to do.

No other human performance requires speech to the same extent as action. In all other performances speech plays a subordinate role, as a means of communication or a mere accompaniment to something that could also be achieved in silence. It is true that speech is extremely useful as a means of communication and information, but as such it could be replaced by a sign language, which then might prove to be even more useful and expedient to convey certain meanings, as in mathematics and other scientific disciplines or in certain forms of teamwork. Thus, it is also true that man's capacity to act, and especially to act in concert, is extremely useful for purposes of self-defense or of pursuit of interests; but if nothing more were at stake here than to use action as a means to an end, it is obvious that the same end could be much more easily attained in mute violence, so that action seems a not very efficient substitute for violence, just as speech, from the viewpoint of sheer utility, seems an awkward substitute for sign language.

In acting and speaking, men show who they are, reveal actively their unique personal identities and thus make their appearance in the human world, while their physical identities appear without any activity of their own in the unique shape of the body and sound of the voice. This disclosure of "who" in contradistinction to "what" somebody is—his qualities, gifts, talents, and shortcomings, which he may display or hide—is implicit in everything somebody says and does. It can be hidden only in complete silence and perfect passivity, but its disclosure can almost never be achieved as a wilful purpose, as though one possessed and could dispose of this "who" in the same manner he has and can dispose of his qualities. On the contrary, it is more than likely that the "who," which appears so clearly and unmistakably to others, remains hidden from the person himself, like the *daimōn* in Greek religion which accompanies each man throughout his life, always looking over his shoulder from behind and thus visible only to those he encounters.

This revelatory quality of speech and action comes to the fore where people are *with* others and neither for nor against them— that is, in sheer human togetherness. Although nobody knows whom he reveals when he discloses himself in deed or word, he must be willing to risk the disclosure, and this neither the doer of good works, who must be without self and preserve complete anonymity, nor the criminal, who must hide himself from others, can take upon themselves. Both are lonely figures, the one being for, the other against, all men; they, therefore, remain outside the pale of human intercourse and are, politically, marginal figures who usually enter the historical scene in times of corruption, dis-integration, and political bankruptcy. Because of its inherent tendency to disclose the agent together with the act, action needs for its full appearance the shining brightness we once called glory, and which is possible only in the public realm.

Without the disclosure of the agent in the act, action loses its specific character and becomes one form of achievement among others. It is then indeed no less a means to an end than making is a means to produce an object. This happens whenever human togetherness is lost, that is, when people are only for or against other people, as for instance in modern warfare, where men go into action and use means of violence in order to achieve certain objectives for their own side and against the enemy. In these instances, which of course have always existed, speech becomes indeed "mere talk," simply one more means toward the end, whether it serves to deceive the enemy or to dazzle everybody with propaganda; here words reveal nothing, disclosure comes only from the deed itself, and this achievement, like all other achievements, cannot disclose the "who," the unique and distinct identity of the agent.

In these instances action has lost the quality through which it transcends mere productive activity, which, from the humble making of use objects to the inspired creation of art works, has no more meaning than is revealed in the finished product and does not intend to show more than is plainly visible at the end of the production process. Action without a name, a "who" attached to it, is meaningless, whereas an art work retains its relevance whether or not we know the master's name. The monuments to the "Unknown Soldier" after World War I bear testimony to the then still existing need for glorification, for finding a "who," an identifiable somebody whom four years of mass slaughter should have revealed. The frustration of this wish and the unwillingness to resign oneself to the brutal fact that the agent of the war was actually nobody inspired the erection of the monuments to the "unknown," to all those whom the war had failed to make known and had robbed thereby, not of their achievement, but of their human dignity.[5]

25

THE WEB OF RELATIONSHIPS AND THE ENACTED STORIES

The manifestation of who the speaker and doer unexchangeably is, though it is plainly visible, retains a curious intangibility that confounds all efforts toward unequivocal verbal expression. The moment we want to say *who* somebody is, our very vocabulary leads us astray into saying *what* he is; we get entangled in a description of qualities he necessarily shares with others like him; we begin to describe a type or a "character" in the old meaning of the word, with the result that his specific uniqueness escapes us.

This frustration has the closest affinity with the well-known philosophic impossibility to arrive at a definition of man, all definitions being determinations or interpretations of *what* man is, of qualities, therefore, which he could possibly share with other living beings, whereas his specific difference would be found in a determination of what kind of a "who" he is. Yet apart from this philosophic perplexity, the impossibility, as it were, to solidify in words the living essence of the person as it shows itself in the flux of action and speech, has great bearing upon the whole realm of human affairs, where we exist primarily as acting and speaking beings. It excludes in principle our ever being able to handle these affairs as we handle things whose nature is at our

5. William Faulkner's *A Fable* (1954) surpasses almost all of World War I literature in perceptiveness and clarity because its hero is the Unknown Soldier.

disposal because we can name them. The point is that the manifestation of the "who" comes to pass in the same manner as the notoriously unreliable manifestations of ancient oracles, which, according to Heraclitus, "neither reveal nor hide in words, but give manifest signs."[6] This is a basic factor in the equally notorious uncertainty not only of all political matters, but of all affairs that go on between men directly, without the intermediary, stabilizing, and solidifying influence of things.[7]

This is only the first of many frustrations by which action, and consequently the togetherness and intercourse of men, are ridden. It is perhaps the most fundamental of those we shall deal with, in so far as it does not rise out of comparisons with more reliable and productive activities, such as fabrication or contemplation or cognition or even labor, but indicates something that frustrates action in terms of its own purposes. What is at stake is the revelatory character without which action and speech would lose all human relevance.

Action and speech go on between men, as they are directed toward them, and they retain their agent-revealing capacity even if their content is exclusively "objective," concerned with the matters of the world of things in which men move, which physically lies between them and out of which arise their specific, objective, worldly interests. These interests constitute, in the word's most literal significance, something which *inter-est*, which lies between people and therefore can relate and bind them together. Most action and speech is concerned with this in-between, which varies with each group of people, so that most words and deeds are *about* some worldly objective reality in addition to being a disclosure of the acting and speaking agent. Since this disclosure of the subject is an integral part of all, even the most "objective" intercourse, the physical, worldly in-between along with its inter-

6. *Oute legei oute kryptei alla sēmainei* (Diels, *Fragmente der Vorsokratiker* [4th ed., 1922], frag. B93).

7. Socrates used the same word as Heraclitus, *sēmainein* ("to show and give signs"), for the manifestation of his *daimonion* (Xenophon *Memorabilia* i. 1. 2, 4). If we are to trust Xenophon, Socrates likened his *daimonion* to the oracles and insisted that both should be used only for human affairs, where nothing is certain, and not for problems of the arts and crafts, where everything is predictable (*ibid.* 7–9).

ests is overlaid and, as it were, overgrown with an altogether different in-between which consists of deeds and words and owes
its origin exclusively to men's acting and speaking directly *to* one
another. This second, subjective in-between is not tangible, since
there are no tangible objects into which it could solidify; the
process of acting and speaking can leave behind no such results
and end products. But for all its intangibility, this in-between is
no less real than the world of things we visibly have in common.
We call this reality the "web" of human relationships, indicating
by the metaphor its somewhat intangible quality.

To be sure, this web is no less bound to the objective world of
things than speech is to the existence of a living body, but the relationship is not like that of a façade or, in Marxian terminology,
of an essentially superfluous superstructure affixed to the useful
structure of the building itself. The basic error of all materialism
in politics—and this materialism is not Marxian and not even
modern in origin, but as old as our history of political theory[8]—
is to overlook the inevitability with which men disclose themselves as subjects, as distinct and unique persons, even when they
wholly concentrate upon reaching an altogether worldly, material
object. To dispense with this disclosure, if indeed it could ever be
done, would mean to transform men into something they are not;
to deny, on the other hand, that this disclosure is real and has
consequences of its own is simply unrealistic.

The realm of human affairs, strictly speaking, consists of the

8. Materialism in political theory is at least as old as the Platonic-Aristotelian
assumption that political communities (*poleis*)—and not only family life or the
coexistence of several households (*oikiai*)—owe their existence to material necessity. (For Plato see *Republic* 369, where the *polis*' origin is seen in our wants and
lack of self-sufficiency. For Aristotle, who here as elsewhere is closer to current
Greek opinion than Plato, see *Politics* 1252b29: "The *polis* comes into existence
for the sake of living, but remains in existence for the sake of living well.") The
Aristotelian concept of *sympheron*, which we later encounter in Cicero's *utilitas*,
must be understood in this context. Both, in turn, are forerunners of the later
interest theory which is fully developed as early as Bodin—as kings rule over
peoples, Interest rules over kings. In the modern development, Marx is outstanding not because of his materialism, but because he is the only political thinker who
was consistent enough to base his theory of material interest on a demonstrably
material human activity, on laboring—that is, on the metabolism of the human
body with matter.

web of human relationships which exists wherever men live to-gether. The disclosure of the "who" through speech, and the setting of a new beginning through action, always fall into an already existing web where their immediate consequences can be felt. Together they start a new process which eventually emerges as the unique life story of the newcomer, affecting uniquely the life stories of all those with whom he comes into contact. It is because of this already existing web of human relationships, with its innumerable, conflicting wills and intentions, that action al-most never achieves its purpose; but it is also because of this medium, in which action alone is real, that it "produces" stories with or without intention as naturally as fabrication produces tangible things. These stories may then be recorded in documents and monuments, they may be visible in use objects or art works, they may be told and retold and worked into all kinds of material. They themselves, in their living reality, are of an altogether dif-ferent nature than these reifications. They tell us more about their subjects, the "hero" in the center of each story, than any product of human hands ever tells us about the master who produced it, and yet they are not products, properly speaking. Although everybody started his life by inserting himself into the human world through action and speech, nobody is the author or producer of his own life story. In other words, the stories, the results of action and speech, reveal an agent, but this agent is not an author or producer. Somebody began it and is its subject in the twofold sense of the word, namely, its actor and sufferer, but nobody is its author.

That every individual life between birth and death can even-tually be told as a story with beginning and end is the prepolitical and prehistorical condition of history, the great story without beginning and end. But the reason why each human life tells its story and why history ultimately becomes the storybook of man-kind, with many actors and speakers and yet without any tangible authors, is that both are the outcome of action. For the great unknown in history, that has baffled the philosophy of history in the modern age, arises not only when one considers history as a whole and finds that its subject, mankind, is an abstraction which never can become an active agent; the same unknown has baffled political philosophy from its beginning in antiquity and contributed to the general contempt in which philosophers since Plato have held the realm of human affairs. The perplexity is that in any series of events that together form a story with a unique meaning we can at best isolate the agent who set the whole process into motion; and although this agent frequently remains the subject, the "hero" of the story, we never can point unequivocally to him as the author of its eventual outcome.

It is for this reason that Plato thought that human affairs (*ta tōn anthrōpōn pragmata*), the outcome of action (*praxis*), should not be treated with great seriousness; the actions of men appear like the gestures of puppets led by an invisible hand behind the scene, so that man seems to be a kind of plaything of a god.[9] It is noteworthy that Plato, who had no inkling of the modern concept of history, should have been the first to invent the metaphor of an actor behind the scenes who, behind the backs of acting men, pulls the strings and is responsible for the story. The Platonic god is but a symbol for the fact that real stories, in distinction from those we invent, have no author; as such, he is the true forerunner of Providence, the "invisible hand," Nature, the "world spirit," class interest, and the like, with which Christian and modern philosophers of history tried to solve the perplexing problem that although history owes its existence to men, it is still obviously not "made" by them. (Nothing in fact indicates more clearly the political nature of history—its being a story of action and deeds rather than of trends and forces or ideas—than the introduction of an invisible actor behind the scenes whom we find in all philosophies of history, which for this reason alone can be recognized as political philosophies in disguise. By the same token, the simple fact that Adam Smith needed an "invisible hand" to guide economic dealings on the exchange market shows plainly that more than sheer economic activity is involved in exchange and that "economic man," when he makes his appearance on the market, is an acting being and neither exclusively a producer nor a trader and barterer.)

The invisible actor behind the scenes is an invention arising from a mental perplexity but corresponding to no real experience.

9. *Laws* 803 and 644.

Through it, the story resulting from action is misconstrued as a fictional story, where indeed an author pulls the strings and directs the play. The fictional story reveals a maker just as every work of art clearly indicates that it was made by somebody; this does not belong to the character of the story itself but only to the mode in which it came into existence. The distinction between a real and a fictional story is precisely that the latter was "made up" and the former not made at all. The real story in which we are engaged as long as we live has no visible or invisible maker because it is not made. The only "somebody" it reveals is its hero, and it is the only medium in which the originally intangible manifestation of a uniquely distinct "who" can become tangible *ex post facto* through action and speech. *Who* somebody is or was we can know only by knowing the story of which he is himself the hero—his biography, in other words; everything else we know of him, including the work he may have produced and left behind, tells us only *what* he is or was. Thus, although we know much less of Socrates, who did not write a single line and left no work behind, than of Plato or Aristotle, we know much better and more intimately who he was, because we know his story, than we know who Aristotle was, about whose opinions we are so much better informed.

The hero the story discloses needs no heroic qualities; the word "hero" originally, that is, in Homer, was no more than a name given each free man who participated in the Trojan enterprise[10] and about whom a story could be told. The connotation of courage, which we now feel to be an indispensable quality of the hero, is in fact already present in a willingness to act and speak at all, to insert one's self into the world and begin a story of one's own. And this courage is not necessarily or even primarily related to a willingness to suffer the consequences; courage and even boldness are already present in leaving one's private hiding place and showing who one is, in disclosing and exposing one's self. The extent of this original courage, without which action and speech and

10. In Homer, the word *hērōs* has certainly a connotation of distinction, but of no other than every free man was capable. Nowhere does it appear in the later meaning of "half-god," which perhaps arose out of a deification of the ancient epic heroes.

therefore, according to the Greeks, freedom, would not be possible at all, is not less great and may even be greater if the "hero" happens to be a coward.

The specific content as well as the general meaning of action and speech may take various forms of reification in art works which glorify a deed or an accomplishment and, by transformation and condensation, show some extraordinary event in its full significance. However, the specific revelatory quality of action and speech, the implicit manifestation of the agent and speaker, is so indissolubly tied to the living flux of acting and speaking that it can be represented and "reified" only through a kind of repetition, the imitation or *mimēsis*, which according to Aristotle prevails in all arts but is actually appropriate only to the *drama*, whose very name (from the Greek verb *dran*, "to act") indicates that play-acting actually is an imitation of acting.[11] But the imitative element lies not only in the art of the actor, but, as Aristotle rightly claims, in the making or writing of the play, at least to the extent that the drama comes fully to life only when it is enacted in the theater. Only the actors and speakers who re-enact the story's plot can convey the full meaning, not so much of the story itself, but of the "heroes" who reveal themselves in it.[12] In terms of Greek tragedy, this would mean that the story's direct as well as its universal meaning is revealed by the chorus, which does not imitate[13] and whose comments are pure poetry, whereas the intangible identities of the agents in the story, since they escape all

11. Aristotle already mentions that the word *drama* was chosen because *drōntes* ("acting people") are imitated (*Poetics* 1448a28). From the treatise itself, it is obvious that Aristotle's model for "imitation" in art is taken from the drama, and the generalization of the concept to make it applicable to all arts seems rather awkward.

12. Aristotle therefore usually speaks not of an imitation of action (*praxis*) but of the agents (*prattontes*) (see *Poetics* 1448a1 ff., 1448b25, 1449b24 ff.). He is not consistent, however, in this use (cf. 1451a29, 1447a28). The decisive point is that tragedy does not deal with the qualities of men, their *poiotēs*, but with whatever happened with respect to them, with their actions and life and good or ill fortune (1450a15–18). The content of tragedy, therefore, is not what we would call character but action or the plot.

13. That the chorus "imitates less" is mentioned in the Ps. Aristotelian *Problemata* (918b28).

generalization and therefore all reification, can be conveyed only through an imitation of their acting. This is also why the theater is the political art par excellence; only there is the political sphere of human life transposed into art. By the same token, it is the only art whose sole subject is man in his relationship to others.

26

THE FRAILTY OF HUMAN AFFAIRS

Action, as distinguished from fabrication, is never possible in isolation; to be isolated is to be deprived of the capacity to act. Action and speech need the surrounding presence of others no less than fabrication needs the surrounding presence of nature for its material, and of a world in which to place the finished product. Fabrication is surrounded by and in constant contact with the world: action and speech are surrounded by and in constant contact with the web of the acts and words of other men. The popular belief in a "strong man" who, isolated against others, owes his strength to his being alone is either sheer superstition, based on the delusion that we can "make" something in the realm of human affairs—"make" institutions or laws, for instance, as we make tables and chairs, or make men "better" or "worse"[14]—or it is conscious despair of all action, political and non-political, coupled with the utopian hope that it may be possible to treat men as one treats other "material."[15] The strength the individual needs for every process of production becomes altogether worthless when action is at stake, regardless of whether this strength is intellectual or a matter of purely material force. History is full of ex-

14. Plato already reproached Pericles because he did not "make the citizen better" and because the Athenians were even worse at the end of his career than before (*Gorgias* 515).

15. Recent political history is full of examples indicating that the term "human material" is no harmless metaphor, and the same is true for a whole host of modern scientific experiments in social engineering, biochemistry, brain surgery, etc., all of which tend to treat and change human material like other matter. This mechanistic approach is typical of the modern age; antiquity, when it pursued similar aims, was inclined to think of men in terms of savage animals who need be tamed and domesticated. The only possible achievement in either case is to kill man, not indeed necessarily as a living organism, but *qua* man.

amples of the impotence of the strong and superior man who does not know how to enlist the help, the co-acting of his fellow men. His failure is frequently blamed upon the fatal inferiority of the many and the resentment every outstanding person inspires in those who are mediocre. Yet true as such observations are bound to be, they do not touch the heart of the matter.

In order to illustrate what is at stake here we may remember that Greek and Latin, unlike the modern languages, contain two altogether different and yet interrelated words with which to designate the verb "to act." To the two Greek verbs *archein* ("to begin," "to lead," finally "to rule") and *prattein* ("to pass through," "to achieve," "to finish") correspond the two Latin verbs *agere* ("to set into motion," "to lead") and *gerere* (whose original meaning is "to bear").[16] Here it seems as though each action were divided into two parts, the beginning made by a single person and the achievement in which many join by "bearing" and "finishing" the enterprise, by seeing it through. Not only are the words interrelated in a similar manner, the history of their usage is very similar too. In both cases the word that originally designated only the second part of action, its achievement—*prattein* and *gerere*—became the accepted word for action in general, whereas the words designating the beginning of action became specialized in meaning, at least in political language. *Archein* came to mean chiefly "to rule" and "to lead" when it was specifically used, and *agere* came to mean "to lead" rather than "to set into motion."

Thus the role of the beginner and leader, who was a *primus inter pares* (in the case of Homer, a king among kings), changed into that of a ruler; the original interdependence of action, the dependence of the beginner and leader upon others for help and the dependence of his followers upon him for an occasion to act themselves, split into two altogether different functions: the function of giving commands, which became the prerogative of the ruler, and the function of executing them, which became the duty of his subjects. This ruler is alone, isolated against others by his force, just as the beginner was isolated through his initiative at

16. For *archein* and *prattein* see especially their use in Homer (cf. C. Capelle, *Wörterbuch des Homeros und der Homeriden* [1889]).

the start, before he had found others to join him. Yet the strength of the beginner and leader shows itself only in his initiative and the risk he takes, not in the actual achievement. In the case of the successful ruler, he may claim for himself what actually is the achievement of many—something that Agamemnon, who was a king but no ruler, would never have been permitted. Through this claim, the ruler monopolizes, so to speak, the strength of those without whose help he would never be able to achieve anything. Thus, the delusion of extraordinary strength arises and with it the fallacy of the strong man who is powerful because he is alone.

Because the actor always moves among and in relation to other acting beings, he is never merely a "doer" but always and at the same time a sufferer. To do and to suffer are like opposite sides of the same coin, and the story that an act starts is composed of its consequent deeds and sufferings. These consequences are boundless, because action, though it may proceed from nowhere, so to speak, acts into a medium where every reaction becomes a chain reaction and where every process is the cause of new processes. Since action acts upon beings who are capable of their own actions, reaction, apart from being a response, is always a new action that strikes out on its own and affects others. Thus action and reaction among men never move in a closed circle and can never be reliably confined to two partners. This boundlessness is characteristic not of political action alone, in the narrower sense of the word, as though the boundlessness of human interrelatedness were only the result of the boundless multitude of people involved, which could be escaped by resigning oneself to action within a limited, graspable framework of circumstances; the smallest act in the most limited circumstances bears the seed of the same boundlessness, because one deed, and sometimes one word, suffices to change every constellation.

Action, moreover, no matter what its specific content, always establishes relationships and therefore has an inherent tendency to force open all limitations and cut across all boundaries.[17] Limita-

17. It is interesting to note that Montesquieu, whose concern was not with laws but with the actions their spirit would inspire, defines laws as *rapports* sub-

tions and boundaries exist within the realm of human affairs, but they never offer a framework that can reliably withstand the on-slaught with which each new generation must insert itself. The frailty of human institutions and laws and, generally, of all matters pertaining to men's living together, arises from the human condi-tion of natality and is quite independent of the frailty of human nature. The fences inclosing private property and insuring the limitations of each household, the territorial boundaries which pro-tect and make possible the physical identity of a people, and the laws which protect and make possible its political existence, are of such great importance to the stability of human affairs precisely because no such limiting and protecting principles rise out of the activities going on in the realm of human affairs itself. The limita-tions of the law are never entirely reliable safeguards against ac-tion from within the body politic, just as the boundaries of the territory are never entirely reliable safeguards against action from without. The boundlessness of action is only the other side of its tremendous capacity for establishing relationships, that is, its specific productivity; this is why the old virtue of moderation, of keeping within bounds, is indeed one of the political virtues par excellence, just as the political temptation par excellence is indeed *hubris* (as the Greeks, fully experienced in the potentialities of action, knew so well) and not the will to power, as we are inclined to believe.

Yet while the various limitations and boundaries we find in every body politic may offer some protection against the inherent boundlessness of action, they are altogether helpless to offset its second outstanding character: its inherent unpredictability. This is not simply a question of inability to foretell all the logical con-sequences of a particular act, in which case an electronic com-puter would be able to foretell the future, but arises directly out of the story which, as the result of action, begins and establishes

sisting between different beings (*Esprit des lois*, Book I, ch. 1; cf. Book XXVI, ch. 1). This definition is surprising because laws had always been defined in terms of boundaries and limitations. The reason for it is that Montesquieu was less interested in what he called the "nature of government"—whether it was a re-public or a monarchy, for instance—than in its "principle . . . by which it is made to act, . . . the human passions which set it in motion" (Book III, ch. 1).

itself as soon as the fleeting moment of the deed is past. The trouble is that whatever the character and content of the subsequent story may be, whether it is played in private or public life, whether it involves many or few actors, its full meaning can reveal itself only when it has ended. In contradistinction to fabrication, where the light by which to judge the finished product is provided by the image or model perceived beforehand by the craftsman's eye, the light that illuminates processes of action, and therefore all historical processes, appears only at their end, frequently when all the participants are dead. Action reveals itself fully only to the storyteller, that is, to the backward glance of the historian, who indeed always knows better what it was all about than the participants. All accounts told by the actors themselves, though they may in rare cases give an entirely trustworthy statement of intentions, aims, and motives, become mere useful source material in the historian's hands and can never match his story in significance and truthfulness. What the storyteller narrates must necessarily be hidden from the actor himself, at least as long as he is in the act or caught in its consequences, because to him the meaningfulness of his act is not in the story that follows. Even though stories are the inevitable results of action, it is not the actor but the storyteller who perceives and "makes" the story.

27

THE GREEK SOLUTION

This unpredictability of outcome is closely related to the revelatory character of action and speech, in which one discloses one's self without ever either knowing himself or being able to calculate beforehand whom he reveals. The ancient saying that nobody can be called *eudaimōn* before he is dead may point to the issue at stake, if we could hear its original meaning after two and a half thousand years of hackneyed repetition; not even its Latin translation, proverbial and trite already in Rome—*nemo ante mortem beatus esse dici potest*—conveys this meaning, although it may have inspired the practice of the Catholic Church to beatify her saints only after they have long been safely dead. For *eudaimonia* means neither happiness nor beatitude; it cannot be translated and perhaps cannot even be explained. It has the connotation of blessedness, but without any religious overtones, and it means literally something like the well-being of the *daimōn* who accompanies each man throughout life, who is his distinct identity, but appears and is visible only to others.[18] Unlike happiness, therefore, which is a passing mood, and unlike good fortune, which one may have at certain periods of life and lack in others, *eudaimonia*, like life itself, is a lasting state of being which is neither subject to change nor capable of effecting change. To be *eudaimōn* and to have been *eudaimōn*, according to Aristotle, are the same, just as to "live well" (*eu dzēn*) and to have "lived well" are the same as long as life lasts; they are not states or activities which change a person's quality, such as learning and having learned, which indicate two altogether different attributes of the same person at different moments.[19]

This unchangeable identity of the person, though disclosing itself intangibly in act and speech, becomes tangible only in the story of the actor's and speaker's life; but as such it can be known, that is, grasped as a palpable entity only after it has come to its end. In other words, human essence—not human nature in general (which does not exist) nor the sum total of qualities and shortcomings in the individual, but the essence of who somebody is—can come into being only when life departs, leaving behind nothing but a story. Therefore whoever consciously aims at being "essential," at leaving behind a story and an identity which will win "immortal fame," must not only risk his life but expressly choose, as Achilles did, a short life and premature death. Only a man who does not survive his one supreme act remains the indisputable master of his identity and possible greatness, because he withdraws into death from the possible consequences and con-

18. For this interpretation of *daimōn* and *eudaimonia*, see Sophocles *Oedipus Rex* 1186 ff., especially the verses: *Tis gar, tis anēr pleon / tas eudaimonias pherei / ē tosouton hoson dokein / kai doxant' apoklinai* ("For which, which man [can] bear more *eudaimonia* than he grasps from appearance and deflects in its appearance?"). It is against this inevitable distortion that the chorus asserts its own knowledge: these others see, they "have" Oedipus' *daimōn* before their eyes as an example; the misery of the mortals is their blindness toward their own *daimōn*.

19. Aristotle *Metaphysics* 1048b23 ff.

tinuation of what he began. What gives the story of Achilles its paradigmatic significance is that it shows in a nutshell that *eudaimonia* can be bought only at the price of life and that one can make sure of it only by foregoing the continuity of living in which we disclose ourselves piecemeal, by summing up all of one's life in a single deed, so that the story of the act comes to its end together with life itself. Even Achilles, it is true, remains dependent upon the storyteller, poet, or historian, without whom everything he did remains futile; but he is the only "hero," and therefore the hero par excellence, who delivers into the narrator's hands the full significance of his deed, so that it is as though he had not merely enacted the story of his life but at the same time also "made" it.

No doubt this concept of action is highly individualistic, as we would say today.[20] It stresses the urge toward self-disclosure at the expense of all other factors and therefore remains relatively untouched by the predicament of unpredictability. As such it became the prototype of action for Greek antiquity and influenced, in the form of the so-called agonal spirit, the passionate drive to show one's self in measuring up against others that underlies the concept of politics prevalent in the city-states. An outstanding symptom of this prevailing influence is that the Greeks, in distinction from all later developments, did not count legislating among the political activities. In their opinion, the lawmaker was like the builder of the city wall, someone who had to do and finish his work before political activity could begin. He therefore was treated like any other craftsman or architect and could be called from abroad and commissioned without having to be a citizen, whereas the right to *politeuesthai*, to engage in the numerous activities which eventually went on in the *polis*, was entirely restricted to citizens. To them, the laws, like the wall around the city, were not results of action but products of making. Before men began to act, a definite space had to be secured and a structure built where all subsequent actions could take place, the space

20. The fact that the Greek word for "every one" (*hekastos*) is derived from *hekas* ("far off") seems to indicate how deep-rooted this "individualism" must have been.

being the public realm of the *polis* and its structure the law; legislator and architect belonged in the same category.[21] But these tangible entities themselves were not the content of politics (not Athens, but the Athenians, were the *polis*[22]), and they did not command the same loyalty we know from the Roman type of patriotism.

Though it is true that Plato and Aristotle elevated lawmaking and city-building to the highest rank in political life, this does not indicate that they enlarged the fundamental Greek experiences of action and politics to comprehend what later turned out to be the political genius of Rome: legislation and foundation. The Socratic school, on the contrary, turned to these activities, which to the Greeks were prepolitical, because they wished to turn against politics and against action. To them, legislating and the execution of decisions by vote are the most legitimate political activities because in them men "act like craftsmen": the result of their action is a tangible product, and its process has a clearly recognizable end.[23] This is no longer or, rather, not yet action (*praxis*), properly speaking, but making (*poiēsis*), which they prefer because of its greater reliability. It is as though they had said that if men only renounce their capacity for action, with its futility, boundlessness, and uncertainty of outcome, there could be a remedy for the frailty of human affairs.

How this remedy can destroy the very substance of human relationships is perhaps best illustrated in one of the rare instances

21. See, for instance, Aristotle *Nicomachean Ethics* 1141b25. There is no more elemental difference between Greece and Rome than their respective attitudes toward territory and law. In Rome, the foundation of the city and the establishment of its laws remained the great and decisive act to which all later deeds and accomplishments had to be related in order to acquire political validity and legitimation.

22. See M. F. Schachermeyr, "La formation de la cité Grecque," *Diogenes*, No. 4 (1953), who compares the Greek usage with that of Babylon, where the notion of "the Babylonians" could be expressed only by saying: the people of the territory of the city of Babylon.

23. "For [the legislators] alone act like craftsmen [*cheirotechnoi*]" because their act has a tangible end, an *eschaton*, which is the decree passed in the assembly (*psēphisma*) (*Nicomachean Ethics* 1141b29).

where Aristotle draws an example of acting from the sphere of private life, in the relationship between the benefactor and his recipient. With that candid absence of moralizing that is the mark of Greek, though not of Roman, antiquity, he states first as a matter of fact that the benefactor always loves those he has helped more than he is loved by them. He then goes on to explain that this is only natural, since the benefactor has done a work, an *ergon*, while the recipient has only endured his beneficence. The benefactor, according to Aristotle, loves his "work," the life of the recipient which he has "made," as the poet loves his poems, and he reminds his readers that the poet's love for his work is hardly less passionate than a mother's love for her children.[24] This explanation shows clearly that he thinks of acting in terms of making, and of its result, the relationship between men, in terms of an accomplished "work" (his emphatic attempts to distinguish between action and fabrication, *praxis* and *poiēsis*, notwithstanding).[25] In this instance, it is perfectly obvious how this interpretation, though it may serve to explain psychologically the phenomenon of ingratitude on the assumption that both benefactor and recipient agree about an interpretation of action in terms of making, actually spoils the action itself and its true result, the relationship it should have established. The example of the legislator is less plausible for us only because the Greek notion of the task and role of the legislator in the public realm is so utterly alien to our own. In any event, work, such as the activity of the legislator in Greek understanding, can become the content of action only on condition that further action is not desirable or possible; and action can result in an end product only on condition that its own authentic, non-tangible, and always utterly fragile meaning is destroyed.

The original, prephilosophic Greek remedy for this frailty had been the foundation of the *polis*. The *polis*, as it grew out of and remained rooted in the Greek pre-*polis* experience and estimate of what makes it worthwhile for men to live together (*syzēn*),

24. *Ibid.* 1168a13 ff.

25. *Ibid.* 1140.

namely, the "sharing of words and deeds,"[26] had a twofold function. First, it was intended to enable men to do permanently, albeit under certain restrictions, what otherwise had been possible only as an extraordinary and infrequent enterprise for which they had to leave their households. The *polis* was supposed to multiply the occasions to win "immortal fame," that is, to multiply the chances for everybody to distinguish himself, to show in deed and word who he was in his unique distinctness. One, if not the chief, reason for the incredible development of gift and genius in Athens, as well as for the hardly less surprising swift decline of the city-state, was precisely that from beginning to end its foremost aim was to make the extraordinary an ordinary occurrence of everyday life. The second function of the *polis*, again closely connected with the hazards of action as experienced before its coming into being, was to offer a remedy for the futility of action and speech; for the chances that a deed deserving fame would not be forgotten, that it actually would become "immortal," were not very good. Homer was not only a shining example of the poet's political function, and therefore the "educator of all Hellas"; the very fact that so great an enterprise as the Trojan War could have been forgotten without a poet to immortalize it several hundred years later offered only too good an example of what could happen to human greatness if it had nothing but poets to rely on for its permanence.

We are not concerned here with the historical causes for the rise of the Greek city-state; what the Greeks themselves thought of it and its *raison d'être*, they have made unmistakably clear. The *polis*—if we trust the famous words of Pericles in the Funeral Oration—gives a guaranty that those who forced every sea and land to become the scene of their daring will not remain without witness and will need neither Homer nor anyone else who knows how to turn words to praise them; without assistance from others, those who acted will be able to establish together the everlasting remembrance of their good and bad deeds, to inspire admiration in the present and in future ages.[27] In other words, men's life together in the form of the *polis* seemed to assure that the most

26. *Logōn kai pragmatōn koinōnein*, as Aristotle once put it (*ibid.* 1126b12).
27. Thucydides ii. 41.

futile of human activities, action and speech, and the least tangible and most ephemeral of man-made "products," the deeds and stories which are their outcome, would become imperishable. The organization of the *polis*, physically secured by the wall around the city and physiognomically guaranteed by its laws—lest the succeeding generations change its identity beyond recognition— is a kind of organized remembrance. It assures the mortal actor that his passing existence and fleeting greatness will never lack the reality that comes from being seen, being heard, and, generally, appearing before an audience of fellow men, who outside the *polis* could attend only the short duration of the performance and therefore needed Homer and "others of his craft" in order to be presented to those who were not there.

According to this self-interpretation, the political realm rises directly out of acting together, the "sharing of words and deeds." Thus action not only has the most intimate relationship to the public part of the world common to us all, but is the one activity which constitutes it. It is as though the wall of the *polis* and the boundaries of the law were drawn around an already existing public space which, however, without such stabilizing protection could not endure, could not survive the moment of action and speech itself. Not historically, of course, but speaking metaphorically and theoretically, it is as though the men who returned from the Trojan War had wished to make permanent the space of action which had arisen from their deeds and sufferings, to prevent its perishing with their dispersal and return to their isolated homesteads.

The *polis*, properly speaking, is not the city-state in its physical location; it is the organization of the people as it arises out of acting and speaking together, and its true space lies between people living together for this purpose, no matter where they happen to be. "Wherever you go, you will be a *polis*": these famous words became not merely the watchword of Greek colonization, they expressed the conviction that action and speech create a space between the participants which can find its proper location almost any time and anywhere. It is the space of appearance in the widest sense of the word, namely, the space where I appear to others as others appear to me, where men exist not merely like other living or inanimate things but make their appearance explicitly.

This space does not always exist, and although all men are capable of deed and word, most of them—like the slave, the foreigner, and the barbarian in antiquity, like the laborer or craftsman prior to the modern age, the jobholder or businessman in our world—do not live in it. No man, moreover, can live in it all the time. To be deprived of it means to be deprived of reality, which, humanly and politically speaking, is the same as appearance. To men the reality of the world is guaranteed by the presence of others, by its appearing to all; "for what appears to all, this we call Being,"[28] and whatever lacks this appearance comes and passes away like a dream, intimately and exclusively our own but without reality.[29]

28

POWER AND THE SPACE OF APPEARANCE

The space of appearance comes into being wherever men are together in the manner of speech and action, and therefore predates and precedes all formal constitution of the public realm and the various forms of government, that is, the various forms in which the public realm can be organized. Its peculiarity is that, unlike the spaces which are the work of our hands, it does not survive the actuality of the movement which brought it into being, but disappears not only with the dispersal of men—as in the case of great catastrophes when the body politic of a people is destroyed—but with the disappearance or arrest of the activities themselves. Wherever people gather together, it is potentially there, but only potentially, not necessarily and not forever. That civilizations can rise and fall, that mighty empires and great cultures can decline and pass away without external catastrophes—and more often than not such external "causes" are preceded by a

28. Aristotle *Nicomachean Ethics* 1172b36 ff.

29. Heraclitus' statement that the world is one and common to those who are awake, but that everybody who is asleep turns away to his own (*Diels, op. cit.*, B89), says essentially the same as Aristotle's remark just quoted.

less visible internal decay that invites disaster—is due to this peculiarity of the public realm, which, because it ultimately resides on action and speech, never altogether loses its potential character. What first undermines and then kills political communities is loss of power and final impotence; and power cannot be stored up and kept in reserve for emergencies, like the instruments of violence, but exists only in its actualization. Where power is not actualized, it passes away, and history is full of examples that the greatest material riches cannot compensate for this loss. Power is actualized only where word and deed have not parted company, where words are not empty and deeds not brutal, where words are not used to veil intentions but to disclose realities, and deeds are not used to violate and destroy but to establish relations and create new realities.

Power is what keeps the public realm, the potential space of appearance between acting and speaking men, in existence. The word itself, its Greek equivalent *dynamis*, like the Latin *potentia* with its various modern dervatives or the German *Macht* (which derives from *mögen* and *möglich*, not from *machen*), indicates its "potential" character. Power is always, as we would say, a power potential and not an unchangeable, measurable, and reliable entity like force or strength. While strength is the natural quality of an individual seen in isolation, power springs up between men when they act together and vanishes the moment they disperse. Because of this peculiarity, which power shares with all potentialities that can only be actualized but never fully materialized, power is to an astonishing degree independent of material factors, either of numbers or means. A comparatively small but well-organized group of men can rule almost indefinitely over large and populous empires, and it is not infrequent in history that small and poor countries get the better of great and rich nations. (The story of David and Goliath is only metaphorically true; the power of a few can be greater than the power of many, but in a contest between two men not power but strength decides, and cleverness, that is, brain power, contributes materially to the outcome on the same level as muscular force.) Popular revolt against materially strong rulers, on the other hand, may engender an almost irresistible power even if it foregoes the use of violence in the face of materially vastly superior forces. To call this "passive resistance" is certainly an ironic idea; it is one of the most active and efficient ways of action ever devised, because it cannot be countered by fighting, where there may be defeat or victory, but only by mass slaughter in which even the victor is defeated, cheated of his prize, since nobody can rule over dead men.

The only indispensable material factor in the generation of power is the living together of people. Only where men live so close together that the potentialities of action are always present can power remain with them, and the foundation of cities, which as city-states have remained paradigmatic for all Western political organization, is therefore indeed the most important material prerequisite for power. What keeps people together after the fleeting moment of action has passed (what we today call "organization") and what, at the same time, they keep alive through remaining together is power. And whoever, for whatever reasons, isolates himself and does not partake in such being together, forfeits power and becomes impotent, no matter how great his strength and how valid his reasons.

If power were more than this potentiality in being together, if it could be possessed like strength or applied like force instead of being dependent upon the unreliable and only temporary agreement of many wills and intentions, omnipotence would be a concrete human possibility. For power, like action, is boundless; it has no physical limitation in human nature, in the bodily existence of man, like strength. Its only limitation is the existence of other people, but this limitation is not accidental, because human power corresponds to the condition of plurality to begin with. For the same reason, power can be divided without decreasing it, and the interplay of powers with their checks and balances is even liable to generate more power, so long, at least, as the interplay is alive and has not resulted in a stalemate. Strength, on the contrary, is indivisible, and while it, too, is checked and balanced by the presence of others, the interplay of plurality in this case spells a definite limitation on the strength of the individual, which is kept in bounds and may be overpowered by the power potential of the many. An identification of the strength necessary for the production of things with the power necessary for action is conceivable only as the divine attribute of one god. Omnipotence therefore is never an attribute of gods in polytheism, no matter how superior the strength of the gods may be to the forces of men. Conversely, aspiration toward omnipotence always implies—apart from its utopian *hubris*—the destruction of plurality.

Under the conditions of human life, the only alternative to power is not strength—which is helpless against power—but force, which indeed one man alone can exert against his fellow men and of which one or a few can possess a monopoly by acquiring the means of violence. But while violence can destroy power, it can never become a substitute for it. From this results the by no means infrequent political combination of force and powerlessness, an array of impotent forces that spend themselves, often spectacularly and vehemently but in utter futility, leaving behind neither monuments nor stories, hardly enough memory to enter into history at all. In historical experience and traditional theory, this combination, even if it is not recognized as such, is known as tyranny, and the time-honored fear of this form of government is not exclusively inspired by its cruelty, which—as the long series of benevolent tyrants and enlightened despots attests—is not among its inevitable features, but by the impotence and futility to which it condemns the rulers as well as the ruled.

More important is a discovery made, as far as I know, only by Montesquieu, the last political thinker to concern himself seriously with the problem of forms of government. Montesquieu realized that the outstanding characteristic of tyranny was that it rested on isolation—on the isolation of the tyrant from his subjects and the isolation of the subjects from each other through mutual fear and suspicion—and hence that tyranny was not one form of government among others but contradicted the essential human condition of plurality, the acting and speaking together, which is the condition of all forms of political organization. Tyranny prevents the development of power, not only in a particular segment of the public realm but in its entirety; it generates, in other words, impotence as naturally as other bodies politic generate power. This, in Montesquieu's interpretation, makes it necessary to assign it a special position in the theory of political bodies: it alone is unable to develop enough power to remain at all in the space of appearance, the public realm; on the contrary, it develops the germs of its own destruction the moment it comes into existence.[30]

Violence, curiously enough, can destroy power more easily than it can destroy strength, and while a tyranny is always characterized by the impotence of its subjects, who have lost their human capacity to act and speak together, it is not necessarily characterized by weakness and sterility; on the contrary, the crafts and arts may flourish under these conditions if the ruler is "benevolent" enough to leave his subjects alone in their isolation. Strength, on the other hand, nature's gift to the individual which cannot be shared with others, can cope with violence more successfully than with power—either heroically, by consenting to fight and die, or stoically, by accepting suffering and challenging all affliction through self-sufficiency and withdrawal from the world; in either case, the integrity of the individual and his strength remain intact. Strength can actually be ruined only by power and is therefore always in danger from the combined force of the many. Power corrupts indeed when the weak band together in order to ruin the strong, but not before. The will to power, as the modern age from Hobbes to Nietzsche understood it in glorification or denunciation, far from being a characteristic of the strong, is, like envy and greed, among the vices of the weak, and possibly even their most dangerous one.

If tyranny can be described as the always abortive attempt to substitute violence for power, ochlocracy, or mob rule, which is its exact counterpart, can be characterized by the much more promising attempt to substitute power for strength. Power indeed can ruin all strength and we know that where the main public realm is society, there is always the danger that, through a perverted form of "acting together"—by pull and pressure and the tricks of cliques—those are brought to the fore who know nothing and can do nothing. The vehement yearning for violence, so char-

---

30. In the words of Montesquieu, who ignores the difference between tyranny and despotism: "Le principe du gouvernement despotique se corrompt sans cesse, parcequ'il est corrompu par sa nature. Les autres gouvernements périssent, parceque des accidents particuliers en violent le principe: celui-ci périt par son vice intérieur, lorsque quelques causes accidentelles n'empêchent point son principe de se corrompre" (*op. cit.*, Book V,III, ch. 10).

acteristic of some of the best modern creative artists, thinkers, scholars, and craftsmen, is a natural reaction of those whom society has tried to cheat of their strength.[31]

Power preserves the public realm and the space of appearance, and as such it is also the lifeblood of the human artifice, which, unless it is the scene of action and speech, of the web of human affairs and relationships and the stories engendered by them, lacks its ultimate *raison d'être*. Without being talked about by men and without housing them, the world would not be a human artifice but a heap of unrelated things to which each isolated individual was at liberty to add one more object; without the human artifice to house them, human affairs would be as floating, as futile and vain, as the wanderings of nomad tribes. The melancholy wisdom of *Ecclesiastes*—"Vanity of vanities; all is vanity. . . . There is no new thing under the sun, . . . there is no remembrance of former things; neither shall there be any remembrance of things that are to come with those that shall come after"—does not necessarily arise from specifically religious experience; but it is certainly unavoidable wherever and whenever trust in the world as a place fit for human appearance, for action and speech, is gone. Without action to bring into the play of the world the new beginning of which each man is capable by virtue of being born, "there is no new thing under the sun"; without speech to materialize and memorialize, however tentatively, the "new things" that appear and shine forth, "there is no remembrance"; without the enduring permanence of a human artifact, there cannot "be any remembrance of things that are to come with those that shall come after." And without power, the space of appearance brought forth through action and speech in public will fade away as rapidly as the living deed and the living word.

Perhaps nothing in our history has been so short-lived as trust in power, nothing more lasting than the Platonic and Christian distrust of the splendor attending its space of appearance, nothing

31. The extent to which Nietzsche's glorification of the will to power was inspired by such experiences of the modern intellectual may be surmised from the following side remark: "Denn die Ohnmacht gegen Menschen, nicht die Ohnmacht gegen die Natur, erzeugt die desperateste Verbitterung gegen das Dasein" (*Wille zur Macht*, No. 55).

—finally in the modern age—more common than the conviction that "power corrupts." The words of Pericles, as Thucydides reports them, are perhaps unique in their supreme confidence that men can enact *and* save their greatness at the same time and, as it were, by one and the same gesture, and that the performance as such will be enough to generate *dynamis* and not need the transforming reification of *homo faber* to keep it in reality.[32] Pericles' speech, though it certainly corresponded to and articulated the innermost convictions of the people of Athens, has always been read with the sad wisdom of hindsight by men who knew that his words were spoken at the beginning of the end. Yet short-lived as this faith in *dynamis* (and consequently in politics) may have been —and it had already come to an end when the first political philosophies were formulated—its bare existence has sufficed to elevate action to the highest rank in the hierarchy of the *vita activa* and to single out speech as the decisive distinction between human and animal life, both of which bestowed upon politics a dignity which even today has not altogether disappeared.

What is outstandingly clear in Pericles' formulations—and, incidentally, no less transparent in Homer's poems—is that the innermost meaning of the acted deed and the spoken word is independent of victory and defeat and must remain untouched by any eventual outcome, by their consequences for better or worse. Unlike human behavior—which the Greeks, like all civilized people, judged according to "moral standards," taking into account motives and intentions on the one hand and aims and consequences on the other—action can be judged only by the criterion of greatness because it is in its nature to break through the commonly accepted and reach into the extraordinary, where whatever is true in common and everyday life no longer applies because everything that exists is unique and *sui generis*.[33] Thucydides, or

32. In the above-mentioned paragraph in the Funeral Oration (n. 27) Pericles deliberately contrasts the *dynamis* of the *polis* with the craftsmanship of the poets.

33. The reason why Aristotle in his *Poetics* finds that greatness (*megethos*) is a prerequisite of the dramatic plot is that the drama imitates acting and acting is judged by greatness, by its distinction from the commonplace (1450b25). The same, incidentally, is true for the beautiful, which resides in greatness and *taxis*, the joining together of the parts (1450b34 ff.).

Pericles, knew full well that he had broken with the normal standards for everyday behavior when he found the glory of Athens in having left behind "everywhere everlasting remembrance [*mnē-meia aidia*] of their good and their evil deeds." The art of politics teaches men how to bring forth what is great and radiant—*ta megala kai lampra*, in the words of Democritus; as long as the *polis* is there to inspire men to dare the extraordinary, all things are safe; if it perishes, everything is lost.[34] Motives and aims, no matter how pure or how grandiose, are never unique; like psychological qualities, they are typical, characteristic of different types of persons. Greatness, therefore, or the specific meaning of each deed, can lie only in the performance itself and neither in its motivation nor its achievement.

It is this insistence on the living deed and the spoken word as the greatest achievements of which human beings are capable that was conceptualized in Aristotle's notion of *energeia* ("actuality"), with which he designated all activities that do not pursue an end (are *ateleis*) and leave no work behind (no *par' autas erga*), but exhaust their full meaning in the performance itself.[35] It is from the experience of this full actuality that the paradoxical "end in itself" derives its original meaning; for in these instances of action and speech[36] the end (*telos*) is not pursued but lies in the activity itself which therefore becomes an *entelecheia*, and the work is not what follows and extinguishes the process but is imbedded in it; the performance is the work, is *energeia*.[37] Aristotle, in his political philosophy, is still well aware of what is at stake in politics, namely, no less than the *ergon tou anthrōpou*[38] (the "work of man" *qua*

---

34. See fragment B157 of Democritus in Diels, *op. cit.*

35. For the concept of *energeia* see *Nicomachean Ethics* 1094a1–5; *Physics* 201b31; *On the Soul* 417a16, 431a6. The examples most frequently used are seeing and flute-playing.

36. It is of no importance in our context that Aristotle saw the highest possibility of "actuality" not in action and speech, but in contemplation and thought, in *theōria* and *nous*.

37. The two Aristotelian concepts, *energeia* and *entelecheia*, are closely interrelated (*energeia . . . synteinei pros tēn entelecheian*): full actuality (*energeia*) effects and produces nothing besides itself, and full reality (*entelecheia*) has no other end besides itself (see *Metaphysics* 1050a22–35).

38. *Nicomachean Ethics* 1097b22.

man), and if he defined this "work" as "to live well" (*eu zēn*), he clearly meant that "work" here is no work product but exists only in sheer actuality. This specifically human achievement lies altogether outside the category of means and ends; the "work of man" is no end because the means to achieve it—the virtues, or *aretai*—are not qualities which may or may not be actualized, but are themselves "actualities." In other words, the means to achieve the end would already be the end; and this "end," conversely, cannot be considered a means in some other respect, because there is nothing higher to attain than this actuality itself.

It is like a feeble echo of the prephilosophical Greek experience of action and speech as sheer actuality to read time and again in political philosophy since Democritus and Plato that politics is a *technē*, belongs among the arts, and can be likened to such activities as healing or navigation, where, as in the performance of the dancer or play-actor, the "product" is identical with the performing act itself. But we may gauge what has happened to action and speech, which are only in actuality, and therefore the highest activities in the political realm, when we hear what modern society, with the peculiar and uncompromising consistency that characterized it in its early stages, had to say about them. For this all-important degradation of action and speech is implied when Adam Smith classifies all occupations which rest essentially on performance—such as the military profession, "churchmen, lawyers, physicians and opera-singers"—together with "menial services," the lowest and most unproductive "labour."[39] It was precisely these occupations—healing, flute-playing, play-acting—which furnished ancient thinking with examples for the highest and greatest activities of man.

29

Homo Faber AND THE SPACE
OF APPEARANCE

The root of the ancient estimation of politics is the conviction that man *qua* man, each individual in his unique distinctness, appears and confirms himself in speech and action, and that these activi-

39. *Wealth of Nations* (Everyman's ed.), II, 295.

ties, despite their material futility, possess an enduring quality of their own because they create their own remembrance.[40] The public realm, the space within the world which men need in order to appear at all, is therefore more specifically "the work of man" than is the work of his hands or the labor of his body.

The conviction that the greatest that man can achieve is his own appearance and actualization is by no means a matter of course. Against it stands the conviction of *homo faber* that a man's products may be more—and not only more lasting—than he is himself, as well as the *animal laborans'* firm belief that life is the highest of all goods. Both, therefore, are, strictly speaking, unpolitical, and will incline to denounce action and speech as idleness, idle busybodyness and idle talk, and generally will judge public activities in terms of their usefulness to supposedly higher ends—to make the world more useful and more beautiful in the case of *homo faber*, to make life easier and longer in the case of the *animal laborans*. This, however, is not to say that they are free to dispense with a public realm altogether, for without a space of appearance and without trusting in action and speech as a mode of being together, neither the reality of one's self, of one's own identity, nor the reality of the surrounding world can be established beyond doubt. The human sense of reality demands that men actualize the sheer passive givenness of their being, not in order to change it but in order to make articulate and call into full existence what otherwise they would have to suffer passively anyhow.[41] This actualization resides and comes to pass in those activities that exist only in sheer actuality.

The only character of the world by which to gauge its reality is its being common to us all, and common sense occupies such a high rank in the hierarchy of political qualities because it is the one sense that fits into reality as a whole our five strictly individual senses and the strictly particular data they perceive. It is by virtue

40. This is a decisive feature of the Greek, though perhaps not of the Roman, concept of "virtue": where *aretē* is, oblivion cannot occur (cf. Aristotle *Nicomachean Ethics* 1100b12–17).

41. This is the meaning of the last sentence of the Dante quotation at the head of this chapter; the sentence, though quite clear and simple in the Latin original, defies translation (*De monarchia* i. 13).

of common sense that the other sense perceptions are known to disclose reality and are not merely felt as irritations of our nerves or resistance sensations of our bodies. A noticeable decrease in common sense in any given community and a noticeable increase in superstition and gullibility are therefore almost infallible signs of alienation from the world.

This alienation—the atrophy of the space of appearance and the withering of common sense—is, of course, carried to a much greater extreme in the case of a laboring society than in the case of a society of producers. In his isolation, not only undisturbed by others but also not seen and heard and confirmed by them, *homo faber* is together not only with the product he makes but also with the world of things to which he will add his own products; in this, albeit indirect, way, he is still together with others who made the world and who also are fabricators of things. We have already mentioned the exchange market on which the craftsmen meet their peers and which represents to them a common public realm in so far as each of them has contributed something to it. Yet while the public realm as exchange market corresponds most adequately to the activity of fabrication, exchange itself already belongs in the field of action and is by no means a mere prolongation of production; it is even less a mere function of automatic processes, as the buying of food and other means of consumption is necessarily incidental to laboring. Marx's contention that economic laws are like natural laws, that they are not made by man to regulate the free acts of exchange but are functions of the productive conditions of society as a whole, is correct only in a laboring society, where all activities are leveled down to the human body's metabolism with nature and where no exchange exists but only consumption.

However, the people who meet on the exchange market are primarily not persons but producers of products, and what they show there is never themselves, not even their skills and qualities as in the "conspicuous production" of the Middle Ages, but their products. The impulse that drives the fabricator to the public market place is the desire for products, not for people, and the power that holds this market together and in existence is not the potentiality which springs up between people when they come together in action and speech, but a combined "power of exchange" (Adam Smith) which each of the participants acquired in isolation. It is this lack of relatedness to others and this primary concern with exchangeable commodities which Marx denounced as the dehumanization and self-alienation of commercial society, which indeed excludes men *qua* men and demands, in striking reversal of the ancient relationship between private and public, that men show themselves only in the privacy of their families or the intimacy of their friends.

The frustration of the human person inherent in a community of producers and even more in commercial society is perhaps best illustrated by the phenomenon of genius, in which, from the Renaissance to the end of the nineteenth century, the modern age saw its highest ideal. (Creative genius as the quintessential expression of human greatness was quite unknown to antiquity or the Middle Ages.) It is only with the beginning of our century that great artists in surprising unanimity have protested against being called "geniuses" and have insisted on craftmanship, competence, and the close relationships between art and handicraft. This protest, to be sure, is partly no more than a reaction against the vulgarization and commercialization of the notion of genius; but it is also due to the more recent rise of a laboring society, for which productivity or creativity is no ideal and which lacks all experiences from which the very notion of greatness can spring. What is important in our context is that the work of genius, as distinguished from the product of the craftsman, appears to have absorbed those elements of distinctness and uniqueness which find their immediate expression only in action and speech. The modern age's obsession with the unique signature of each artist, its unprecedented sensitivity to style, shows a preoccupation with those features by which the artist transcends his skill and workmanship in a way similar to the way each person's uniqueness transcends the sum total of his qualities. Because of this transcendence, which indeed distinguishes the great work of art from all other products of human hands, the phenomenon of the creative genius seemed like the highest legitimation for the conviction of *homo faber* that a man's products may be more and essentially greater than himself.

However, the great reverence the modern age so willingly paid to genius, so frequently bordering on idolatry, could hardly change the elementary fact that the essence of who somebody is cannot be reified by himself. When it appears "objectively"—in the style of an art work or in ordinary handwriting—it manifests the identity of a person and therefore serves to identify authorship, but it remains mute itself and escapes us if we try to interpret it as the mirror of a living person. In other words, the idolization of genius harbors the same degradation of the human person as the other tenets prevalent in commercial society.

It is an indispensable element of human pride to believe that who somebody is transcends in greatness and importance anything he can do and produce. "Let physicians and confectioners and the servants of the great houses be judged by what they have done, and even by what they have meant to do; the great people themselves are judged by what they are."[42] Only the vulgar will condescend to derive their pride from what they have done; they will, by this condescension, become the "slaves and prisoners" of their own faculties and will find out, should anything more be left in them than sheer stupid vanity, that to be one's own slave and prisoner is no less bitter and perhaps even more shameful than to be the servant of somebody else. It is not the glory but the predicament of the creative genius that in his case the superiority of man to his work seems indeed inverted, so that he, the living creator, finds himself in competition with his creations which he outlives, although they may survive him eventually. The saving grace of all really great gifts is that the persons who bear their burden remain superior to what they have done, at least as long as the source of creativity is alive; for this source springs indeed from *who* they are and remains outside the actual work process as well as independent of *what* they may achieve. That the predicament of genius is nevertheless a real one becomes quite apparent in the case of the *literati*, where the inverted order between man and his product is in fact consummated; what is so outrageous in their case, and incidentally incites popular hatred even more than spurious intellectual superiority, is that even their worst product is likely to be better than they are themselves. It is the hallmark of the "intellectual" that he remains quite undisturbed

42. I use here Isak Dinesen's wonderful story "The Dreamers," in *Seven Gothic Tales* (Modern Library ed.), especially pp. 340 ff.

by "the terrible humiliation" under which the true artist or writer labors, which is "to feel that he becomes the son of his work," in which he is condemned to see himself "as in a mirror, limited, such and such."[43]

30

THE LABOR MOVEMENT

The activity of work, for which isolation from others is a necessary prerequisite, although it may not be able to establish an autonomous public realm in which men *qua* men can appear, still is connected with this space of appearances in many ways; at the very least, it remains related to the tangible world of things it produced. Workmanship, therefore, may be an unpolitical way of life, but it certainly is not an antipolitical one. Yet this precisely is the case of laboring, an activity in which man is neither together with the world nor with other people, but alone with his body, facing the naked necessity to keep himself alive.[44] To be sure, he too lives in the presence of and together with others, but this togetherness has none of the distinctive marks of true plurality. It does not consist in the purposeful combination of different skills and callings as in the case of workmanship (let alone in the relationships between unique persons), but exists in the multiplication of specimens which are fundamentally all alike because they are what they are as mere living organisms.

43. The full text of the aphorism of Paul Valéry from which the quotations are taken reads as follows: "*Créateur créé*. Qui vient d'achever un long ouvrage le voit former enfin un être qu'il n'avait pas voulu, qu'il n'a pas conçu, précisément puisqu'il l'a enfanté, et ressent cette terrible humiliation de se sentir devenir le fils de son œuvre, de lui emprunter des traits irrécusables, une ressemblance, des manies, une borne, un miroir; et ce qu'il a de pire dans un miroir, s'y voir limité, tel et tel" (*Tel quel* II, 149).

44. The loneliness of the laborer *qua* laborer is usually overlooked in the literature on the subject because social conditions and the organization of labor demand the simultaneous presence of many laborers for any given task and break down all barriers of isolation. However, M. Halbwachs (*La classe ouvrière et les niveaux de vie* [1913]) is aware of the phenomenon: "L'ouvrier est celui qui dans et par son travail ne se trouve en rapport qu'avec de la matière, et non avec des hommes" and finds in this inherent lack of contact the reason why, for so many centuries, this whole class was put outside society (p. 118).

It is indeed in the nature of laboring to bring men together in the form of a labor gang where any number of individuals "labor together as though they were one,"[45] and in this sense togetherness may permeate laboring even more intimately than any other activity.[46] But this "collective nature of labor,"[47] far from establishing a recognizable, identifiable reality for each member of the labor gang, requires on the contrary the actual loss of all awareness of individuality and identity; and it is for this reason that all those "values" which derive from laboring, beyond its obvious function in the life process, are entirely "social" and essentially not different from the additional pleasure derived from eating and drinking in company. The sociability arising out of those activities which spring from the human body's metabolism with nature rest not on equality but on sameness, and from this viewpoint it is perfectly true that "by nature a philosopher is not in genius and disposition half so different from a street porter as a mastiff is from a greyhound." This remark of Adam Smith, which Marx quoted with great delight,[48] indeed fits a consumers' society much

45. Viktor von Weizsäcker, the German psychiatrist, describes the relationship between laborers during their labor as follows: "Es ist zunächst bemerkenswert, dass die zwei Arbeiter sich zusammen verhalten, als ob sie einer wären. . . . Wir haben hier einen Fall von Kollektivbildung vor uns, der in der annähernden Identität oder Einswerdung der zwei Individuen besteht. Man kann auch sagen, dass zwei Personen durch Verschmelzung eine einzige dritte geworden seien; aber die Regeln, nach der diese dritte arbeitet, unterscheiden sich in nichts von der Arbeit einer einzigen Person" ("Zum Begriff der Arbeit," in *Festschrift für Alfred Weber* [1948], pp. 739–40).

46. This seems to be the reason why, etymologically, "Arbeit und Gemeinschaft für den Menschen älterer geschichtlicher Stufen grosse Inhaltsflächen gemeinsam [haben]" (for the relation between labor and community see Jost Trier, "Arbeit und Gemeinschaft," *Studium Generale*, Vol. III, No. 11 [November, 1950]).

47. See R. P. Genelli ("Facteur humain ou facteur social du travail," *Revue française du travail*, Vol. VII, Nos. 1–3 [January–March, 1952]), who believes that a "new solution of the labor problem" should be found which would take into account the "collective nature of labor" and therefore provide not for the individual laborer but for him as member of his group. This "new" solution is of course the one prevailing in modern society.

48. Adam Smith, *op. cit.*, I, 15, and Marx, *Das Elend der Philosophie* (Stuttgart, 1885), p. 125: Adam Smith "hat sehr wohl gesehen, dass 'in Wirklichkeit die

better than the gathering of people on the exchange market, which brings to light the skills and qualities of the producers and thus always provides some basis for distinction.

The sameness prevailing in a society resting on labor and consumption and expressed in its conformity is intimately connected with the somatic experience of laboring together, where the biological rhythm of labor unites the group of laborers to the point that each may feel that he is no longer an individual but actually one with all others. To be sure, this eases labor's toil and trouble in much the same way as marching together eases the effort of walking for each soldier. It is therefore quite true that for the *animal laborans* "labor's sense and value depend entirely upon the social conditions," that is, upon the extent to which the labor and consumption process is permitted to function smoothly and easily, independent of "professional attitudes properly speaking";[49] the trouble is only that the best "social conditions" are those under which it is possible to lose one's identity. This unitedness of many into one is basically antipolitical; it is the very opposite of the togetherness prevailing in political or commercial communities, which—to take the Aristotelian example—consist not of an association (*koinōnia*) between two physicians, but between a physi-

---

Verschiedenheit der natürlichen Anlagen zwischen den Individuen weit geringer ist als wir glauben.' . . . Ursprünglich unterscheidet sich ein Lastträger weniger von einem Philosophen als ein Kettenhund von einem Windhund. Es ist die Arbeitsteilung, welche einen Abgrund zwischen beiden aufgetan hat." Marx uses the term "division of labor" indiscriminately for professional specialization and for the dividing of the labor process itself, but means here of course the former. Professional specialization is indeed a form of distinction, and the craftsman or professional worker, even if he is helped by others, works essentially in isolation. He meets others *qua* worker only when it comes to the exchange of products. In the true division of labor, the laborer cannot accomplish anything in isolation; his effort is only part and function of the effort of all laborers among whom the task is divided. But these other laborers *qua* laborers are not different from him, they are all the same. Thus, it is not the relatively recent division of labor, but the age-old professional specialization which "opened the gulf" between the street porter and the philosopher.

49. Alain Touraine, *L'évolution du travail ouvrier aux usines Renault* (1955), p. 177.

cian and a farmer, "and in general between people who are different and unequal."[50]

The equality attending the public realm is necessarily an equality of unequals who stand in need of being "equalized" in certain respects and for specific purposes. As such, the equalizing factor arises not from human "nature" but from outside, just as money—to continue the Aristotelian example—is needed as an outside factor to equate the unequal activities of physician and farmer. Political equality, therefore, is the very opposite of our equality before death, which as the common fate of all men arises out of the human condition, or of equality before God, at least in its Christian interpretation, where we are confronted with an equality of sinfulness inherent in human nature. In these instances, no equalizer is needed because sameness prevails anyhow; by the same token, however, the actual experience of this sameness, the experience of life and death, occurs not only in isolation but in utter loneliness, where no true communication, let alone association and community, is possible. From the viewpoint of the world and the public realm, life and death and everything attesting to sameness are non-worldly, antipolitical, truly transcendent experiences.

The incapacity of the *animal laborans* for distinction and hence for action and speech seems to be confirmed by the striking absence of serious slave rebellions in ancient and modern times.[51] No less striking, however, is the sudden and frequently extraordinarily productive role which the labor movements have played in modern politics. From the revolutions of 1848 to the Hungarian revolution of 1956, the European working class, by virtue of being the only organized and hence the leading section of the people, has written one of the most glorious and probably the most promising chapter of recent history. However, although the line between political and economic demands, between political organizations and trade unions, was blurred enough, the two should not

50. *Nicomachean Ethics* 1133a16.

51. The decisive point is that modern rebellions and revolutions always asked for freedom and justice for all, whereas in antiquity "slaves never raised the demand of freedom as an inalienable right for all men, and there never was an attempt to achieve abolition of slavery as such through combined action" (W. L. Westermann, "Sklaverei," in Pauly-Wissowa, Suppl. VI, p. 981).

be confused. The trade unions, defending and fighting for the interests of the working class, are responsible for its eventual incorporation into modern society, especially for an extraordinary increase in economic security, social prestige, and political power. The trade unions were never revolutionary in the sense that they desired a transformation of society together with a transformation of the political institutions in which this society was represented, and the political parties of the working class have been interest parties most of the time, in no way different from the parties which represented other social classes. A distinction appeared only in those rare and yet decisive moments when during the process of a revolution it suddenly turned out that these people, if not led by official party programs and ideologies, had their own ideas about the possibilities of democratic government under modern conditions. In other words, the dividing line between the two is not a matter of extreme social and economic demands but solely of the proposition of a new form of government.

What is so easily overlooked by the modern historian who faces the rise of totalitarian systems, especially when he deals with developments in the Soviet Union, is that just as the modern masses and their leaders succeeded, at least temporarily, in bringing forth in totalitarianism an authentic, albeit all-destructive, new form of government, thus the people's revolutions, for more than a hundred years now, have come forth, albeit never successfully, with another new form of government: the system of people's councils to take the place of the Continental party system, which, one is tempted to say, was discredited even before it came into existence.[52] The historical destinies of the two trends in the work-

52. It is important to keep in mind the sharp difference in substance and political function between the Continental party system and both the British and American systems. It is a decisive, though little noticed, fact in the development of European revolutions that the slogan of Councils (Soviets, *Räte*, etc.) was never raised by the parties and movements which took an active hand in organizing them, but always sprang from spontaneous rebellions; as such, the councils were neither properly understood nor particularly welcomed by the ideologists of the various movements who wanted to use the revolution in order to impose a preconceived form of government on the people. The famous slogan of the Kronstadt rebellion, which was one of the decisive turning points of the Russian Revolution,

ing class, the trade-union movement and the people's political aspirations, could not be more at variance: the trade unions, that is, the working class in so far as it is but one of the classes of modern society, have gone from victory to victory, while at the same time the political labor movement has been defeated each time it dared to put forth its own demands, as distinguished from party programs and economic reforms. If the tragedy of the Hungarian revolution achieved nothing more than that it showed the world that, all defeats and all appearances notwithstanding, this political élan has not yet died, its sacrifices were not in vain.

This apparently flagrant discrepancy between historical fact—the political productivity of the working class—and the phenomenal data obtained from an analysis of the laboring activity is likely to disappear upon closer inspection of the labor movement's development and substance. The chief difference between slave labor and modern, free labor is not that the laborer possesses personal freedom—freedom of movement, economic activity, and personal inviolability—but that he is admitted to the political realm and fully emancipated as a citizen. The turning point in the history of labor came with the abolition of property qualifications for the right to vote. Up to this time the status of free labor had been very similar to the status of the constantly increasing emancipated slave population in antiquity; these men were free, being assimilated to the status of resident aliens, but not citizens. In contrast to ancient slave emancipations, where as a rule the slave ceased to be a laborer when he ceased to be a slave, and where, therefore, slavery remained the social condition of laboring no matter how many slaves were emancipated, the modern emancipation of labor was intended to elevate the laboring activity itself, and this was achieved long before the laborer as a person was granted personal and civil rights.

However, one of the important side effects of the actual emanci-

---

was: Soviets without Communism; and this at the time implied: Soviets without parties.

The thesis that the totalitarian regimes confront us with a new form of government is explained at some length in my article, "Ideology and Terror: A Novel Form of Government," *Review of Politics* (July, 1953). A more detailed analysis of the Hungarian revolution and the council system can be found in a recent article, "Totalitarian Imperialism," *Journal of Politics* (February, 1958).

pation of laborers was that a whole new segment of the population was more or less suddenly admitted to the public realm, that is, *appeared* in public,[53] and this without at the same time being admitted to society, without playing any leading role in the all-important economic activities of this society, and without, therefore, being absorbed by the social realm and, as it were, spirited away from the public. The decisive role of mere appearance, of distinguishing oneself and being conspicuous in the realm of human affairs is perhaps nowhere better illustrated than in the fact that laborers, when they entered the scene of history, felt it necessary to adopt a costume of their own, the *sans-culotte*, from which, during the French Revolution, they even derived their name.[54] By this costume they won a distinction of their own, and the distinction was directed against all others.

The very pathos of the labor movement in its early stages—and it is still in its early stages in all countries where capitalism has not reached its full development, in Eastern Europe, for example, but also in Italy or Spain and even in France—stemmed from its fight against society as a whole. The enormous power potential these movements acquired in a relatively short time and often under very

53. An anecdote, reported by Seneca from imperial Rome, may illustrate how dangerous mere appearance in public was thought to be. At that time a proposition was laid before the senate to have slaves dress uniformly in public so that they could immediately be distinguished from free citizens. The proposition was turned down as too dangerous, since the slaves would now be able to recognize each other and become aware of their potential power. Modern interpreters were of course inclined to conclude from this incident that the number of slaves at the time must have been very great, yet this conclusion turned out to be quite erroneous. What the sound political instinct of the Romans judged to be dangerous was appearance as such, quite independent from the number of people involved (see Westermann, *op. cit.*, p. 1000).

54. A. Soboul ("Problèmes de travail en l'an II," *Journal de psychologie normale et pathologique*, Vol. LII, No. 1 [January–March, 1955]) describes very well how the workers made their first appearance on the historical scene: "Les travailleurs ne sont pas désignés par leur fonction sociale, mais simplement par leur costume. Les ouvriers adoptèrent le pantalon boutonné à la veste, et ce costume devint une caractéristique du peuple: des sans-culottes . . . 'en parlant des *sans-culottes*, déclare Petion à la Convention, le 10 avril 1793, on n'entend pas tous les citoyens, les nobles et les aristocrates exceptés, mais on entend des hommes qui n'ont pas, pour les distinguer de ceux qui ont.' "

adverse circumstances sprang from the fact that despite all the talk and theory they were the only group on the political scene which not only defended its economic interests but fought a full-fledged political battle. In other words, when the labor movement appeared on the public scene, it was the only organization in which men acted and spoke *qua* men—and not *qua* members of society.

For this political and revolutionary role of the labor movement, which in all probability is nearing its end, it is decisive that the economic activity of its members was incidental and that its force of attraction was never restricted to the ranks of the working class. If for a time it almost looked as if the movement would succeed in founding, at least within its own ranks, a new public space with new political standards, the spring of these attempts was not labor—neither the laboring activity itself nor the always utopian rebellion against life's necessity—but those injustices and hypocrisies which have disappeared with the transformation of a class society into a mass society and with the substitution of a guaranteed annual wage for daily or weekly pay.

The workers today are no longer outside of society; they are its members, and they are jobholders like everybody else. The political significance of the labor movement is now the same as that of any other pressure group; the time is past when, as for nearly a hundred years, it could represent the people as a whole—if we understand by *le peuple* the actual political body, distinguished as such from the population as well as from society.[55] (In the Hungarian revolution the workers were in no way distinguished from the rest of the people; what from 1848 to 1918 had been almost a monopoly of the working class—the notion of a parliamentary system based on councils instead of parties—had now become the unanimous demand of the whole people.) The labor movement, equivocal in its content and aims from the beginning, lost this representation and hence its political role at once wherever the working class became an integral part of society, a social and economic power of its own as in the most developed economies of the Western world, or

55. Originally, the term *le peuple*, which became current at the end of the eighteenth century, designated simply those who had no property. As we mentioned before, such a class of completely destitute people was not known prior to the modern age.

where it "succeeded" in transforming the whole population into a labor society as in Russia and as may happen elsewhere even under non-totalitarian conditions. Under circumstances where even the exchange market is being abolished, the withering of the public realm, so conspicuous throughout the modern age, may well find its consummation.

31

THE TRADITIONAL SUBSTITUTION OF MAKING FOR ACTING

The modern age, in its early concern with tangible products and demonstrable profits or its later obsession with smooth functioning and sociability, was not the first to denounce the idle uselessness of action and speech in particular and of politics in general.[56] Exasperation with the threefold frustration of action—the unpredictability of its outcome, the irreversibility of the process, and the anonymity of its authors—is almost as old as recorded history. It has always been a great temptation, for men of action no less than for men of thought, to find a substitute for action in the hope that the realm of human affairs may escape the haphazardness and moral irresponsibility inherent in a plurality of agents. The remarkable monotony of the proposed solutions throughout our recorded history testifies to the elemental simplicity of the matter. Generally speaking, they always amount to seeking shelter from action's calamities in an activity where one man, isolated from all others, remains master of his doings from beginning to end. This attempt to replace acting with making is manifest in the whole body of argument against "democracy," which, the more consistently and better reasoned it is, will turn into an argument against the essentials of politics.

The calamities of action all arise from the human condition of plurality, which is the condition *sine qua non* for that space of appearance which is the public realm. Hence the attempt to do away with this plurality is always tantamount to the abolition of the public realm itself. The most obvious salvation from the dangers of

56. The classic author on this matter is still Adam Smith, to whom the only legitimate function of government is "the defence of the rich against the poor, or of those who have some property against those who have none at all" (*op. cit.*, II, 198 ff.; for the quotation see II, 203).

plurality is mon-archy, or one-man-rule, in its many varieties, from outright tyranny of one against all to benevolent despotism and to those forms of democracy in which the many form a collective body so that the people "is many in one" and constitute themselves as a "monarch."[57] Plato's solution of the philosopher-king, whose "wisdom" solves the perplexities of action as though they were solvable problems of cognition, is only one—and by no means the least tyrannical—variety of one-man rule. The trouble with these forms of government is not that they are cruel, which often they are not, but rather that they work too well. Tyrants, if they know their business, may well be "kindly and mild in everything," like Peisistratus, whose rule even in antiquity was compared to "the Golden Age of Cronos";[58] their measures may sound very "un-tyrannical" and beneficial to modern ears, especially when we hear that the only—albeit unsuccessful—attempt to abolish slavery in antiquity was made by Periandros, tyrant of Corinth.[59] But they all have in common the banishment of the citizens from the public realm and the insistence that they mind their private business while only "the ruler should attend to public affairs."[60] This, to be sure,

57. This is the Aristotelian interpretation of tyranny in the form of a democracy (*Politics* 1292a16 ff.). Kingship, however, does not belong among the tyrannical forms of government, nor can it be defined as one-man rule or monarchy. While the terms "tyranny" and "monarchy" could be used interchangeably, the words "tyrant" and *basileus* ("king") are used as opposites (see, for instance, Aristotle *Nicomachean Ethics* 1160b3; Plato *Republic* 576D). Generally speaking, one-man rule is praised in antiquity only for household matters or for warfare, and it is usually in some military or "economic" context that the famous line from the *Iliad, ouk agathon polykoiraniē; heis koiranos estō, heis basileus*—"the rule by many is not good; one should be master, one be king" (ii. 204)—is quoted. (Aristotle, who applies Homer's saying in his *Metaphysics* [1076a3 ff.] to political community life [*politeuesthai*] in a metaphorical sense, is an exception. In *Politics* 1292a13, where he quotes the Homeric line again, he takes a stand against the many having the power "not as individuals, but collectively," and states that this is only a disguised form of one-man rule, or tyranny.) Conversely, the rule of the many, later called *polyarkhia*, is used disparagingly to mean confusion of command in warfare (see, for instance, Thucydides vi. 72; cf. Xenophon *Anabasis* vi. 1. 18).

58. Aristotle *Athenian Constitution* xvi. 2, 7.

59. See Fritz Heichelheim, *Wirtschaftsgeschichte des Altertums* (1938), I, 258.

60. Aristotle (*Athenian Constitution* xv. 5) reports this of Peisistratus.

was tantamount to furthering private industry and industriousness, but the citizens could see in this policy nothing but the attempt to deprive them of the time necessary for participation in common matters. It is the obvious short-range advantages of tyranny, the advantages of stability, security, and productivity, that one should beware, if only because they pave the way to an inevitable loss of power, even though the actual disaster may occur in a relatively distant future.

Escape from the frailty of human affairs into the solidity of quiet and order has in fact so much to recommend it that the greater part of political philosophy since Plato could easily be interpreted as various attempts to find theoretical foundations and practical ways for an escape from politics altogether. The hallmark of all such escapes is the concept of rule, that is, the notion that men can lawfully and politically live together only when some are entitled to command and the others forced to obey. The commonplace notion already to be found in Plato and Aristotle that every political community consists of those who rule and those who are ruled (on which assumption in turn are based the current definitions of forms of government—rule by one or monarchy, rule by few or oligarchy, rule by many or democracy) rests on a suspicion of action rather than on a contempt for men, and arose from the earnest desire to find a substitute for action rather than from any irresponsible or tyrannical will to power.

Theoretically, the most brief and most fundamental version of the escape from action into rule occurs in the *Statesman*, where Plato opens a gulf between the two modes of action, *archein* and *prattein* ("beginning" and "achieving"), which according to Greek understanding were interconnected. The problem, as Plato saw it, was to make sure that the beginner would remain the complete master of what he had begun, not needing the help of others to carry it through. In the realm of action, this isolated mastership can be achieved only if the others are no longer needed to join the enterprise of their own accord, with their own motives and aims, but are used to execute orders, and if, on the other hand, the beginner who took the initiative does not permit himself to get involved in the action itself. To begin (*archein*) and to act (*prattein*) thus can become two altogether different activities, and the beginner has become a ruler (an *archōn* in the twofold sense of the word) who "does not have to act at all (*prattein*), but rules (*archein*) over those who are capable of execution." Under these circumstances, the essence of politics is "to know how to begin and to rule in the gravest matters with regard to timeliness and untimeliness"; action as such is entirely eliminated and has become the mere "execution of orders."[61] Plato was the first to introduce the division between those who know and do not act and those who act and do not know, instead of the old articulation of action into beginning and achieving, so that knowing what to do and doing it became two altogether different performances.

Since Plato himself immediately identified the dividing line between thought and action with the gulf which separates the rulers from those over whom they rule, it is obvious that the experiences on which the Platonic division rests are those of the household, where nothing would ever be done if the master did not know what to do and did not give orders to the slaves who executed them without knowing. Here indeed, he who knows does not have to do and he who does needs no thought or knowledge. Plato was still quite aware that he proposed a revolutionary transformation of the *polis* when he applied to its administration the currently recognized maxims for a well-ordered household.[62] (It is a common error to interpret Plato as though he wanted to abolish the family and the household; he wanted, on the contrary, to extend this type of life until one family embraced every citizen. In other words, he wanted to eliminate from the household community its private character, and it is for this purpose that he recommended the abolition of private property and individual marital status.)[63] According to Greek understanding, the relationship between ruling and being ruled,

61. *Statesman* 305.

62. It is the decisive contention of the *Statesman* that no difference existed between the constitution of a large household and that of the *polis* (see 259), so that the same science would cover political and "economic" or household matters.

63. This is particularly manifest in those passages of the fifth book of the *Republic* in which Plato describes how the fear lest one attack his own son, brother, or father would further general peace in his utopian republic. Because of the community of women, nobody would know who his blood relatives were (see esp. 463C and 465B).

between command and obedience, was by definition identical with the relationship between master and slaves and therefore precluded all possibility of action. Plato's contention, therefore, that the rules of behavior in public matters should be derived from the master-slave relationship in a well-ordered household actually meant that action should not play any part in human affairs.

It is obvious that Plato's scheme offers much greater chances for a permanent order in human affairs than the tyrant's efforts to eliminate everybody but himself from the public realm. Although each citizen would retain some part in the handling of public affairs, they would indeed "act" like one man without even the possibility of internal dissension, let alone factional strife: through rule, "the many become one in every respect" except bodily appearance.[64] Historically, the concept of rule, though originating in the household and family realm, has played its most decisive part in the organization of public matters and is for us invariably connected with politics. This should not make us overlook the fact that for Plato it was a much more general category. He saw in it the chief device for ordering and judging human affairs in every respect. This is not only evident from his insistence that the city-state must be considered to be "man writ large" and from his construction of a psychological order which actually follows the public order of his utopian city, but is even more manifest in the grandiose consistency with which he introduced the principle of domination into the intercourse of man with himself. The supreme criterion of fitness for ruling others is, in Plato and in the aristocratic tradition of the West, the capacity to rule one's self. Just as the philosopher-king commands the city, the soul commands the body and reason commands the passions. In Plato himself, the legitimacy of this tyranny in everything pertaining to man, his conduct toward himself no less than his conduct toward others, is still firmly rooted in the equivocal significance of the word *archein*, which means both beginning and ruling; it is decisive for Plato, as he says expressly at the end of the *Laws*, that only the beginning (*archē*) is entitled to rule (*archein*). In the tradition of Platonic thought, this original, linguistically predetermined identity of ruling and beginning had the consequence that all beginning was understood as the

64. *Republic* 443E.

legitimation for rulership, until, finally, the element of beginning disappeared altogether from the concept of rulership. With it the most elementary and authentic understanding of human freedom disappeared from political philosophy.

The Platonic separation of knowing and doing has remained at the root of all theories of domination which are not mere justifications of an irreducible and irresponsible will to power. By sheer force of conceptualization and philosophical clarification, the Platonic identification of knowledge with command and rulership and of action with obedience and execution overruled all earlier experiences and articulations in the political realm and became authoritative for the whole tradition of political thought, even after the roots of experience from which Plato derived his concepts had long been forgotten. Apart from the unique Platonic mixture of depth and beauty, whose weight was bound to carry his thoughts through the centuries, the reason for the longevity of this particular part of his work is that he strengthened his substitution of rulership for action through an even more plausible interpretation in terms of making and fabrication. It is indeed true—and Plato, who had taken the key word of his philosophy, the term "idea," from experiences in the realm of fabrication, must have been the first to notice it—that the division between knowing and doing, so alien to the realm of action, whose validity and meaningfulness are destroyed the moment thought and action part company, is an everyday experience in fabrication, whose processes obviously fall into two parts: first, perceiving the image or shape (*eidos*) of the product-to-be, and then organizing the means and starting the execution.

The Platonic wish to substitute making for acting in order to bestow upon the realm of human affairs the solidity inherent in work and fabrication becomes most apparent where it touches the very center of his philosophy, the doctrine of ideas. When Plato was not concerned with political philosophy (as in the *Symposium* and elsewhere), he describes the ideas as what "shines forth most" (*ekphanestaton*) and therefore as variations of the beautiful. Only in the *Republic* were the ideas transformed into standards, measurements, and rules of behavior, all of which are variations or derivations of the idea of the "good" in the Greek sense of the word, that is, of the "good for" or of fitness.[65] This transformation was necessary to apply the doctrine of ideas to politics, and it is essentially for a political purpose, the purpose of eliminating the character of frailty from human affairs, that Plato found it necessary to declare the good, and not the beautiful, to be the highest idea. But this idea of the good is not the highest idea of the philosopher, who wishes to contemplate the true essence of Being and therefore leaves the dark cave of human affairs for the bright sky of ideas; even in the *Republic*, the philosopher is still defined as a lover of beauty, not of goodness. The good is the highest idea of the philosopher-*king*, who wishes to be the ruler of human affairs because he must spend his life among men and cannot dwell forever under the sky of ideas. It is only when he returns to the dark cave of human affairs to live once more with his fellow men that he needs the ideas for guidance as standards and rules by which to measure and under which to subsume the varied multitude of human deeds and words with the same absolute, "objective" certainty with which the craftsman can be guided in making and the layman in judging individual beds by using the unwavering ever-present model, the "idea" of bed in general.[66]

Technically, the greatest advantage of this transformation and application of the doctrine of ideas to the political realm lay in the elimination of the personal element in the Platonic notion of ideal

65. The word *ekphanestaton* occurs in the *Phaedrus* (250) as the chief quality of the beautiful. In the *Republic* (518) a similar quality is claimed for the idea of the good, which is called *phanotaton*. Both words derive from *phainesthai* ("to appear" and "shine forth"), and in both cases the superlative is used. Obviously, the quality of shining brightness applies to the beautiful much more than to the good.

66. Werner Jaeger's statement (*Paideia* [1945], II, 416 n.), "The idea that there is a supreme art of measurement and that the philosopher's knowledge of value (*phronēsis*) is the ability to measure, runs through all Plato's work right down to the end," is true only for Plato's political philosophy, where the idea of the good replaces the idea of the beautiful. The parable of the Cave, as told in the *Republic*, is the very center of Plato's political philosophy, but the doctrine of ideas as presented there must be understood as its application to politics, not as the original, purely philosophical development, which we cannot discuss here. Jaeger's characterization of the "philosopher's knowledge of values" as *phronēsis* indicates, in fact, the political and non-philosophical nature of this knowledge; for the very word *phronēsis* characterizes in Plato and Aristotle the insight of the statesman rather than the vision of the philosopher.

rulership. Plato knew quite well that his favorite analogies taken from household life, such as the master-slave or the shepherd-flock relationship, would demand a quasi-divine quality in the ruler of men to distinguish him as sharply from his subjects as the slaves are distinguished from the master or the sheep from the shepherd.[67] The construction of the public space in the image of a fabricated object, on the contrary, carried with it only the implication of ordinary mastership, experience in the art of politics as in all other arts, where the compelling factor lies not in the person of the artist or craftsman but in the impersonal object of his art or craft. In the *Republic*, the philosopher-king applies the ideas as the craftsman applies his rules and standards; he "makes" his City as the sculptor makes a statue;[68] and in the final Platonic work these same ideas have even become laws which need only be executed.[69]

Within this frame of reference, the emergence of a utopian political system which could be construed in accordance with a model by somebody who has mastered the techniques of human affairs becomes almost a matter of course; Plato, who was the first to design a blueprint for the making of political bodies, has remained the inspiration of all later utopias. And although none of these utopias ever came to play any noticeable role in history—for in the few instances where utopian schemes were realized, they broke down quickly under the weight of reality, not so much the reality of exterior circumstances as of the real human relationships they could not control—they were among the most efficient vehicles to conserve and develop a tradition of political thinking in

67. In the *Statesman*, where Plato chiefly pursues this line of thought, he concludes ironically: Looking for someone who would be as fit to rule over man as the shepherd is to rule over his flock, we found "a god instead of a mortal man" (275).

68. *Republic* 420.

69. It may be interesting to note the following development in Plato's political theory: In the *Republic*, his division between rulers and ruled is guided by the relationship between expert and layman; in the *Statesman*, he takes his bearings from the relation between knowing and doing; and in the *Laws*, the execution of unchangeable laws is all that is left to the statesman or necessary for the functioning of the public realm. What is most striking in this development is the progressive shrinkage of faculties needed for the mastering of politics.

which, consciously or unconsciously, the concept of action was interpreted in terms of making and fabrication.

One thing, however, is noteworthy in the development of this tradition. It is true that violence, without which no fabrication could ever come to pass, has always played an important role in political schemes and thinking based upon an interpretation of action in terms of making; but up to the modern age, this element of violence remained strictly instrumental, a means that needed an end to justify and limit it, so that glorifications of violence as such are entirely absent from the tradition of political thought prior to the modern age. Generally speaking, they were impossible as long as contemplation and reason were supposed to be the highest capacities of man, because under this assumption all articulations of the *vita activa*, fabrication no less than action and let alone labor, remained themselves secondary and instrumental. Within the narrower sphere of political theory, the consequence was that the notion of rule and the concomitant questions of legitimacy and rightful authority played a much more decisive role than the understanding and interpretations of action itself. Only the modern age's conviction that man can know only what he makes, that his allegedly higher capacities depend upon making and that he therefore is primarily *homo faber* and not an *animal rationale*, brought forth the much older implications of violence inherent in all interpretations of the realm of human affairs as a sphere of making. This has been particularly striking in the series of revolutions, characteristic of the modern age, all of which—with the exception of the American Revolution—show the same combination of the old Roman enthusiasm for the foundation of a new body politic with the glorification of violence as the only means for "making" it. Marx's dictum that "violence is the midwife of every old society pregnant with a new one," that is, of all change in history and politics,[70] only sums up the conviction of the whole modern age and draws the consequences of its innermost belief that history is "made" by men as nature is "made" by God.

70. The quote is from *Capital* (Modern Library ed.), p. 824. Other passages in Marx show that he does not restrict his remark to the manifestation of social or economic forces. For example: "In actual history it is notorious that conquest, enslavement, robbery, murder, briefly violence, play the great part" (*ibid.*, 785).

How persistent and successful the transformation of action into a mode of making has been is easily attested by the whole terminology of political theory and political thought, which indeed makes it almost impossible to discuss these matters without using the category of means and ends and thinking in terms of instrumentality. Perhaps even more convincing is the unanimity with which popular proverbs in all modern languages advise us that "he who wants an end must also want the means" and that "you can't make an omelette without breaking eggs." We are perhaps the first generation which has become fully aware of the murderous consequences inherent in a line of thought that forces one to admit that all means, provided that they are efficient, are permissible and justified to pursue something defined as an end. However, in order to escape these beaten paths of thought it is not enough to add some qualifications, such as that not all means are permissible or that under certain circumstances means may be more important than ends; these qualifications either take for granted a moral system which, as the very exhortations demonstrate, can hardly be taken for granted, or they are overpowered by the very language and analogies they use. For to make a statement about ends that do not justify all means is to speak in paradoxes, the definition of an end being precisely the justification of the means; and paradoxes always indicate perplexities, they do not solve them and hence are never convincing. As long as we believe that we deal with ends and means in the political realm, we shall not be able to prevent anybody's using all means to pursue recognized ends.

The substitution of making for acting and the concomitant degradation of politics into a means to obtain an allegedly "higher" end—in antiquity the protection of the good men from the rule of the bad in general, and the safety of the philosopher in particular,[71] in the Middle Ages the salvation of souls, in the modern age the productivity and progress of society—is as old as the tradition of political philosophy. It is true that only the modern age defined man primarily as *homo faber*, a toolmaker and producer of things,

71. Compare Plato's statement that the wish of the philosopher to become a ruler of men can spring only from the fear of being ruled by those who are worse (*Republic* 347) with Augustine's statement that the function of government is to enable "the good" to live more quietly among "the bad" (*Epistolae* 153. 6).

and therefore could overcome the deep-seated contempt and suspicion in which the tradition had held the whole sphere of fabrication. Yet, this same tradition, in so far as it also had turned against action—less openly, to be sure, but no less effectively—had been forced to interpret acting in terms of making, and thereby, its suspicion and contempt notwithstanding, had introduced into political philosophy certain trends and patterns of thought upon which the modern age could fall back. In this respect, the modern age did not reverse the tradition but rather liberated it from the "prejudices" which had prevented it from declaring openly that the work of the craftsman should rank higher than the "idle" opinions and doings which constitute the realm of human affairs. The point is that Plato and, to a lesser degree, Aristotle, who thought craftsmen not even worthy of full-fledged citizenship, were the first to propose handling political matters and ruling political bodies in the mode of fabrication. This seeming contradiction clearly indicates the depth of the authentic perplexities inherent in the human capacity for action and the strength of the temptation to eliminate its risks and dangers by introducing into the web of human relationships the much more reliable and solid categories inherent in activities with which we confront nature and build the world of the human artifice.

32

THE PROCESS CHARACTER OF ACTION

The instrumentalization of action and the degradation of politics into a means for something else has of course never really succeeded in eliminating action, in preventing its being one of the decisive human experiences, or in destroying the realm of human affairs altogether. We saw before that in our world the seeming elimination of labor, as the painful effort to which all human life is bound, had first of all the consequence that work is now performed in the mode of laboring, and the products of work, objects for use, are consumed as though they were mere consumer goods. Similarly, the attempt to eliminate action because of its uncertainty and to save human affairs from their frailty by dealing with them as though they were or could become the planned products of human making has first of all resulted in channeling the human capacity for action, for beginning new and spontaneous processes which without men never would come into existence, into an attitude toward nature which up to the latest stage of the modern age had been one of exploring natural laws and fabricating objects out of natural material. To what an extent we have begun to act into nature, in the literal sense of the word, is perhaps best illustrated by a recent, casual remark of a scientist who quite seriously suggested that "basic research is when I am doing what I don't know what I am doing."[72]

This started harmlessly enough with the experiment in which men were no longer content to observe, to register, and contemplate whatever nature was willing to yield in her own appearance, but began to prescribe conditions and to provoke natural processes. What then developed into an ever-increasing skill in unchaining elemental processes, which, without the interference of men, would have lain dormant and perhaps never have come to pass, has finally ended in a veritable art of "making" nature, that is, of creating "natural" processes which without men would never exist and which earthly nature by herself seems incapable of accomplishing, although similar or identical processes may be commonplace phenomena in the universe surrounding the earth. Through the introduction of the experiment, in which we prescribed man-thought conditions to natural processes and forced them to fall into man-made patterns, we eventually learned how to "repeat the process that goes on in the sun," that is, how to win from natural processes on the earth those energies which without us develop only in the universe.

The very fact that natural sciences have become exclusively sciences of process and, in their last stage, sciences of potentially irreversible, irremediable "processes of no return" is a clear indication that, whatever the brain power necessary to start them, the actual underlying human capacity which alone could bring about this development is no "theoretical" capacity, neither contemplation nor reason, but the human ability to act—to start new unprecedented processes whose outcome remains uncertain and unpre-

72. Quoted from an interview with Wernher von Braun, as reported in the *New York Times*, December 16, 1957.

dictable whether they are let loose in the human or the natural realm.

In this aspect of action—all-important to the modern age, to its enormous enlargement of human capabilities as well as to its unprecedented concept and consciousness of history—processes are started whose outcome is unpredictable, so that uncertainty rather than frailty becomes the decisive character of human affairs. This property of action had escaped the attention of antiquity, by and large, and had, to say the least, hardly found adequate articulation in ancient philosophy, to which the very concept of history as we know it is altogether alien. The central concept of the two entirely new sciences of the modern age, natural science no less than historical, is the concept of process, and the actual human experience underlying it is action. Only because we are capable of acting, of starting processes of our own, can we conceive of both nature and history as systems of processes. It is true that this character of modern thinking first came to the fore in the science of history, which, since Vico, has been consciously presented as a "new science," while the natural sciences needed several centuries before they were forced by the very results of their triumphal achievements to exchange an obsolete conceptual framework for a vocabulary that is strikingly similar to the one used in the historical sciences.

However that may be, only under certain historical circumstances does frailty appear to be the chief characteristic of human affairs. The Greeks measured them against the ever-presence or eternal recurrence of all natural things, and the chief Greek concern was to measure up to and become worthy of an immortality which surrounds men but which mortals do not possess. To people who are not possessed by this concern with immortality, the realm of human affairs is bound to show an altogether different, even somehow contradictory aspect, namely, an extraordinary resiliency whose force of persistence and continuity in time is far superior to the stable durability of the solid world of things. Whereas men have always been capable of destroying whatever was the product of human hands and have become capable today even of the potential destruction of what man did not make—the earth and earthly nature—men never have been and never will be able to undo or even to control reliably any of the processes they start through action. Not even oblivion and confusion, which can cover up so efficiently the origin and the responsibility for every single deed, are able to undo a deed or prevent its consequences. And this incapacity to undo what has been done is matched by an almost equally complete incapacity to foretell the consequences of any deed or even to have reliable knowledge of its motives.[73]

While the strength of the production process is entirely absorbed in and exhausted by the end product, the strength of the action process is never exhausted in a single deed but, on the contrary, can grow while its consequences multiply; what endures in the realm of human affairs are these processes, and their endurance is as unlimited, as independent of the perishability of material and the mortality of men as the endurance of humanity itself. The reason why we are never able to foretell with certainty the outcome and end of any action is simply that action has no end. The process of a single deed can quite literally endure throughout time until mankind itself has come to an end.

That deeds possess such an enormous capacity for endurance, superior to every other man-made product, could be a matter of pride if men were able to bear its burden, the burden of irreversibility and unpredictability, from which the action process draws its very strength. That this is impossible, men have always known. They have known that he who acts never quite knows what he is doing, that he always becomes "guilty" of consequences he never intended or even foresaw, that no matter how disastrous and unexpected the consequences of his deed he can never undo it, that the process he starts is never consummated unequivocally in one single deed or event, and that its very meaning never discloses itself to the actor but only to the backward glance of the historian who himself does not act. All this is reason enough to turn away with despair from the realm of human affairs and to hold in contempt the human capacity for freedom, which, by producing the web of human relationships, seems to entangle its producer to such an ex-

73. "Man weiss die Herkunft nicht, man weiss die Folgen nicht . . . [der Wert der Handlung ist] unbekannt," as Nietzsche once put it (*Wille zur Macht*, No. 291), hardly aware that he only echoed the age-old suspicion of the philosopher against action.

tent that he appears much more the victim and the sufferer than the author and doer of what he has done. Nowhere, in other words, neither in labor, subject to the necessity of life, nor in fabrication, dependent upon given material, does man appear to be less free than in those capacities whose very essence is freedom and in that realm which owes its existence to nobody and nothing but man.

It is in accordance with the great tradition of Western thought to think along these lines: to accuse freedom of luring man into necessity, to condemn action, the spontaneous beginning of something new, because its results fall into a predetermined net of relationships, invariably dragging the agent with them, who seems to forfeit his freedom the very moment he makes use of it. The only salvation from this kind of freedom seems to lie in non-acting, in abstention from the whole realm of human affairs as the only means to safeguard one's sovereignty and integrity as a person. If we leave aside the disastrous consequences of these recommendations (which materialized into a consistent system of human behavior only in Stoicism), their basic error seems to lie in that identification of sovereignty with freedom which has always been taken for granted by political as well as philosophic thought. If it were true that sovereignty and freedom are the same, then indeed no man could be free, because sovereignty, the ideal of uncompromising self-sufficiency and mastership, is contradictory to the very condition of plurality. No man can be sovereign because not one man, but men, inhabit the earth—and not, as the tradition since Plato holds, because of man's limited strength, which makes him depend upon the help of others. All the recommendations the tradition has to offer to overcome the condition of non-sovereignty and win an untouchable integrity of the human person amount to a compensation for the intrinsic "weakness" of plurality. Yet, if these recommendations were followed and this attempt to overcome the consequences of plurality were successful, the result would be not so much sovereign domination of one's self as arbitrary domination of all others, or, as in Stoicism, the exchange of the real world for an imaginary one where these others would simply not exist.

In other words, the issue here is not strength or weakness in the sense of self-sufficiency. In polytheist systems, for instance, even a god, no matter how powerful, cannot be sovereign; only under the assumption of one god ("One is one and all alone and evermore shall be so") can sovereignty and freedom be the same. Under all other circumstances, sovereignty is possible only in imagination, paid for by the price of reality. Just as Epicureanism rests on the illusion of happiness when one is roasted alive in the Phaleric Bull, Stoicism rests on the illusion of freedom when one is enslaved. Both illusions testify to the psychological power of imagination, but this power can exert itself only as long as the reality of the world and the living, where one is and appears to be either happy or unhappy, either free or slave, are eliminated to such an extent that they are not even admitted as spectators to the spectacle of self-delusion.

If we look upon freedom with the eyes of the tradition, identifying freedom with sovereignty, the simultaneous presence of freedom and non-sovereignty, of being able to begin something new and of not being able to control or even foretell its consequences, seems almost to force us to the conclusion that human existence is absurd.[74] In view of human reality and its phenomenal evidence, it is indeed as spurious to deny human freedom to act because the actor never remains the master of his acts as it is to maintain that human sovereignty is possible because of the incontestable fact of human freedom.[75] The question which then arises is whether our notion that freedom and non-sovereignty are mutually exclusive is

74. This "existentialist" conclusion is much less due to an authentic revision of traditional concepts and standards than it appears to be; actually, it still operates within the tradition and with traditional concepts, though in a certain spirit of rebellion. The most consistent result of this rebellion is therefore a return to "religious values" which, however, have no root any longer in authentic religious experiences or faith, but are like all modern spiritual "values," exchange values, obtained in this case for the discarded "values" of despair.

75. Where human pride is still intact, it is tragedy rather than absurdity which is taken to be the hallmark of human existence. Its greatest representative is Kant, to whom the spontaneity of acting, and the concomitant faculties of practical reason, including force of judgment, remain the outstanding qualities of man, even though his action falls into the determinism of natural laws and his judgment cannot penetrate the secret of absolute reality (the *Ding an sich*). Kant had the courage to acquit man from the consequences of his deed, insisting solely on the purity of his motives, and this saved him from losing faith in man and his potential greatness.

not defeated by reality, or to put it another way, whether the capacity for action does not harbor within itself certain potentialities which enable it to survive the disabilities of non-sovereignty.

33

IRREVERSIBILITY AND THE
POWER TO FORGIVE

We have seen that the *animal laborans* could be redeemed from its predicament of imprisonment in the ever-recurring cycle of the life process, of being forever subject to the necessity of labor and consumption, only through the mobilization of another human capacity, the capacity for making, fabricating, and producing of *homo faber*, who as a toolmaker not only eases the pain and trouble of laboring but also erects a world of durability. The redemption of life, which is sustained by labor, is worldliness, which is sustained by fabrication. We saw furthermore that *homo faber* could be redeemed from his predicament of meaninglessness, the "devaluation of all values," and the impossibility of finding valid standards in a world determined by the category of means and ends, only through the interrelated faculties of action and speech, which produce meaningful stories as naturally as fabrication produces use objects. If it were not outside the scope of these considerations, one could add the predicament of thought to these instances; for thought, too, is unable to "think itself" out of the predicaments which the very activity of thinking engenders. What in each of these instances saves man—man *qua animal laborans*, *qua homo faber*, *qua* thinker— is something altogether different; it comes from the outside—not, to be sure, outside of man, but outside of each of the respective activities. From the viewpoint of the *animal laborans*, it is like a miracle that it is also a being which knows of and inhabits a world; from the viewpoint of *homo faber*, it is like a miracle, like the revelation of divinity, that meaning should have a place in this world.

The case of action and action's predicaments is altogether different. Here, the remedy against the irreversibility and unpredictability of the process started by acting does not arise out of another and possibly higher faculty, but is one of the potentialities

[*236*]

of action itself. The possible redemption from the predicament of irreversibility—of being unable to undo what one has done though one did not, and could not, have known what he was doing—is the faculty of forgiving. The remedy for unpredictability, for the chaotic uncertainty of the future, is contained in the faculty to make and keep promises. The two faculties belong together in so far as one of them, forgiving, serves to undo the deeds of the past, whose "sins" hang like Damocles' sword over every new generation; and the other, binding oneself through promises, serves to set up in the ocean of uncertainty, which the future is by definition, islands of security without which not even continuity, let alone durability of any kind, would be possible in the relationships between men.

Without being forgiven, released from the consequences of what we have done, our capacity to act would, as it were, be confined to one single deed from which we could never recover; we would remain the victims of its consequences forever, not unlike the sorcerer's apprentice who lacked the magic formula to break the spell. Without being bound to the fulfilment of promises, we would never be able to keep our identities; we would be condemned to wander helplessly and without direction in the darkness of each man's lonely heart, caught in its contradictions and equivocalities —a darkness which only the light shed over the public realm through the presence of others, who confirm the identity between the one who promises and the one who fulfils, can dispel. Both faculties, therefore, depend on plurality, on the presence and acting of others, for no one can forgive himself and no one can feel bound by a promise made only to himself; forgiving and promising enacted in solitude or isolation remain without reality and can signify no more than a role played before one's self.

Since these faculties correspond so closely to the human condition of plurality, their role in politics establishes a diametrically different set of guiding principles from the "moral" standards inherent in the Platonic notion of rule. For Platonic rulership, whose legitimacy rested upon the domination of the self, draws its guiding principles—those which at the same time justify and limit power over others—from a relationship established between me and myself, so that the right and wrong of relationships with others are determined by attitudes toward one's self, until the whole of the public realm is seen in the image of "man writ large," of the right order between man's individual capacities of mind, soul, and body. The moral code, on the other hand, inferred from the faculties of forgiving and of making promises, rests on experiences which nobody could ever have with himself, which, on the contrary, are entirely based on the presence of others. And just as the extent and modes of self-rule justify and determine rule over others—how one rules himself, he will rule others—thus the extent and modes of being forgiven and being promised determine the extent and modes in which one may be able to forgive himself or keep promises concerned only with himself.

Because the remedies against the enormous strength and resiliency inherent in action processes can function only under the condition of plurality, it is very dangerous to use this faculty in any but the realm of human affairs. Modern natural science and technology, which no longer observe or take material from or imitate processes of nature but seem actually to act into it, seem, by the same token, to have carried irreversibility and human unpredictability into the natural realm, where no remedy can be found to undo what has been done. Similarly, it seems that one of the great dangers of acting in the mode of making and within its categorical framework of means and ends lies in the concomitant self-deprivation of the remedies inherent only in action, so that one is bound not only to *do* with the means of violence necessary for all fabrication, but also to *undo* what he has done as he undoes an unsuccessful object, by means of destruction. Nothing appears more manifest in these attempts than the greatness of human power, whose source lies in the capacity to act, and which without action's inherent remedies inevitably begins to overpower and destroy not man himself but the conditions under which life was given to him.

The discoverer of the role of forgiveness in the realm of human affairs was Jesus of Nazareth. The fact that he made this discovery in a religious context and articulated it in religious language is no reason to take it any less seriously in a strictly secular sense. It has been in the nature of our tradition of political thought (and for reasons we cannot explore here) to be highly selective and to exclude from articulate conceptualization a great variety of authentic political experiences, among which we need not be surprised to find some of an even elementary nature. Certain aspects of the teaching of Jesus of Nazareth which are not primarily related to the Christian religious message but sprang from experiences in the small and closely knit community of his followers, bent on challenging the public authorities in Israel, certainly belong among them, even though they have been neglected because of their allegedly exclusively religious nature. The only rudimentary sign of an awareness that forgiveness may be the necessary corrective for the inevitable damages resulting from action may be seen in the Roman principle to spare the vanquished (*parcere subiectis*)—a wisdom entirely unknown to the Greeks—or in the right to commute the death sentence, probably also of Roman origin, which is the prerogative of nearly all Western heads of state.

It is decisive in our context that Jesus maintains against the "scribes and pharisees" first that it is not true that only God has the power to forgive,[76] and second that this power does not derive from God—as though God, not men, would forgive through the medium of human beings—but on the contrary must be mobilized by men toward each other before they can hope to be forgiven by God also. Jesus' formulation is even more radical. Man in the gospel is not supposed to forgive because God forgives and he must do "likewise," but "if ye from your hearts forgive," God shall do "likewise."[77] The reason for the insistence on a duty to forgive is clearly "for they know not what they do" and it does not apply to the extremity of crime and willed evil, for then it would not have been necessary to teach: "And if he trespass

76. This is stated emphatically in Luke 5:21–24 (cf. Matt. 9:4–6 or Mark 12:7–10), where Jesus performs a miracle to prove that "the Son of man hath power upon earth to forgive sins," the emphasis being on "upon earth." It is his insistence on the "power to forgive," even more than his performance of miracles, that shocks the people, so that "they that sat at meat with him began to say within themselves, Who is this that forgives sins also?" (Luke 7:49).

77. Matt. 18:35; cf. Mark 11:25; "And when ye stand praying, forgive, . . . that your Father also which is in heaven may forgive you your trespasses." Or: "If ye forgive men their trespasses, your heavenly Father will also forgive you: But if ye forgive not men their trespasses, neither will your Father forgive your trespasses" (Matt. 6:14–15). In all these instances, the power to forgive is primarily a human power: God forgives "us our debts, as we forgive our debtors."

against thee seven times a day, and seven times in a day turn again to thee, saying, I repent; thou shalt forgive him."[78] Crime and willed evil are rare, even rarer perhaps than good deeds; according to Jesus, they will be taken care of by God in the Last Judgment, which plays no role whatsoever in life on earth, and the Last Judgment is not characterized by forgiveness but by just retribution (*apodounai*).[79] But trespassing is an everyday occurrence which is in the very nature of action's constant establishment of new relationships within a web of relations, and it needs forgiving, dismissing, in order to make it possible for life to go on by constantly releasing men from what they have done unknowingly.[80] Only through this constant mutual release from what they do can men remain free agents, only by constant willingness to change their minds and start again can they be trusted with so great a power as that to begin something new.

In this respect, forgiveness is the exact opposite of vengeance, which acts in the form of re-acting against an original trespassing, whereby far from putting an end to the consequences of the first misdeed, everybody remains bound to the process, permitting the chain reaction contained in every action to take its unhindered

78. Luke 17:3–4. It is important to keep in mind that the three key words of the text—*aphienai*, *metanoein*, and *hamartanein*—carry certain connotations even in New Testament Greek which the translations fail to render fully. The original meaning of *aphienai* is "dismiss" and "release" rather than "forgive"; *metanoein* means "change of mind" and—since it serves also to render the Hebrew *shuv*—"return," "trace back one's steps," rather than "repentance" with its psychological emotional overtones; what is required is: change your mind and "sin no more," which is almost the opposite of doing penance. *Hamartanein*, finally, is indeed very well rendered by "trespassing" in so far as it means rather "to miss," "fail and go astray," than "to sin" (see Heinrich Ebeling, *Griechisch-deutsches Wörterbuch zum Neuen Testamente* [1923]). The verse which I quote in the standard translation could also be rendered as follows: "And if he trespass against thee . . . and . . . turn again to thee, saying, *I changed my mind*; thou shalt *release* him."

79. Matt. 16:27.

80. This interpretation seems justified by the context (Luke 17:1–5): Jesus introduces his words by pointing to the inevitability of "offenses" (*skandala*) which are unforgivable, at least on earth; for "woe unto him, through whom they come! It were better for him that a millstone were hanged about his neck, and he cast into the sea"; and then continues by teaching forgiveness for "trespassing" (*hamartanein*).

course. In contrast to revenge, which is the natural, automatic re-action to transgression and which because of the irreversibility of the action process can be expected and even calculated, the act of forgiving can never be predicted; it is the only reaction that acts in an unexpected way and thus retains, though being a reaction, some-thing of the original character of action. Forgiving, in other words, is the only reaction which does not merely re-act but acts anew and unexpectedly, unconditioned by the act which provoked it and therefore freeing from its consequences both the one who forgives and the one who is forgiven. The freedom contained in Jesus' teach-ings of forgiveness is the freedom from vengeance, which incloses both doer and sufferer in the relentless automatism of the action process, which by itself need never come to an end.

The alternative to forgiveness, but by no means its opposite, is punishment, and both have in common that they attempt to put an end to something that without interference could go on endlessly. It is therefore quite significant, a structural element in the realm of human affairs, that men are unable to forgive what they cannot punish and that they are unable to punish what has turned out to be unforgivable. This is the true hallmark of those offenses which, since Kant, we call "radical evil" and about whose nature so little is known, even to us who have been exposed to one of their rare outbursts on the public scene. All we know is that we can neither punish nor forgive such offenses and that they therefore transcend the realm of human affairs and the potentialities of human power, both of which they radically destroy wherever they make their appearance. Here, where the deed itself dispossesses us of all power, we can indeed only repeat with Jesus: "It were better for him that a millstone were hanged about his neck, and he cast into the sea."

Perhaps the most plausible argument that forgiving and acting are as closely connected as destroying and making comes from that aspect of forgiveness where the undoing of what was done seems to show the same revelatory character as the deed itself. Forgiving and the relationship it establishes is always an eminently personal (though not necessarily individual or private) affair in which *what* was done is forgiven for the sake of *who* did it. This, too, was clearly recognized by Jesus ("Her sins which are many are forgiven; for she loved much: but to whom little is forgiven, the same loveth little"), and it is the reason for the current conviction that only love has the power to forgive. For love, although it is one of the rarest occurrences in human lives,[81] indeed possesses an unequaled power of self-revelation and an unequaled clarity of vision for the disclosure of *who*, precisely because it is unconcerned to the point of total unworldliness with *what* the loved person may be, with his qualities and shortcomings no less than with his achievements, failings, and transgressions. Love, by reason of its passion, destroys the in-between which relates us to and separates us from others. As long as its spell lasts, the only in-between which can insert itself between two lovers is the child, love's own product. The child, this in-between to which the lovers now are related and which they hold in common, is representative of the world in that it also separates them; it is an indication that they will insert a new world into the existing world.[82] Through the child, it is as though the lovers return to the world from which their love had expelled them. But this new worldliness, the possible result and the only possibly happy ending of a love affair, is, in a sense, the end of love, which must either overcome the partners anew or be transformed into another mode of belonging together. Love, by its very nature, is unworldly, and it is for this reason rather than its rarity that it is not only apolitical but antipolitical, perhaps the most powerful of all antipolitical human forces.

If it were true, therefore, as Chrsitianity assumed, that only love can forgive because only love is fully receptive to *who* somebody

81. The common prejudice that love is as common as "romance" may be due to the fact that we all learned about it first through poetry. But the poets fool us; they are the only ones to whom love is not only a crucial, but an indispensable experience, which entitles them to mistake it for a universal one.

82. This world-creating faculty of love is not the same as fertility, upon which most creation myths are based. The following mythological tale, on the contrary, draws its imagery clearly from the experience of love: the sky is seen as a gigantic goddess who still bends down upon the earth god, from whom she is being separated by the air god who was born between them and is now lifting her up. Thus a world space composed of air comes into being and inserts itself between earth and sky. See H. A. Frankfort, *The Intellectual Adventure of Ancient Man* (Chicago, 1946), p. 18, and Mircea Eliade, *Traité d'Histoire des Religions* (Paris, 1953), p. 212.

is, to the point of being always willing to forgive him whatever he may have done, forgiving would have to remain altogether outside our considerations. Yet what love is in its own, narrowly circumscribed sphere, respect is in the larger domain of human affairs. Respect, not unlike the Aristotelian *philia politikē*, is a kind of "friendship" without intimacy and without closeness; it is a regard for the person from the distance which the space of the world puts between us, and this regard is independent of qualities which we may admire or of achievements which we may highly esteem. Thus, the modern loss of respect, or rather the conviction that respect is due only where we admire or esteem, constitutes a clear symptom of the increasing depersonalization of public and social life. Respect, at any rate, because it concerns only the person, is quite sufficient to prompt forgiving of what a person did, for the sake of the person. But the fact that the same *who*, revealed in action and speech, remains also the subject of forgiving is the deepest reason why nobody can forgive himself; here, as in action and speech generally, we are dependent upon others, to whom we appear in a distinctness which we ourselves are unable to perceive. Closed within ourselves, we would never be able to forgive ourselves any failing or transgression because we would lack the experience of the person for the sake of whom one can forgive.

34

UNPREDICTABILITY AND THE POWER OF PROMISE

In contrast to forgiving, which—perhaps because of its religious context, perhaps because of the connection with love attending its discovery—has always been deemed unrealistic and inadmissible in the public realm, the power of stabilization inherent in the faculty of making promises has been known throughout our tradition. We may trace it back to the Roman legal system, the inviolability of agreements and treaties (*pacta sunt servanda*); or we may see its discoverer in Abraham, the man from Ur, whose whole story, as the Bible tells it, shows such a passionate drive toward making covenants that it is as though he departed from his country for no other reason than to try out the power of mutual promise in the

The Human Condition

wilderness of the world, until eventually God himself agreed to make a Covenant with him. At any rate, the great variety of contract theories since the Romans attests to the fact that the power of making promises has occupied the center of political thought over the centuries.

The unpredictability which the act of making promises at least partially dispels is of a twofold nature: it arises simultaneously out of the "darkness of the human heart," that is, the basic unreliability of men who never can guarantee today who they will be tomorrow, and out of the impossibility of foretelling the consequences of an act within a community of equals where everybody has the same capacity to act. Man's inability to rely upon himself or to have complete faith in himself (which is the same thing) is the price human beings pay for freedom; and the impossibility of remaining unique masters of what they do, of knowing its consequences and relying upon the future, is the price they pay for plurality and reality, for the joy of inhabiting together with others a world whose reality is guaranteed for each by the presence of all.

The function of the faculty of promising is to master this twofold darkness of human affairs and is, as such, the only alternative to a mastery which relies on domination of one's self and rule over others; it corresponds exactly to the existence of a freedom which was given under the condition of non-sovereignty. The danger and the advantage inherent in all bodies politic that rely on contracts and treaties is that they, unlike those that rely on rule and sovereignty, leave the unpredictability of human affairs and the unreliability of men as they are, using them merely as the medium, as it were, into which certain islands of predictability are thrown and in which certain guideposts of reliability are erected. The moment promises lose their character as isolated islands of certainty in an ocean of uncertainty, that is, when this faculty is misused to cover the whole ground of the future and to map out a path secured in all directions, they lose their binding power and the whole enterprise becomes self-defeating.

We mentioned before the power generated when people gather together and "act in concert," which disappears the moment they depart. The force that keeps them together, as distinguished from the space of appearances in which they gather and the power which

[244]

keeps this public space in existence, is the force of mutual promise or contract. Sovereignty, which is always spurious if claimed by an isolated single entity, be it the individual entity of the person or the collective entity of a nation, assumes, in the case of many men mutually bound by promises, a certain limited reality. The sovereignty resides in the resulting, limited independence from the incalculability of the future, and its limits are the same as those inherent in the faculty itself of making and keeping promises. The sovereignty of a body of people bound and kept together, not by an identical will which somehow magically inspires them all, but by an agreed purpose for which alone the promises are valid and binding, shows itself quite clearly in its unquestioned superiority over those who are completely free, unbound by any promises and unkept by any purpose. This superiority derives from the capacity to dispose of the future as though it were the present, that is, the enormous and truly miraculous enlargement of the very dimension in which power can be effective. Nietzsche, in his extraordinary sensibility to moral phenomena, and despite his modern prejudice to see the source of all power in the will power of the isolated individual, saw in the faculty of promises (the "memory of the will," as he called it) the very distinction which marks off human from animal life.[83] If sovereignty is in the realm of action and human affairs what mastership is in the realm of making and the world of things, then their chief distinction is that the one can only be achieved by the many bound together, whereas the other is conceivable only in isolation.

In so far as morality is more than the sum total of *mores*, of customs and standards of behavior solidified through tradition and valid on the ground of agreements, both of which change with time, it has, at least politically, no more to support itself than the good will to counter the enormous risks of action by readiness to forgive and to be forgiven, to make promises and to keep them.

83. Nietzsche saw with unequaled clarity the connection between human sovereignty and the faculty of making promises, which led him to a unique insight into the relatedness of human pride and human conscience. Unfortunately, both insights remained unrelated with and without effect upon his chief concept, the "will to power," and therefore are frequently overlooked even by Nietzsche scholars. They are to be found in the first two aphorisms of the second treatise in *Zur Genealogie der Moral*.

These moral precepts are the only ones that are not applied to action from the outside, from some supposedly higher faculty or from experiences outside action's own reach. They arise, on the contrary, directly out of the will to live together with others in the mode of acting and speaking, and thus they are like control mechanisms built into the very faculty to start new and unending processes. If without action and speech, without the articulation of natality, we would be doomed to swing forever in the ever-recurring cycle of becoming, then without the faculty to undo what we have done and to control at least partially the processes we have let loose, we would be the victims of an automatic necessity bearing all the marks of the inexorable laws which, according to the natural sciences before our time, were supposed to constitute the outstanding characteristic of natural processes. We have seen before that to mortal beings this natural fatality, though it swings in itself and may be eternal, can only spell doom. If it were true that fatality is the inalienable mark of historical processes, then it would indeed be equally true that everything done in history is doomed.

And to a certain extent this is true. If left to themselves, human affairs can only follow the law of mortality, which is the most certain and the only reliable law of a life spent between birth and death. It is the faculty of action that interferes with this law because it interrupts the inexorable automatic course of daily life, which in its turn, as we saw, interrupted and interfered with the cycle of the biological life process. The life span of man running toward death would inevitably carry everything human to ruin and destruction if it were not for the faculty of interrupting it and beginning something new, a faculty which is inherent in action like an ever-present reminder that men, though they must die, are not born in order to die but in order to begin. Yet just as, from the standpoint of nature, the rectilinear movement of man's life-span between birth and death looks like a peculiar deviation from the common natural rule of cyclical movement, thus action, seen from the viewpoint of the automatic processes which seem to determine the course of the world, looks like a miracle. In the language of natural science, it is the "infinite improbability which occurs regularly." Action is, in fact, the one miracle-working faculty of man, as Jesus

of Nazareth, whose insights into this faculty can be compared in
their originality and unprecedentedness with Socrates' insights into
the possibilities of thought, must have known very well when he
likened the power to forgive to the more general power of performing miracles, putting both on the same level and within the
reach of man.[84]

The miracle that saves the world, the realm of human affairs,
from its normal, "natural" ruin is ultimately the fact of natality, in
which the faculty of action is ontologically rooted. It is, in other
words, the birth of new men and the new beginning, the action they
are capable of by virtue of being born. Only the full experience of
this capacity can bestow upon human affairs faith and hope, those
two essential characteristics of human existence which Greek
antiquity ignored altogether, discounting the keeping of faith as a
very uncommon and not too important virtue and counting hope
among the evils of illusion in Pandora's box. It is this faith in and
hope for the world that found perhaps its most glorious and most
succinct expression in the few words with which the Gospels
announced their "glad tidings": "A child has been born unto us."

84. Cf. the quotations given in n. 77. Jesus himself saw the human root of this
power to perform miracles in faith—which we leave out of our considerations.
In our context, the only point that matters is that the power to perform miracles is
not considered to be divine—faith will move mountains and faith will forgive; the
one is no less a miracle than the other, and the reply of the apostles when Jesus
demanded of them to forgive seven times in a day was: "Lord, increase our
faith."

CHAPTER VI

The *Vita Activa*
and the Modern Age

Er hat den archimedischen Punkt gefunden, hat ihn aber gegen sich ausgenutzt, offenbar hat er ihn nur unter dieser Bedingung finden dürfen.

(He found the Archimedean point, but he used it against himself; it seems that he was permitted to find it only under this condition.)

FRANZ KAFKA

35

WORLD ALIENATION

Three great events stand at the threshold of the modern age and determine its character: the discovery of America and the ensuing exploration of the whole earth; the Reformation, which by expropriating ecclesiastical and monastic possessions started the two-fold process of individual expropriation and the accumulation of social wealth; the invention of the telescope and the development of a new science that considers the nature of the earth from the viewpoint of the universe. These cannot be called modern events as we know them since the French Revolution, and although they cannot be explained by any chain of causality, because no event can, they are still happening in an unbroken continuity, in which precedents exist and predecessors can be named. None of them exhibits the peculiar character of an explosion of undercurrents which, having gathered their force in the dark, suddenly erupt. The names we connect with them, Galileo Galilei and Martin Luther and the great seafarers, explorers, and adventurers in the age of discovery, still belong to a premodern world. Moreover, the strange pathos of novelty, the almost violent insistence of nearly

The Vita Activa *and the Modern Age*

all the great authors, scientists, and philosophers since the seventeenth century that they saw things never seen before, thought thoughts never thought before, can be found in none of them, not even in Galileo.[1] These precursors are not revolutionists, and their motives and intentions are still securely rooted in tradition.

In the eyes of their contemporaries, the most spectacular of these events must have been the discoveries of unheard-of continents and undreamed-of oceans; the most disturbing might have been the Reformation's irremediable split of Western Christianity, with its inherent challenge to orthodoxy as such and its immediate threat to the tranquillity of men's souls; certainly the least noticed was the addition of a new implement to man's already large arsenal of tools, useless except to look at the stars, even though it was the first purely scientific instrument ever devised. However, if we could measure the momentum of history as we measure natural processes, we might find that what originally had the least noticeable impact, man's first tentative steps toward the discovery of the universe, has constantly increased in momentousness as well as

1. The term *scienza nuova* seems to occur for the first time in the work of the sixteenth-century Italian mathematician Niccolò Tartaglia, who designed the new science of ballistics which he claimed to have discovered because he was the first to apply geometrical reasoning to the motion of projectiles. (I owe this information to Professor Alexandre Koyré.) Of greater relevance in our context is that Galileo, in the *Sidereus Nuncius* (1610), insists on the "absolute novelty" of his discoveries, but this certainly is a far cry from Hobbes's claim that political philosophy was "no older than my own book *De Cive*" (*English Works*, ed. Molesworth [1839], I, ix) or Descartes' conviction that no philosopher before him had succeeded in philosophy ("Lettre au traducteur pouvant servir de préface" for *Les principes de la philosophie*). From the seventeenth century on, the insistence on absolute novelty and the rejection of the whole tradition became commonplace. Karl Jaspers (*Descartes und die Philosophie* [2d ed.; 1948], pp. 61 ff.) stresses the difference between Renaissance philosophy, where "Drang nach Geltung der originalen Persönlichkeit ... das Neusein als Auszeichnung verlangte," and modern science, where "sich das Wort 'neu' als sachliches Wertpraedikat verbreitet." In the same context, he shows how different in significance the claim to novelty is in science and philosophy. Descartes certainly presented his philosophy as a scientist may present a new scientific discovery. Thus, he writes as follows about his "considérations": "Je ne mérite point plus de gloire de les avoir trouvées, que ferait un passant d'avoir rencontré par bonheur à ses pieds quelque riche trésor, que la diligence de plusieurs aurait inutilement cherché longtemps auparavant" (*La recherche de la vérité* [Pléiade ed.], p. 669).

[*249*]

speed until it has eclipsed not only the enlargement of the earth's surface, which found its final limitation only in the limitations of the globe itself, but also the still apparently limitless economic accumulation process.

But these are mere speculations. As a matter of fact, the discovery of the earth, the mapping of her lands and the chartering of her waters, took many centuries and has only now begun to come to an end. Only now has man taken full possession of his mortal dwelling place and gathered the infinite horizons, which were temptingly and forbiddingly open to all previous ages, into a globe whose majestic outlines and detailed surface he knows as he knows the lines in the palm of his hand. Precisely when the immensity of available space on earth was discovered, the famous shrinkage of the globe began, until eventually in our world (which, though the result of the modern age, is by no means identical with the modern age's world) each man is as much an inhabitant of the earth as he is an inhabitant of his country. Men now live in an earth-wide continuous whole where even the notion of distance, still inherent in the most perfectly unbroken contiguity of parts, has yielded before the onslaught of speed. Speed has conquered space; and though this conquering process finds its limit at the unconquerable boundary of the simultaneous presence of one body at two different places, it has made distance meaningless, for no significant part of a human life—years, months, or even weeks—is any longer necessary to reach any point on the earth.

Nothing, to be sure, could have been more alien to the purpose of the explorers and circumnavigators of the early modern age than this closing-in process; they went to enlarge the earth, not shrink her into a ball, and when they submitted to the call of the distant, they had no intention of abolishing distance. Only the wisdom of hindsight sees the obvious, that nothing can remain immense if it can be measured, that every survey brings together distant parts and therefore establishes closeness where distance ruled before. Thus the maps and navigation charts of the early stages of the modern age anticipated the technical inventions through which all earthly space has become small and close at hand. Prior to the shrinkage of space and the abolition of distance through railroads, steamships, and airplanes, there is the infinitely greater and more effective shrinkage which comes about through the surveying capacity of the human mind, whose use of numbers, symbols, and models can condense and scale earthly physical distance down to the size of the human body's natural sense and understanding. Before we knew how to circle the earth, how to circumscribe the sphere of human habitation in days and hours, we had brought the globe into our living rooms to be touched by our hands and swirled before our eyes.

There is another aspect of this matter which, as we shall see, will be of greater importance in our context. It is in the nature of the human surveying capacity that it can function only if man disentangles himself from all involvement in and concern with the close at hand and withdraws himself to a distance from everything near him. The greater the distance between himself and his surroundings, world or earth, the more he will be able to survey and to measure and the less will worldly, earth-bound space be left to him. The fact that the decisive shrinkage of the earth was the consequence of the invention of the airplane, that is, of leaving the surface of the earth altogether, is like a symbol for the general phenomenon that any decrease of terrestrial distance can be won only at the price of putting a decisive distance between man and earth, of alienating man from his immediate earthly surroundings.

The fact that the Reformation, an altogether different event, eventually confronts us with a similar phenomenon of alienation, which Max Weber even identified, under the name of "innerworldly asceticism," as the innermost spring of the new capitalist mentality, may be one of the many coincidences that make it so difficult for the historian not to believe in ghosts, demons, and *Zeitgeists*. What is so striking and disturbing is the similarity in utmost divergence. For this innerworldly alienation has nothing to do, either in intent or content, with the alienation from the earth inherent in the discovery and taking possession of the earth. Moreover, the innerworldly alienation whose historical factuality Max Weber demonstrated in his famous essay is not only present in the new morality that grew out of Luther's and Calvin's attempts to restore the uncompromising otherworldliness of the Christian faith; it is equally present, albeit on an altogether different level, in the expropriation of the peasantry, which was the unforeseen consequence of the expropriation of church property and, as such, the greatest single factor in the breakdown of the feudal system.[2] It is, of course, idle to speculate on what the course of our economy would have been without this event, whose impact propelled Western mankind into a development in which all property was destroyed in the process of its appropriation, all things devoured in the process of their production, and the stability of the world undermined in a constant process of change. Yet, such speculations are meaningful to the extent that they remind us that history is a story of events and not of forces or ideas with predictable courses. They are idle and even dangerous when used as arguments against reality and when meant to point to positive potentialities and alternatives, because their number is not only indefinite by definition but they also lack the tangible unexpectedness of the event, and compensate for it by mere plausibility. Thus, they remain sheer phantoms no matter in how pedestrian a manner they may be presented.

In order not to underestimate the momentum this process has reached after centuries of almost unhindered development, it may be well to reflect on the so-called "economic miracle" of postwar Germany, a miracle only if seen in an outdated frame of reference. The German example shows very clearly that under modern conditions the expropriation of people, the destruction of objects, and the devastation of cities will turn out to be a radical stimulant for a process, not of mere recovery, but of quicker and more efficient accumulation of wealth—if only the country is modern enough to respond in terms of the production process. In Germany, outright destruction took the place of the relentless process of depreciation of all worldly things, which is the hallmark of the waste economy

2. This is not to·deny the greatness of Max Weber's discovery of the enormous power that comes from an otherworldliness directed toward the world (see "Protestant Ethics and the Spirit of Capitalism," in *Religionssoziologie* [1920], Vol. I). Weber finds the Protestant work ethos preceded by certain traits of monastic ethics, and one can indeed see a first germ of these attitudes in Augustine's famous distinction between *uti* and *frui*, between the things of this world which one may use but not enjoy and those of the world to come which may be enjoyed for their own sake. The increase in power of man over the things of this world springs in either case from the distance which man puts between himself and the world, that is, from world alienation.

in which we now live. The result is almost the same: a booming prosperity which, as postwar Germany illustrates, feeds not on the abundance of material goods or on anything stable and given but on the process of production and consumption itself. Under modern conditions, not destruction but conservation spells ruin because the very durability of conserved objects is the greatest impediment to the turnover process, whose constant gain in speed is the only constancy left wherever it has taken hold.[3]

We saw before that property, as distinguished from wealth and appropriation, indicates the privately owned share of a common world and therefore is the most elementary political condition for man's worldliness. By the same token, expropriation and world alienation coincide, and the modern age, very much against the intentions of all the actors in the play, began by alienating certain strata of the population from the world. We tend to overlook the central importance of this alienation for the modern age because we usually stress its secular character and identify the term secularity with worldliness. Yet secularization as a tangible historical event means no more than separation of Church and State, of religion and politics, and this, from a religious viewpoint, implies a return to the early Christian attitude of "Render unto Caesar the things that are Caesar's and unto God the things that are God's" rather than a loss of faith and transcendence or a new and emphatic interest in the things of this world.

Modern loss of faith is not religious in origin—it cannot be traced to the Reformation and Counter Reformation, the two great religious movements of the modern age—and its scope is by no means restricted to the religious sphere. Moreover, even if we admitted that the modern age began with a sudden, inexplicable

3. The reason most frequently given for the surprising recovery of Germany —that she did not have to carry the burden of a military budget—is inconclusive on two accounts: first, Germany had to pay for a number of years the costs of occupation, which amounted to a sum almost equal to a full-fledged military budget, and second, war production is held in other economies to be the greatest single factor in the postwar prosperity. Moreover, the point I wish to make could be equally well illustrated by the common and yet quite uncanny phenomenom that prosperity is closely connected with the "useless" production of means of destruction, of goods produced to be wasted either by using them up in destruction or—and this is the more common case—by destroying them because they soon become obsolete.

eclipse of transcendence, of belief in a hereafter, it would by no means follow that this loss threw man back upon the world. The historical evidence, on the contrary, shows that modern men were not thrown back upon this world but upon themselves. One of the most persistent trends in modern philosophy since Descartes and perhaps its most original contribution to philosophy has been an exclusive concern with the self, as distinguished from the soul or person or man in general, an attempt to reduce all experiences, with the world as well as with other human beings, to experiences between man and himself. The greatness of Max Weber's discovery about the origins of capitalism lay precisely in his demonstration that an enormous, strictly mundane activity is possible without any care for or enjoyment of the world whatever, an activity whose deepest motivation, on the contrary, is worry and care about the self. World alienation, and not self-alienation as Marx thought,[4] has been the hallmark of the modern age.

Expropriation, the deprivation for certain groups of their place

4. There are several indications in the writings of the young Marx that he was not altogether unaware of the implications of world alienation in capitalist economy. Thus, in the early article of 1842, "Debatten über das Holzdiebstahlsgesetz" (see *Marx-Engels Gesamtausgabe* [Berlin, 1932], Part 1, Vol. I, pp. 266 ff.), he criticizes a law against theft not only because the formal opposition of owner and thief leaves "human needs" out of account—the fact that the thief who uses the wood needs it more urgently than the owner who sells it—and therefore dehumanizes men by equating wood-user and wood-seller as wood proprietors, but also that the wood itself is deprived of its nature. A law which regards men only as property-owners considers things only as properties and properties only as exchange objects, not as use things. That things are denatured when they are used for exchange was probably suggested to Marx by Aristotle, who pointed out that though a shoe may be wanted for either usage or exchange, it is against the nature of a shoe to be exchanged, "for a shoe is not made to be an object of barter" (*Politics* 1257a8). (Incidentally the influence of Aristotle on the style of Marx's thought seems to me almost as characteristic and decisive as the influence of Hegel's philosophy.) However, such occasional considerations play a minor role in his work, which remained firmly rooted in the modern age's extreme subjectivism. In his ideal society, where men will produce as human beings, world alienation is even more present than it was before; for then they will be able to objectify (*vergegenständlichen*) their individuality, their peculiarity, to confirm and actualize their true being: "Unsere Produktionen wären ebensoviele Spiegel, woraus unser Wesen sich entgegen leuchtete" ("Aus den Exzerptheften" [1844–45], in *Gesamtausgabe*, Part 1, Vol. III, pp. 546–47).

in the world and their naked exposure to the exigencies of life, created both the original accumulation of wealth and the possibility of transforming this wealth into capital through labor. These together constituted the conditions for the rise of a capitalist economy. That this development, started by expropriation and fed upon it, would result in an enormous increase in human productivity was manifest from the beginning, centuries before the industrial revolution. The new laboring class, which literally lived from hand to mouth, stood not only directly under the compelling urgency of life's necessity[5] but was at the same time alienated from all cares and worries which did not immediately follow from the life process itself. What was liberated in the early stages of the first free laboring class in history was the force inherent in "labor power," that is, in the sheer natural abundance of the biological process, which like all natural forces—of procreation no less than of laboring—provides for a generous surplus over and beyond the reproduction of young to balance the old. What distinguishes this development at the beginning of the modern age from similar occurrences in the past is that expropriation and wealth accumulation did not simply result in new property or lead to a new redistribution of wealth, but were fed back into the process to generate further expropriations, greater productivity, and more appropriation.

In other words, the liberation of labor power as a natural process did not remain restricted to certain classes of society, and appropriation did not come to an end with the satisfaction of wants and desires; capital accumulation, therefore, did not lead to the stagnation we know so well from rich empires prior to the modern age, but spread throughout the society and initiated a steadily increasing flow of wealth. But this process, which indeed is the "life process of society," as Marx used to call it, and whose wealth-producing capacity can be compared only with the fertility of natural processes where the creation of one man and one woman would suffice to produce by multiplication any given number of human beings, remains bound to the principle of world alienation from

5. This of course is markedly different from present conditions, where the day laborer has already become a weekly wage-earner; in a probably not very distant future the guaranteed annual wage will do away with these early conditions altogether.

which it sprang; the process can continue only provided that no worldly durability and stability is permitted to interfere, only as long as all worldly things, all end products of the production process, are fed back into it at an ever-increasing speed. In other words, the process of wealth accumulation, as we know it, stimulated by the life process and in turn stimulating human life, is possible only if the world and the very worldliness of man are sacrificed.

The first stage of this alienation was marked by its cruelty, the misery and material wretchedness it meant for a steadily increasing number of "labouring poor," whom expropriation deprived of the twofold protection of family and property, that is, of a family-owned private share in the world, which until the modern age had housed the individual life process and the laboring activity subject to its necessities. The second stage was reached when society became the subject of the new life process, as the family had been its subject before. Membership in a social class replaced the protection previously offered by membership in a family, and social solidarity became a very efficient substitute for the earlier, natural solidarity ruling the family unit. Moreover, society as a whole, the "collective subject" of the life process, by no means remained an intangible entity, the "communist fiction" needed by classical economics; just as the family unit had been identified with a privately owned piece of the world, its property, society was identified with a tangible, albeit collectively owned, piece of property, the territory of the nation-state, which until its decline in the twentieth century offered all classes a substitute for the privately owned home of which the class of the poor had been deprived.

The organic theories of nationalism, especially in its Central European version, all rest on an identification of the nation and the relationships between its members with the family and family relationships. Because society becomes the substitute for the family, "blood and soil" is supposed to rule the relationships between its members; homogeneity of population and its rootedness in the soil of a given territory become the requisites for the nation-state everywhere. However, while this development undoubtedly mitigated cruelty and misery, it hardly influenced the process of expropriation and world alienation, since collective ownership, strictly speaking, is a contradiction in terms.

The Vita Activa *and the Modern Age*

The decline of the European nation-state system; the economic and geographic shrinkage of the earth, so that prosperity and depression tend to become world-wide phenomena; the transformation of mankind, which until our own time was an abstract notion or a guiding principle for humanists only, into a really existing entity whose members at the most distant points of the globe need less time to meet than the members of a nation needed a generation ago—these mark the beginnings of the last stage in this development. Just as the family and its property were replaced by class membership and national territory, so mankind now begins to replace nationally bound societies, and the earth replaces the limited state territory. But whatever the future may bring, the process of world alienation, started by expropriation and characterized by an ever-increasing progress in wealth, can only assume even more radical proportions if it is permitted to follow its own inherent law. For men cannot become citizens of the world as they are citizens of their countries, and social men cannot own collectively as family and household men own their private property. The rise of society brought about the simultaneous decline of the public as well as the private realm. But the eclipse of a common public world, so crucial to the formation of the lonely mass man and so dangerous in the formation of the worldless mentality of modern ideological mass movements, began with the much more tangible loss of a privately owned share in the world.

36

THE DISCOVERY OF THE ARCHIMEDEAN POINT

"Since a babe was born in a manger, it may be doubted whether so great a thing has happened with so little stir." These are the words with which Whitehead introduces Galileo and the discovery of the telescope on the stage of the "modern world."[6] Nothing in these words is an exaggeration. Like the birth in a manger, which spelled not the end of antiquity but the beginning of something so unexpectedly and unpredictably new that neither hope nor fear could have anticipated it, these first tentative glances into the uni-

6. A. N. Whitehead, *Science and the Modern World* (Pelican ed., 1926), p. 12.

verse through an instrument, at once adjusted to human senses and destined to uncover what definitely and forever must lie beyond them, set the stage for an entirely new world and determined the course of other events, which with much greater stir were to usher in the modern age. Except for the numerically small, politically inconsequential milieu of learned men—astronomers, philosophers, and theologians—the telescope created no great excitement; public attention was drawn, rather, to Galileo's dramatic demonstration of the laws of falling bodies, taken to be the beginning of modern natural science (although it may be doubted that by themselves, without being transformed later by Newton into the universal law of gravitation—still one of the most grandiose examples of the modern amalgamation of astronomy and physics—they would ever have led the new science on the path of astrophysics). For what most drastically distinguished the new world view not only from that of antiquity or the Middle Ages, but from the great thirst for direct experience in the Renaissance as well, was the assumption that the same kind of exterior force should be manifest in the fall of terrestrial and the movements of heavenly bodies.

Moreover, the novelty of Galileo's discovery was clouded by its close relationship to antecedents and predecessors. Not the philosophical speculations of Nicholas of Cusa and Giordano Bruno alone, but the mathematically trained imagination of the astronomers, Copernicus and Kepler, had challenged the finite, geocentric world view which men had held since time immemorial. Not Galileo but the philosophers were the first to abolish the dichotomy between one earth and one sky above it, promoting, as they thought, the earth "to the rank of the noble stars" and finding her a home in an eternal and infinite universe.[7] And it seems the astronomers needed no telescope to assert that, contrary to all sense experience, it is not the sun that moves around the earth but the earth that circles the sun. If the historian looks back upon these beginnings with all the wisdom and prejudices of hindsight, he is tempted to conclude that no empirical confirmation was needed to abolish the Ptolemaic system. What was wanted was, rather, the

7. I follow the excellent recent exposition of the interrelated history of philosophic and scientific thought in "the seventeenth century revolution" by Alexandre Koyré (*From the Closed World to the Infinite Universe* [1957], pp. 43 ff.).

speculative courage to follow the ancient and medieval principle of simplicity in nature—even if it led to the denial of all sense experience—and the great boldness of Copernicus' imagination, which lifted him from the earth and enabled him to look down upon her as though he actually were an inhabitant of the sun. And the historian feels justified in his conclusions when he considers that Galileo's discoveries were preceded by a "véritable retour à Archimède" which had been effective since the Renaissance. It certainly is suggestive that Leonardo studied him with passionate interest and that Galileo can be called his disciple.[8]

However, neither the speculations of philosophers nor the imaginings of astronomers has ever constituted an event. Prior to the telescopic discoveries of Galileo, Giordano Bruno's philosophy attracted little attention even among learned men, and without the factual confirmation they bestowed upon the Copernican revolution, not only the theologians but all "sensible men . . . would have pronounced it a wild appeal . . . of an uncontrolled imagination."[9] In the realm of ideas there are only originality and depth, both personal qualities, but no absolute, objective novelty; ideas come and go, they have a permanence, even an immortality of their own, depending upon their inherent power of illumination, which is and endures independently of time and history. Ideas, moreover, as distinguished from events, are never unprecedented, and empirically unconfirmed speculations about the earth's movement around the sun were no more unprecedented than contemporary theories about atoms would be if they had no basis in experiments and no consequences in the factual world.[10] What Galileo did and what nobody had done before was to use the telescope in such a

8. See P.-M. Schuhl, *Machinisme et philosophie* (1947), pp. 28–29.

9. E. A. Burtt, *Metaphysical Foundations of Modern Science* (Anchor ed.), p. 38 (cf. Koyré, *op. cit.*, p. 55, who states that Bruno's influence made itself felt "only after the great telescopic discoveries of Galileo").

10. The first "to save the phenomena by the assumption that the heaven is at rest, but that the earth revolves in an oblique orbit, while also rotating about its own axis" was Aristarchus of Samos in the third century B.C., and the first to conceive of an atomic structure of matter was Democritus of Abdera in the fifth century B.C. A very instructive account of the Greek physical world from the viewpoint of modern science is given by S. Sambursky, *The Physical World of the Greeks* (1956).

way that the secrets of the universe were delivered to human cognition "with the certainty of sense-perception";[11] that is, he put within the grasp of an earth-bound creature and its body-bound senses what had seemed forever beyond his reach, at best open to the uncertainties of speculation and imagination.

This difference in relevance between the Copernican system and Galileo's discoveries was quite clearly understood by the Catholic Church, which raised no objections to the pre-Galilean theory of an immobile sun and a moving earth as long as the astronomers used it as a convenient hypothesis for mathematical purposes; but, as Cardinal Bellarmine pointed out to Galileo, "to prove that the hypothesis . . . saves the appearances is not at all the same thing as to demonstrate the reality of the movement of the earth."[12] How pertinent this remark was could be seen immediately by the sudden change of mood which overtook the learned world after the confirmation of Galileo's discovery. From then on, the enthusiasm with which Giordano Bruno had conceived of an infinite universe, and the pious exultation with which Kepler had contemplated the sun, "the most excellent of all the bodies in the universe whose whole essence is nothing but pure light" and which therefore was to him the most fitting dwelling place of "God and the blessed angels,"[13] or the more sober satisfaction of Nicholas of Cusa of seeing the earth finally at home in the starred sky, were conspicuous by their absence. By "confirming" his predecessors, Galileo established a demonstrable fact where before him there were inspired speculations. The immediate philosophic reaction to this reality was not exultation but the Cartesian doubt by which modern philosophy—that "school of suspicion," as Nietzsche once called

11. Galileo (*op. cit.*) himself stressed this point: "Any one can know with the certainty of sense-perception that the moon is by no means endowed with a smooth and polished surface, etc." (quoted from Koyré, *op. cit.*, p. 89).

12. A similar stand was taken by the Lutheran theologian Osiander of Nuremberg, who wrote in an introduction to Copernicus' posthumous work, *On the Revolutions of Celestial Bodies* (1546): "The hypotheses of this book are not necessarily true or even probable. Only one thing matters. They must lead by computation to results that are in agreement with the observed phenomena." Both quotations are from Philipp Frank, "Philosophical Uses of Science," *Bulletin of Atomic Scientists*, Vol. XIII, No. 4 (April, 1957).

13. Burtt, *op. cit.*, p. 58.

it—was founded, and which ended in the conviction that "only on the firm foundation of unyielding despair can the soul's habitation henceforth be safely built."[14]

For many centuries the consequences of this event, again not unlike the consequences of the Nativity, remained contradictory and inconclusive, and even today the conflict between the event itself and its almost immediate consequences is far from resolved. The rise of the natural sciences is credited with a demonstrable, ever-quickening increase in human knowledge and power; shortly before the modern age European mankind knew less than Archimedes in the third century B.C., while the first fifty years of our century have witnessed more important discoveries than all the centuries of recorded history together. Yet the same phenomenon is blamed with equal right for the hardly less demonstrable increase in human despair or the specifically modern nihilism which has spread to ever larger sections of the population, their most significant aspect perhaps being that they no longer spare the scientists themselves, whose well-founded optimism could still, in the nineteenth century, stand up against the equally justifiable pessimism of thinkers and poets. The modern astrophysical world view, which began with Galileo, and its challenge to the adequacy of the senses to reveal reality, have left us a universe of whose qualities we know no more than the way they affect our measuring instruments, and—in the words of Eddington—"the former have as much resemblance to the latter as a telephone number has to a subscriber."[15] Instead of objective qualities, in other words, we find instruments, and instead of nature or the universe—in the words of Heisenberg—man encounters only himself.[16]

14. Bertrand Russell, "A Free Man's Worship," in *Mysticism and Logic* (1918), p. 46.

15. As quoted by J. W. N. Sullivan, *Limitations of Science* (Mentor ed.), p. 141.

16. The German physicist Werner Heisenberg has expressed this thought in a number of recent publications. For instance: "Wenn man versucht, von der Situation in der modernen Naturwissenschaft ausgehend, sich zu den in Bewegung geratenen Fundamenten vorzutasten, so hat man den Eindruck, ... dass zum erstenmal im Laufe der Geschichte der Mensch auf dieser Erde nur noch sich selbst gegenübersteht ..., dass wir gewissermassen immer nur uns selbst begegnen" (*Das Naturbild der heutigen Physik* [1955], pp. 17–18). Heisenberg's point is that the observed object has no existence independent of the observing

The point, in our context, is that both despair and triumph are inherent in the same event. If we wish to put this into historical perspective, it is as if Galileo's discovery proved in demonstrable fact that both the worst fear and the most presumptuous hope of human speculation, the ancient fear that our senses, our very organs for the reception of reality, might betray us, and the Archimedean wish for a point outside the earth from which to unhinge the world, could only come true together, as though the wish would be granted only provided that we lost reality and the fear was to be consummated only if compensated by the acquisition of supramundane powers. For whatever we do today in physics—whether we release energy processes that ordinarily go on only in the sun, or attempt to initiate in a test tube the processes of cosmic evolution, or penetrate with the help of telescopes the cosmic space to a limit of two and even six billion light years, or build machines for the production and control of energies unknown in the household of earthly nature, or attain speeds in atomic accelerators which approach the speed of light, or produce elements not to be found in nature, or disperse radioactive particles, created by us through the use of cosmic radiation, on the earth—we always handle nature from a point in the universe outside the earth. Without actually standing where Archimedes wished to stand (*dos moi pou stō*), still bound to the earth through the human condition, we have found a way to act on the earth and within terrestrial nature as though we dispose of it from outside, from the Archimedean point. And even at the risk of endangering the natural life process we expose the earth to universal, cosmic forces alien to nature's household.

While these achievements were anticipated by no one, and while most present-day theories flatly contradict those formulated during the first centuries of the modern age, this development itself was possible only because at the beginning the old dichotomy between earth and sky was abolished and a unification of the universe effected, so that from then on nothing occurring in earthly nature

---

subject: "Durch die Art der Beobachtung wird entschieden, welche Züge der Natur bestimmt werden und welche wir durch unsere Beobachtungen verwischen" (*Wandlungen in den Grundlagen der Naturwissenschaft* [1949], p. 67).

was viewed as a mere earthly happening. All events were considered to be subject to a universally valid law in the fullest sense of the word, which means, among other things, valid beyond the reach of human sense experience (even of the sense experiences made with the help of the finest instruments), valid beyond the reach of human memory and the appearance of mankind on earth, valid even beyond the coming into existence of organic life and the earth herself. All laws of the new astrophysical science are formulated from the Archimedean point, and this point probably lies much farther away from the earth and exerts much more power over her than Archimedes or Galileo ever dared to think.

If scientists today point out that we may assume with equal validity that the earth turns around the sun or the sun turns around the earth, that both assumptions are in agreement with observed phenomena and the difference is only a difference of the chosen point of reference, it by no means indicates a return to Cardinal Bellarmine's or Copernicus' position, where the astronomers dealt with mere hypotheses. It rather signifies that we have moved the Archimedean point one step farther away from the earth to a point in the universe where neither earth nor sun are centers of a universal system. It means that we no longer feel bound even to the sun, that we move freely in the universe, choosing our point of reference wherever it may be convenient for a specific purpose. For the actual accomplishments of modern science this change from the earlier heliocentric system to a system without a fixed center is, no doubt, as important as the original shift from the geocentric to the heliocentric world view. Only now have we established ourselves as "universal" beings, as creatures who are terrestrial not by nature and essence but only by the condition of being alive, and who therefore by virtue of reasoning can overcome this condition not in mere speculation but in actual fact. Yet the general relativism that results automatically from the shift from a heliocentric to a centerless world view—conceptualized in Einstein's theory of relativity with its denial that "at a definite present instant all matter is simultaneously real"[17] and the concomitant, implied denial that Being which appears in time and space possesses an absolute reality—was already contained in, or at least

17. Whitehead, *op. cit.*, p. 120.

preceded by, those seventeenth-century theories according to which blue is nothing but a "relation to a seeing eye" and heaviness nothing but a "relation of reciprocal acceleration."[18] The parentage of modern relativism is not in Einstein but in Galileo and Newton.

What ushered in the modern age was not the age-old desire of astronomers for simplicity, harmony, and beauty, which made Copernicus look upon the orbits of the planets from the sun instead of the earth, nor the Renaissance's new-awakened love for the earth and the world, with its rebellion against the rationalism of medieval scholasticism; this love of the world, on the contrary, was the first to fall victim to the modern age's triumphal world alienation. It was rather the discovery, due to the new instrument, that Copernicus' image of "the virile man standing in the sun . . . overlooking the planets"[19] was much more than an image or a gesture, was in fact an indication of the astounding human capacity to think in terms of the universe while remaining on the earth, and the perhaps even more astounding human ability to use cosmic laws as guiding principles for terrestrial action. Compared with the earth alienation underlying the whole development of natural science in the modern age, the withdrawal from terrestrial proximity contained in the discovery of the globe as a whole and the world alienation produced in the twofold process of expropriation and wealth accumulation are of minor significance.

At any event, while world alienation determined the course and the development of modern society, earth alienation became and has remained the hallmark of modern science. Under the sign of earth alienation, every science, not only physical and natural science, so radically changed its innermost content that one may doubt whether prior to the modern age anything like science existed at all. This is perhaps clearest in the development of the new science's most important mental instrument, the devices of modern algebra, by which mathematics "succeeded in freeing itself from

18. Ernst Cassirer's early essay, *Einstein's Theory of Relativity* (Dover Publications, 1953), strongly emphasizes this continuity between twentieth-century and seventeenth-century science.

19. J. Bronowski, in an article "Science and Human Values," points out the great role the metaphor played in the mind of important scientists (see *Nation*, December 29, 1956).

the shackles of spatiality,"[20] that is, from geometry, which, as the name indicates, depends on terrestrial measures and measurements. Modern mathematics freed man from the shackles of earth-bound experience and his power of cognition from the shackles of finitude.

The decisive point here is not that men at the beginning of the modern age still believed with Plato in the mathematical structure of the universe nor that, one generation later, they believed with Descartes that certain knowledge is possible only where the mind plays with its own forms and formulas. What is decisive is the entirely un-Platonic subjection of geometry to algebraic treatment, which discloses the modern ideal of reducing terrestrial sense data and movements to mathematical symbols. Without this non-spatial symbolic language Newton would not have been able to unite astronomy and physics into a single science or, to put it another way, to formulate a law of gravitation where the same equation will cover the movements of heavenly bodies in the sky and the motion of terrestrial bodies on earth. Even then it was clear that modern mathematics, in an already breathtaking development, had discovered the amazing human faculty to grasp in symbols those dimensions and concepts which at most had been thought of as negations and hence limitations of the mind, because their immensity seemed to transcend the minds of mere mortals, whose existence lasts an insignificant time and remains bound to a not too important corner of the universe. Yet even more significant than this possibility—to reckon with entities which could not be "seen" by the eye of the mind—was the fact that the new mental instrument, in this respect even newer and more significant than all the scientific tools it helped to devise, opened the way for an altogether novel mode of meeting and approaching nature in the experiment. In the experiment man realized his newly won freedom from the shackles of earth-bound experience; instead of observing natural phenomena as they were given to him, he placed nature under the conditions of his own mind, that is, under conditions won from a universal, astrophysical viewpoint, a cosmic standpoint outside nature itself.

It is for this reason that mathematics became the leading science of the modern age, and this elevation has nothing to do with Plato,

20. Burtt, *op. cit.*, p. 44.

who deemed mathematics to be the noblest of all sciences, second only to philosophy, which he thought nobody should be permitted to approach without having become familiar first with the mathematical world of ideal forms. For mathematics (that is, geometry) was the proper introduction to that sky of ideas where no mere images (*eidōla*) and shadows, no perishable matter, could any longer interfere with the appearing of eternal being, where these appearances are saved (*sōzein ta phainomena*) and safe, as purified of human sensuality and mortality as of material perishability. Yet mathematical and ideal forms were not the products of the intellect, but given to the eyes of the mind as sense data were given to the organs of the senses; and those who were trained to perceive what was hidden from the eyes of bodily vision and the untrained mind of the many perceived true being, or rather being in its true appearance. With the rise of modernity, mathematics does not simply enlarge its content or reach out into the infinite to become applicable to the immensity of an infinite and infinitely growing, expanding universe, but ceases to be concerned with appearances at all. It is no longer the beginning of philosophy, of the "science" of Being in its true appearance, but becomes instead the science of the structure of the human mind.

When Descartes' analytical geometry treated space and extension, the *res extensa* of nature and the world, so "that its relations, however complicated, must always be expressible in algebraic formulae," mathematics succeeded in reducing and translating all that man is not into patterns which are identical with human, mental structures. When, moreover, the same analytical geometry proved "conversely that numerical truths . . . can be fully represented spatially," a physical science had been evolved which required no principles for its completion beyond those of pure mathematics, and in this science man could move, risk himself into space and be certain that he would not encounter anything but himself, nothing that could not be reduced to patterns present in him.[21] Now the phenomena could be saved only in so far as they could be reduced to a mathematical order, and this mathematical operation does not serve to prepare man's mind for the revelation of true being by directing it to the ideal measures that appear in the

21. *Ibid.*, p. 106.

sensually given data, but serves, on the contrary, to reduce these data to the measure of the human mind, which, given enough distance, being sufficiently remote and uninvolved, can look upon and handle the multitude and variety of the concrete in accordance with its own patterns and symbols. These are no longer ideal forms disclosed to the eye of the mind, but are the results of removing the eyes of the mind, no less than the eyes of the body, from the phenomena, of reducing all appearances through the force inherent in distance.

Under this condition of remoteness, every assemblage of things is transformed into a mere multitude, and every multitude, no matter how disordered, incoherent, and confused, will fall into certain patterns and configurations possessing the same validity and no more significance than the mathematical curve, which, as Leibniz once remarked, can always be found between points thrown at random on a piece of paper. For if "it can be shown that a mathematical web of some kind can be woven about any universe containing several objects . . . then the fact that our universe lends itself to mathematical treatment is not a fact of any great philosophic significance."[22] It certainly is neither a demonstration of an inherent and inherently beautiful order of nature nor does it offer a confirmation of the human mind, of its capacity to surpass the senses in perceptivity or of its adequateness as an organ for the reception of truth.

The modern *reductio scientiae ad mathematicam* has overruled the testimony of nature as witnessed at close range by human senses in the same way that Leibniz overruled the knowledge of the haphazard origin and the chaotic nature of the dot-covered piece of paper. And the feeling of suspicion, outrage, and despair, which was the first, and spiritually is still the most lasting consequence of the discovery that the 'Archimedean point was no vain dream of idle speculation, is not unlike the helpless outrage of a man who,

22. Bertrand Russell, as quoted by J. W. N. Sullivan, *op. cit.*, p. 144. See also Whitehead's distinction between the traditional scientific method of classification and the modern approach of measurement: the former follows objective realities whose principle is found in the otherness of nature; the latter is entirely subjective, independent of qualities, and requires not more than that a multitude of objects be given.

having watched with his own eyes how these dots were thrown arbitrarily and without foresight onto the paper, is shown and forced to admit that all his senses and all his powers of judgment have betrayed him and that what he saw was the evolution of a "geometrical line whose direction is constantly and uniformly defined by one rule."[23]

37

UNIVERSAL VERSUS NATURAL SCIENCE

It took many generations and quite a few centuries before the true meaning of the Copernican revolution and the discovery of the Archimedean point came to light. Only we, and we only for hardly more than a few decades, have come to live in a world thoroughly determined by a science and a technology whose objective truth and practical know-how are derived from cosmic and universal, as distinguished from terrestrial and "natural," laws, and in which a knowledge acquired by selecting a point of reference outside the earth is applied to earthly nature and the human artifice. There is a deep gulf between those before us who knew that the earth re-volves around the sun, that neither the one nor the other is the cen-ter of the universe, and who concluded that man had lost his home as well as his privileged position in creation, and ourselves, who still and probably forever are earth-bound creatures, dependent upon metabolism with a terrestrial nature, and who have found the means to bring about processes of cosmic origin and possibly cosmic dimension. If one wishes to draw a distinctive line between the modern age and the world we have come to live in, he may well find it in the difference between a science which looks upon nature from a universal standpoint and thus acquires complete mastery over her, on one hand, and a truly "universal" science, on the other, which imports cosmic processes into nature even at the ob-vious risk of destroying her and, with her, man's mastership over her.

Foremost in our minds at this moment is of course the enor-mously increased human power of destruction, that we are able to

23. Leibniz, *Discours de métaphysique*, No. 6.

destroy all organic life on earth and shall probably be able one day to destroy even the earth itself. However, no less awesome and no less difficult to come to terms with is the corresponding new creative power, that we can produce new elements never found in nature, that we are able not only to speculate about the relationships between mass and energy and their innermost identity but actually to transform mass into energy or to transform radiation into matter. At the same time, we have begun to populate the space surrounding the earth with man-made stars, creating as it were, in the form of satellites, new heavenly bodies, and we hope that in a not very distant future we shall be able to perform what times before us regarded as the greatest, the deepest, and holiest secret of nature, to create or re-create the miracle of life. I use the word "create" deliberately, to indicate that we are actually doing what all ages before ours thought to be the exclusive prerogative of divine action.

This thought strikes us as blasphemous, and though it is blasphemous in every traditional Western or Eastern philosophic or theological frame of reference, it is no more blasphemous than what we have been doing and what we are aspiring to do. The thought loses its blasphemous character, however, as soon as we understand what Archimedes understood so well, even though he did not know how to reach his point outside the earth, namely, that no matter how we explain the evolution of the earth and nature and man, they must have come into being by some transmundane, "universal" force, whose work must be comprehensible to the point of imitation by somebody who is able to occupy the same location. It is ultimately nothing but this assumed location in the universe outside the earth that enables us to produce processes which do not occur on the earth and play no role in stable matter but are decisive for the coming into being of matter. It is indeed in the very nature of the thing that astrophysics and not geophysics, that "universal" science and not "natural" science, should have been able to penetrate the last secrets of the earth and of nature. From the viewpoint of the universe, the earth is but a special case and can be understood as such, just as in this view there cannot be a decisive distinction between matter and energy, both being "only different forms of the selfsame basic substance."[24]

With Galileo already, certainly since Newton, the word "universal" has begun to acquire a very specific meaning indeed; it means "valid beyond our solar system." And something quite similar has happened to another word of philosophic origin, the word "absolute," which is applied to "absolute time," "absolute space," "absolute motion," or "absolute speed," in each usage meaning a time, a space, a movement, a velocity which is present in the universe and compared to which earth-bound time or space or movement or speed are only "relative." Everything happening on earth has become relative since the earth's relatedness to the universe became the point of reference for all measurements.

Philosophically, it seems that man's ability to take this cosmic, universal standpoint without changing his location is the clearest possible indication of his universal origin, as it were. It is as though we no longer needed theology to tell us that man is not, cannot possibly be, of this world even though he spends his life here; and we may one day be able to look upon the age-old enthusiasm of philosophers for the universal as the first indication, as though they alone possessed a foreboding, that the time would come when men would have to live under the earth's conditions and at the same time be able to look upon and act on her from a point outside. (The trouble is only—or so it seems now—that while man can *do* things from a "universal," absolute standpoint, what the philosophers had never deemed possible, he has lost his capacity to *think* in universal, absolute terms, thus realizing and defeating at the same time the standards and ideals of traditional philosophy. Instead of the old dichotomy between earth and sky we have a new one between man and the universe, or between the capacities of the human mind for understanding and the universal laws which man can discover and handle without true comprehension.) Whatever the rewards and the burdens of this yet uncertain future may turn out to be, one thing is sure: while it may affect greatly, perhaps even radically, the vocabulary and metaphoric content of existing religions, it

24. I follow the presentation given by Werner Heisenberg, "Elementarteile der Materie," in *Vom Atom zum Weltsystem* (1954).

neither abolishes nor removes nor even shifts the unknown that is the region of faith.

While the new science, the science of the Archimedean point, needed centuries and generations to develop its full potentialities, taking roughly two hundred years before it even began to change the world and to establish new conditions for the life of man, it took no more than a few decades, hardly one generation, for the human mind to draw certain conclusions from Galileo's discoveries and the methods and assumption by which they had been accomplished. The human mind changed in a matter of years or decades as radically as the human world in a matter of centuries; and while this change naturally remained restricted to the few who belonged to that strangest of all modern societies, the society of scientists and the republic of letters (the only society which has survived all changes of conviction and conflict without a revolution and without ever forgetting to "honor the man whose beliefs it no longer shares"),[25] this society anticipated in many respects, by sheer force of trained and controlled imagination, the radical change of mind of all modern men which became a politically demonstrable reality only in our own time.[26] Descartes is no less the father of modern

25. Bronowski, *op. cit.*

26. The foundation and early history of the Royal Society is quite suggestive. When it was founded, members had to agree to take no part in matters outside the terms of reference given it by the King, especially to take no part in political or religious strife. One is tempted to conclude that the modern scientific ideal of "objectivity" was born here, which would suggest that its origin is political and not scientific. Furthermore, it seems noteworthy that the scientists found it necessary from the beginning to organize themselves into a society, and the fact that the work done inside the Royal Society turned out to be vastly more important than work done outside it demonstrated how right they were. An organization, whether of scientists who have abjured politics or of politicians, is always a political institution; where men organize they intend to act and to acquire power. No scientific teamwork is pure science, whether its aim is to act upon society and secure its members a certain position within it or—as was and still is to a large extent the case of organized research in the natural sciences—to act together and in concert in order to conquer nature. It is indeed, as Whitehead once remarked, "no accident that an age of science has developed into an age of organisation. Organised thought is the basis of organised action," not, one is tempted to add, because thought is the basis of action but rather because modern science as "the organisation of thought" introduced an element of action into thinking. (See *The Aims of Education* [Mentor ed.], pp. 106–7.)

philosophy than Galileo is the ancestor of modern science, and while it is true that after the seventeenth century, and chiefly because of the development of modern philosophy, science and philosophy parted company more radically than ever before[27]—Newton was almost the last to consider his own endeavors as "experimental philosophy" and to offer his discoveries to the reflection of "astronomers and philosophers,"[28] as Kant was the last philosopher who was also a kind of astronomer and natural scientist[29]—modern philosophy owes its origin and its course more exclusively to specific scientific discoveries than any previous philosophy. That this philosophy, the exact counterpart of a scientific world view long since discarded, has not become obsolete today is not only due to the nature of philosophy, which, wherever it is authentic, possesses the same permanence and durability as art works, but is in this particular case closely related to the eventual evolution of a world where truths for many centuries accessible only to the few have become realities for everybody.

It would be folly indeed to overlook the almost too precise congruity of modern man's world alienation with the subjectivism of modern philosophy, from Descartes and Hobbes to English sensualism, empiricism, and pragmatism, as well as German idealism and materialism up to the recent phenomenological existentialism and logical or epistemological positivism. But it would be equally foolish to believe that what turned the philosopher's mind away from the old metaphysical questions toward a great variety of introspections—introspection into his sensual or cognitive apparatus, into his consciousness, into psychological and logical processes—was an impetus that grew out of an autonomous development of ideas, or, in a variation of the same approach, to believe that our world would have become different if only philosophy had held

27. Karl Jaspers, in his masterful interpretation of Cartesian philosophy, insists on the strange ineptitude of Descartes' "scientific" ideas, his lack of understanding for the spirit of modern science, and his inclination to accept theories uncritically without tangible evidence, which had already surprised Spinoza (*op. cit.*, esp. pp. 50 ff. and 93 ff.).

28. See Newton's *Mathematical Principles of Natural Philosophy*, trans. Motte (1803), II, 314.

29. Among Kant's early publications was an *Allgemeine Naturgeschichte und Theorie des Himmels*.

fast to tradition. As we said before, not ideas but events change the world—the heliocentric system as an idea is as old as Pythagorean speculation and as persistent in our history as Neo-Platonic traditions, without, for that matter, ever having changed the world or the human mind—and the author of the decisive event of the modern age is Galileo rather than Descartes. Descartes himself was quite aware of this, and when he heard of Galileo's trial and his recantation, he was tempted for a moment to burn all his papers, because "if the movement of the earth is false, all the foundations of my philosophy are also false."[30] But Descartes and the philosophers, since they elevated what had happened to the level of uncompromising thought, registered with unequaled precision the enormous shock of the event; they anticipated, at least partially, the very perplexities inherent in the new standpoint of man with which the scientists were too busy to bother until, in our own time, they began to appear in their own work and to interfere with their own inquiries. Since then, the curious discrepancy between the mood of modern philosophy, which from the beginning had been predominantly pessimistic, and the mood of modern science, which until very recently had been so buoyantly optimistic, has been bridged. There seems to be little cheerfulness left in either of them.

38

THE RISE OF THE CARTESIAN DOUBT

Modern philosophy began with Descartes' *de omnibus dubitandum est*, with doubt, but with doubt not as an inherent control of the human mind to guard against deceptions of thought and illusions of sense, not as skepticism against the morals and prejudices of men and times, not even as a critical method in scientific inquiry and philosophic speculation. Cartesian doubt is much more far-reaching in scope and too fundamental in intent to be determined by such concrete contents. In modern philosophy and thought, doubt occupies much the same central position as that occupied for all the centuries before by the Greek *thaumazein*, the wonder at everything that is as it is. Descartes was the first to conceptualize this modern doubting, which after him became the self-evident, in-

30. See Descartes' letter to Mersenne of November, 1633.

audible motor which has moved all thought, the invisible axis around which all thinking has been centered. Just as from Plato and Aristotle to the modern age conceptual philosophy, in its greatest and most authentic representatives, had been the articulation of wonder, so modern philosophy since Descartes has consisted in the articulations and ramifications of doubting.

Cartesian doubt, in its radical and universal significance, was originally the response to a new reality, a reality no less real for its being restricted for centuries to the small and politically insignificant circle of scholars and learned men. The philosophers understood at once that Galileo's discoveries implied no mere challenge to the testimony of the senses and that it was no longer reason, as in Aristarchus and Copernicus, that had "committed such a rape on their senses," in which case men indeed would have needed only to choose between their faculties and to let innate reason become "the mistress of their credulity."[31] It was not reason but a man-made instrument, the telescope, which actually changed the physical world view; it was not contemplation, observation, and speculation which led to the new knowledge, but the active stepping in of *homo faber*, of making and fabricating. In other words, man had been deceived so long as he trusted that reality and truth would reveal themselves to his senses and to his reason if only he remained true to what he saw with the eyes of body and mind. The old opposition of sensual and rational truth, of the inferior truth capacity of the senses and the superior truth capacity of reason, paled beside this challenge, beside the obvious implication that neither truth nor reality is given, that neither of them appears as it is, and that only interference with appearance, doing away with appearances, can hold out a hope for true knowledge.

The extent to which reason and faith in reason depend not upon single sense perceptions, each of which may be an illusion, but upon the unquestioned assumption that the senses as a whole—kept together and ruled over by common sense, the sixth and the highest sense—fit man into the reality which surrounds him, was only now

31. In these words, Galileo expresses his admiration for Copernicus and Aristarchus, whose reason "was able . . . to commit such a rape on their senses, as in despite thereof to make herself mistress of their credulity" (*Dialogues concerning the Two Great Systems of the World*, trans. Salusbury [1661], p. 301).

discovered. If the human eye can betray man to the extent that so many generations of men were deceived into believing that the sun turns around the earth, then the metaphor of the eyes of the mind cannot possibly hold any longer; it was based, albeit implicitly and even when it was used in opposition to the senses, on an ultimate trust in bodily vision. If Being and Appearance part company forever, and this—as Marx once remarked—is indeed the basic assumption of all modern science, then there is nothing left to be taken upon faith; everything must be doubted. It was as though Democritus' early prediction that a victory of the mind over the senses could end only in the mind's defeat had come true, except that now the reading of an instrument seemed to have won a victory over both the mind and the senses.[31a]

The outstanding characteristic of Cartesian doubt is its universality, that nothing, no thought and no experience, can escape it. No one perhaps explored its true dimensions more honestly than Kierkegaard when he leaped—not from reason, as he thought, but from doubt—into belief, thereby carrying doubt into the very heart of modern religion.[32] Its universality spreads from the testimony of the senses to the testimony of reason to the testimony of faith because this doubt resides ultimately in the loss of self-evidence, and all thought had always started from what is evident in and by itself—evident not only for the thinker but for everybody. Cartesian doubt did not simply doubt that human understanding may not be open to every truth or that human vision may not be able to see everything, but that intelligibility to human understanding does not at all constitute a demonstration of truth, just as visibility did not at all constitute proof of reality. This doubt doubts

31a. Democritus, after having stated that "in reality there is no white, or black, or bitter, or sweet," added: "Poor mind, from the senses you take your arguments, and then want to defeat them? Your victory is your defeat" (Diels, *Fragmente der Vorsokratiker* [4th ed., 1922], frag. B125).

32. See *Johannes Climacus oder De omnibus dubitandum est*, one of the earliest manuscripts of Kierkegaard and perhaps still the deepest interpretation of Cartesian doubt. It tells in the form of a spiritual autobiography how he learned about Descartes from Hegel and then regretted not having started his philosophical studies with his works. This little treatise, the Danish edition of the *Collected Works* (Copenhagen, 1909), Vol. IV, is available in a German translation (Darmstadt, 1948).

that such a thing as truth exists at all, and discovers thereby that the traditional concept of truth, whether based on sense perception or on reason or on belief in divine revelation, had rested on the twofold assumption that what truly is will appear of its own accord and that human capabilities are adequate to receive it.[33] That truth reveals itself was the common creed of pagan and Hebrew antiquity, of Christian and secular philosophy. This is the reason why the new, modern philosophy turned with such vehemence—in fact with a violence bordering on hatred—against tradition, making short shrift of the enthusiastic Renaissance revival and rediscovery of antiquity.

The poignancy of Descartes' doubt is fully realized only if one understands that the new discoveries dealt an even more disastrous blow to human confidence in the world and in the universe than is indicated by a clear-cut separation of being and appearance. For here the relationship between these two is no longer static as it was in traditional skepticism, as though appearances simply hide and cover a true being which forever escapes the notice of man. This Being, on the contrary, is tremendously active and energetic: it creates its own appearances, except that these appearances are delusions. Whatever human senses perceive is brought about by invisible, secret forces, and if through certain devices, ingenious instruments, these forces are caught in the act rather than discovered —as an animal is trapped or a thief is caught much against their own will and intentions—it turns out that this tremendously effec-

33. The close relatedness of confidence in the senses and confidence in reason in the traditional concept of truth was clearly recognized by Pascal. According to him: "Ces deux principes de vérité, la raison et les sens, outre qu'ils manquent chacun de sincérité, s'abusent réciproquement l'un et l'autre. Les sens abusent la raison par de fausses apparences; et cette même piperie qu'ils apportent à la raison, ils la reçoivent d'elle à leur tour: elle s'en revanche. Les passions de l'âme troublent les sens, et leur font des impressions fausses. Ils mentent et se trompent à l'envi" (*Pensées* [Pléiades ed., 1950], No. 92, p. 849). Pascal's famous wager that he certainly would risk less by believing what Christianity has to teach about a hereafter than by disbelieving it is sufficient demonstration of the interrelatedness of rational and sensory truth with the truth of divine revelation. To Pascal, as to Descartes, God is *un Dieu caché* (*ibid.*, No. 366, p. 923) who does not reveal himself, but whose existence and even goodness is the only hypothetical guaranty that human life is not a dream (the Cartesian nightmare recurs in Pascal, *ibid.*, No. 380, p. 928) and human knowledge not a divine fraud.

tive Being is of such a nature that its disclosures must be illusions and that conclusions drawn from its appearances must be delusions.

Descartes' philosophy is haunted by two nightmares which in a sense became the nightmares of the whole modern age, not because this age was so deeply influenced by Cartesian philosophy, but because their emergence was almost inescapable once the true implications of the modern world view were understood. These nightmares are very simple and very well known. In the one, reality, the reality of the world as well as of human life, is doubted; if neither the senses nor common sense nor reason can be trusted, then it may well be that all that we take for reality is only a dream. The other concerns the general human condition as it was revealed by the new discoveries and the impossibility for man to trust his senses and his reason; under these circumstances it seems, indeed, much more likely that an evil spirit, a *Dieu trompeur*, wilfully and spitefully betrays man than that God is the ruler of the universe. The consummate devilry of this evil spirit would consist in having created a creature which harbors a notion of truth only to bestow on it such other faculties that it will never be able to reach any truth, never be able to be certain of anything.

Indeed, this last point, the question of certainty, was to become decisive for the whole development of modern morality. What was lost in the modern age, of course, was not the capacity for truth or reality or faith nor the concomitant inevitable acceptance of the testimony of the senses and of reason, but the certainty that formerly went with it. In religion it was not belief in salvation or a hereafter that was immediately lost, but the *certitudo salutis*—and this happened in all Protestant countries where the downfall of the Catholic Church had eliminated the last tradition-bound institution which, wherever its authority remained unchallenged, stood between the impact of modernity and the masses of believers. Just as the immediate consequence of this loss of certainty was a new zeal for making good in this life as though it were only an overlong period of probation,[34] so the loss of certainty of truth ended in a

34. Max Weber, who, despite some errors in detail which by now have been corrected, is still the only historian who raised the question of the modern age with the depth and relevance corresponding to its importance, was also aware that it was not a simple loss of faith that caused the reversal in the estimate of

new, entirely unprecedented zeal for truthfulness—as though man could afford to be a liar only so long as he was certain of the unchallengeable existence of truth and objective reality, which surely would survive and defeat all his lies.[35] The radical change in moral standards occurring in the first century of the modern age was inspired by the needs and ideals of its most important group of men, the new scientists; and the modern cardinal virtues—success, industry, and truthfulness—are at the same time the greatest virtues of modern science.[36]

The learned societies and Royal Academies became the morally influential centers where scientists were organized to find ways and means by which nature could be trapped by experiments and instruments so that she would be forced to yield her secrets. And this gigantic task, to which no single man but only the collective effort of the best minds of mankind could possibly be adequate, prescribed the rules of behavior and the new standards of judgment. Where formerly truth had resided in the kind of "theory" that since the Greeks had meant the contemplative glance of the beholder who was concerned with, and received, the reality opening up before him, the question of success took over and the test of theory became a "practical" one—whether or not it will work. Theory became hypothesis, and the success of the hypothesis became truth. This all-important standard of success, however, does not depend upon practical considerations or the technical developments which may or may not accompany specific scientific discoveries. The criterion of success is inherent in the very essence and progress of modern science quite apart from its applicability. Success here is not at all the empty idol to which it degenerated in

---

work and labor, but the loss of the *certitudo salutis*, of the certainty of salvation. In our context, it would appear that this certainty was only one among the many certainties lost with the arrival of the modern age.

35. It certainly is quite striking that not one of the major religions, with the exception of Zoroastrianism, has ever included lying as such among the mortal sins. Not only is there no commandment: Thou shalt not lie (for the commandment: Thou shalt not bear false witness against thy neighbor, is of course of a different nature), but it seems as though prior to puritan morality nobody ever considered lies to be serious offenses.

36. This is the chief point of Bronowski's article quoted above.

bourgeois society; it was, and in the sciences has been ever since, a veritable triumph of human ingenuity against overwhelming odds.

The Cartesian solution of universal doubt or its salvation from the two interconnected nightmares—that everything is a dream and there is no reality and that not God but an evil spirit rules the world and mocks man—was similar in method and content to the turning away from truth to truthfulness and from reality to reliability. Descartes' conviction that "though our mind is not the measure of things or of truth, it must assuredly be the measure of things that we affirm or deny"[37] echoes what scientists in general and without explicit articulation had discovered: that even if there is no truth, man can be truthful, and even if there is no reliable certainty, man can be reliable. If there was salvation, it had to lie in man himself, and if there was a solution to the questions raised by doubting, it had to come from doubting. If everything has become doubtful, then doubting at least is certain and real. Whatever may be the state of reality and of truth as they are given to the senses and to reason, "nobody can doubt of his doubt and remain uncertain whether he doubts or does not doubt."[38] The famous *cogito ergo sum* ("I think, hence I am") did not spring for Descartes from any self-certainty of thought as such—in which case, indeed, thought would have acquired a new dignity and significance for man—but was a mere generalization of a *dubito ergo sum*.[39] In

37. From a letter of Descartes to Henry More, quoted from Koyré, *op. cit.*, p. 117.

38. In the dialogue *La recherche de la vérité par la lumière naturelle*, where Descartes exposes his fundamental insights without technical formality, the central position of doubting is even more in evidence than in his other works. Thus Eudoxe, who stands for Descartes, explains: "Vous pouvez douter avec raison de toutes les choses dont la connaissance ne vous vient que par l'office des sens; mais pouvez-vous douter de votre doute et rester incertain si vous doutez ou non? . . . vous qui doutez vous êtes, et cela est si vrai que vous n'en pouvez douter davantage" (Pléiade ed., p. 680).

39. "Je doute, donc je suis, ou bien ce qui est la même chose: je pense, donc je suis" (*ibid.*, p. 687). Thought in Descartes has indeed a mere derivative character: "Car s'il est vrai que je doute, comme je n'en puis douter, il est également vrai que je pense; en effet douter est-il autre chose que penser d'une certaine manière?" (*ibid.*, p. 686). The leading idea of this philosophy is by no means that I would not be able to think without being, but that "nous ne saurions douter sans être, et que cela est la première connaissance certaine qu'on peut acquérir" (*Prin-*

other words, from the mere logical certainty that in doubting something I remain aware of a process of doubting in my consciousness,
Descartes concluded that those processes which go on in the mind
of man himself have a certainty of their own, that they can become
the object of investigation in introspection.

39

INTROSPECTION AND THE LOSS OF COMMON SENSE

Introspection, as a matter of fact, not the reflection of man's mind
on the state of his soul or body but the sheer cognitive concern of
consciousness with its own content (and this is the essence of the
Cartesian *cogitatio*, where *cogito* always means *cogito me cogitare*)
must yield certainty, because here nothing is involved except what
the mind has produced itself; nobody is interfering but the producer
of the product, man is confronted with nothing and nobody but
himself. Long before the natural and physical sciences began to
wonder if man is capable of encountering, knowing, and comprehending anything except himself, modern philosophy had made
sure in introspection that man concerns himself only with himself.
Descartes believed that the certainty yielded by his new method of
introspection is the certainty of the I-am.[40] Man, in other words,
carries his certainty, the certainty of his existence, within himself;
the sheer functioning of consciousness, though it cannot possibly
assure a worldly reality given to the senses and to reason, confirms
beyond doubt the reality of sensations and of reasoning, that is, the
reality of processes which go on in the mind. These are not unlike

cipes [Pléiade ed.], Part I, sec. 7). The argument itself is of course not new. One
finds it, for instance, almost word for word in Augustine's *De libero arbitrio* (ch.
3), but without the implication that this is the only certainty against the possibility of a *Dieu trompeur* and, generally, without being the very fundament of a
philosophical system.

40. That the *cogito ergo sum* contains a logical error, that, as Nietzsche pointed
out, it should read: *cogito, ergo cogitationes sunt*, and that therefore the mental
awareness expressed in the *cogito* does not prove that I am, but only that consciousness is, is another matter and need not interest us here (see Nietzsche,
Wille zur Macht, No. 484).

the biological processes that go on in the body and which, when one becomes aware of them, can also convince one of its working reality. In so far as even dreams are real, since they presuppose a dreamer and a dream, the world of consciousness is real enough. The trouble is only that just as it would be impossible to infer from the awareness of bodily processes the actual shape of any body, including one's own, so it is impossible to reach out from the mere consciousness of sensations, in which one senses his senses and in which even the sensed object becomes part of sensation, into reality with its shapes, forms, colors, and constellations. The seen tree may be real enough for the sensation of vision, just as the dreamed tree is real enough for the dreamer as long as the dream lasts, but neither can ever become a real tree.

It is out of these perplexities that Descartes and Leibniz needed to prove, not the existence of God, but his goodness, the one demonstrating that no evil spirit rules the world and mocks man and the other that this world, including man, is the best of all possible worlds. The point about these exclusively modern justifications, known since Leibniz as theodicies, is that the doubt does not concern the existence of a highest being, which, on the contrary, is taken for granted, but concerns his revelation, as given in biblical tradition, and his intentions with respect to man and world, or rather the adequateness of the relationship between man and world. Of these two, the doubt that the Bible or nature contains divine revelation is a matter of course, once it has been shown that revelation as such, the disclosure of reality to the senses and of truth to reason, is no guaranty for either. Doubt of the goodness of God, however, the notion of a *Dieu trompeur*, arose out of the very experience of deception inherent in the acceptance of the new world view, a deception whose poignancy lies in its irremediable repetitiveness, for no knowledge about the heliocentric nature of our planetary system can change the fact that every day the sun is seen circling the earth, rising and setting at its preordained location. Only now, when it appeared as though man, if it had not been for the accident of the telescope, might have been deceived forever, did the ways of God really become wholly inscrutable; the more man learned about the universe, the less he could understand the intentions and purposes for which he should have been created.

The Human Condition

The goodness of the God of the theodicies, therefore, is strictly the quality of a *deus ex machina;* inexplicable goodness is ultimately the only thing that saves reality in Descartes' philosophy (the co-existence of mind and extension, *res cogitans* and *res extensa*), as it saves the prestabilized harmony between man and world in Leibniz.[41]

The very ingenuity of Cartesian introspection, and hence the reason why this philosophy became so all-important to the spiritual and intellectual development of the modern age, lies first in that it had used the nightmare of non-reality as a means of submerging all worldly objects into the stream of consciousness and its processes. The "seen tree" found in consciousness through introspection is no longer the tree given in sight and touch, an entity in itself with an unalterable identical shape of its own. By being processed into an object of consciousness on the same level with a merely remembered or entirely imaginary thing, it becomes part and parcel of this process itself, of that consciousness, that is, which one knows only as an ever-moving stream. Nothing perhaps could prepare our minds better for the eventual dissolution of matter into energy, of objects into a whirl of atomic occurrences, than this dissolution of objective reality into subjective states of mind or, rather, into subjective mental processes. Second, and this was of even greater relevance to the initial stages of the modern age, the Cartesian method of securing certainty against universal doubt corresponded most precisely to the most obvious conclusion to be drawn from the new physical science: though one cannot know truth as something given and disclosed, man can at least know what he makes himself. This, indeed, became the most general and most generally accepted attitude of the modern age, and it is this conviction, rather

41. This quality of God as a *deus ex machina*, as the only possible solution to universal doubt, is especially manifest in Descartes' *Méditations*. Thus, he says in the third meditation: In order to eliminate the cause of doubting, "je dois examiner s'il y a un Dieu . . . ; et si je trouve qu'il y en ait un, je dois aussi examiner s'il peut être trompeur: car sans la connaissance de ces deux vérités, je ne vois pas que je puisse jamais être certain d'aucune chose." And he concludes at the end of the fifth meditation: "Ainsi je reconnais très clairement que la certitude et la vérité de toute science dépend de la seule connaissance du vrai Dieu: en sorte qu'avant que je le connusse, je ne pouvais savoir parfaitement aucune autre chose" (Pléiade ed., pp. 177, 208).

than the doubt underlying it, that propelled one generation after another for more than three hundred years into an ever-quickening pace of discovery and development.

Cartesian reason is entirely based "on the implicit assumption that the mind can only know that which it has itself produced and retains in some sense within itself."[42] Its highest ideal must therefore be mathematical knowledge as the modern age understands it, that is, not the knowledge of ideal forms given outside the mind but of forms produced by a mind which in this particular instance does not even need the stimulation—or, rather, the irritation—of the senses by objects other than itself. This theory is certainly what Whitehead calls it, "the outcome of common-sense in retreat."[43] For common sense, which once had been the one by which all other senses, with their intimately private sensations, were fitted into the common world, just as vision fitted man into the visible world, now became an inner faculty without any world relationship. This sense now was called common merely because it happened to be common to all. What men now have in common is not the world but the structure of their minds, and this they cannot have in common, strictly speaking; their faculty of reasoning can only happen to be the same in everybody.[44] The fact that, given the problem of two plus two we all will come out with the same answer, four, is henceforth the very model of common-sense reasoning.

Reason, in Descartes no less than in Hobbes, becomes "reckoning with consequences," the faculty of deducing and concluding, that is, of a process which man at any moment can let loose within himself. The mind of this man—to remain in the sphere of mathematics—no longer looks upon "two-and-two-are-four" as an equation in which two sides balance in a self-evident harmony, but understands the equation as the expression of a process in which two and two *become* four in order to generate further processes of addi-

42. A. N. Whitehead, *The Concept of Nature* (Ann Arbor ed.), p. 32.

43. *Ibid.*, p. 43. The first to comment on and criticize the absence of common sense in Descartes was Vico (see *De nostri temporis studiorum ratione*, ch. 3).

44. This transformation of common sense into an inner sense is characteristic of the whole modern age; in the German language it is indicated by the difference between the older German word *Gemeinsinn* and the more recent expression *gesunder Menschenverstand* which replaced it.

tion which eventually will lead into the infinite. This faculty the modern age calls common-sense reasoning; it is the playing of the mind with itself, which comes to pass when the mind is shut off from all reality and "senses" only itself. The results of this play are compelling "truths" because the structure of one man's mind is supposed to differ no more from that of another than the shape of his body. Whatever difference there may be is a difference of mental power, which can be tested and measured like horsepower. Here the old definition of man as an *animal rationale* acquires a terrible precision: deprived of the sense through which man's five animal senses are fitted into a world common to all men, human beings are indeed no more than animals who are able to reason, "to reckon with consequences."

The perplexity inherent in the discovery of the Archimedean point was and still is that the point outside the earth was found by an earth-bound creature, who found that he himself lived not only in a different but in a topsy-turvy world the moment he tried to apply his universal world view to his actual surroundings. The Cartesian solution of this perplexity was to move the Archimedean point into man himself,[45] to choose as ultimate point of reference the pattern of the human mind itself, which assures itself of reality and certainty within a framework of mathematical formulas which are its own products. Here the famous *reductio scientiae ad mathematicam* permits replacement of what is sensuously given by a system of mathematical equations where all real relationships are dissolved into logical relations between man-made symbols. It is this replacement which permits modern science to fulfil its "task of *producing*" the phenomena and objects it wishes to observe.[46] And the assumption is that neither God nor an evil spirit can change the fact that two and two equal four.

45. This removal of the Archimedean point into man himself was a conscious operation of Descartes: "Car à partir de ce doute universel, comme à partir d'un point fixe et immobile, je me suis proposé de faire dériver la connaissance de Dieu, de vous-mêmes et de toutes les choses qui existent dans le monde" (*Recherche de la vérité*, p. 680).

46. Frank, *op. cit.*, defines science by its "task of producing desired observable phenomena."

40

THOUGHT AND THE MODERN WORLD VIEW

The Cartesian removal of the Archimedean point into the mind of man, while it enabled man to carry it, as it were, within himself wherever he went and thus freed him from given reality altogether —that is, from the human condition of being an inhabitant of the earth—has perhaps never been as convincing as the universal doubt from which it sprang and which it was supposed to dispel.[47] Today, at any rate, we find in the perplexities confronting natural scientists in the midst of their greatest triumphs the same nightmares which have haunted the philosophers from the beginning of the modern age. This nightmare is present in the fact that a mathematical equation, such as of mass and energy—which originally was destined only to save the phenomena, to be in agreement with observable facts that could also be explained differently, just as the Ptolemaic and Copernican systems originally differed only in simplicity and harmony—actually lends itself to a very real conversion of mass into energy and vice versa, so that the mathematical "conversion" implicit in every equation corresponds to convertibility in reality; it is present in the weird phenomenon that the systems of non-Euclidean mathematics were found without any forethought of applicability or even empirical meaning before they gained their surprising validity in Einstein's theory; and it is even more troubling in the inevitable conclusion that "the possibility of such an application must be held open for all, even the most remote constructions of pure mathematics."[48] If it should be true that a whole universe, or rather any number of utterly different universes will spring into existence and "prove" whatever over-all pattern the

47. Ernst Cassirer's hope that "doubt is overcome by being outdone" and that the theory of relativity would free the human mind from its last "earthly remainder," namely, the anthropomorphism inherent in "the manner in which we make empirical measurements of space and time" (*op. cit.*, pp. 389, 382), has not been fulfilled; on the contrary, doubt not of the validity of scientific statements but of the intelligibility of scientific data has increased during the last decades.

48. *Ibid.*, p. 443.

human mind has constructed, then man may indeed, for a moment, rejoice in a reassertion of the "pre-established harmony between pure mathematics and physics,"[49] between mind and matter, between man and the universe. But it will be difficult to ward off the suspicion that this mathematically preconceived world may be a dream world where every dreamed vision man himself produces has the character of reality only as long as the dream lasts. And his suspicions will be enforced when he must discover that the events and occurrences in the infinitely small, the atom, follow the same laws and regularities as in the infinitely large, the planetary systems.[50] What this seems to indicate is that if we inquire into nature from the standpoint of astronomy we receive planetary systems, while if we carry out our astronomical inquiries from the standpoint of the earth we receive geocentric, terrestrial systems.

In any event, wherever we try to transcend appearance beyond all sensual experience, even instrument-aided, in order to catch the ultimate secrets of Being, which according to our physical world view is so secretive that it never appears and still so tremendously powerful that it produces all appearance, we find that the same patterns rule the macrocosm and the microcosm alike, that we receive the same instrument readings. Here again, we may for a moment rejoice in a refound unity of the universe, only to fall prey to the suspicion that what we have found may have nothing to do with either the macrocosmos or the microcosmos, that we deal only with the patterns of our own mind, the mind which designed the instruments and put nature under its conditions in the experiment—prescribed its laws to nature, in Kant's phrase—in which case it is really as though we were in the hands of an evil spirit who

49. Hermann Minkowski, "Raum und Zeit," in Lorentz, Einstein, and Minkowski, *Das Relativitätsprinzip* (1913); quoted from Cassirer, *op. cit.*, p. 419.

50. And this doubt is not assuaged if another coincidence is added, the coincidence between logic and reality. Logically, it seems evident indeed that "the electrons if they were to explain the sensory qualities of matter could not very well possess these sensory qualities, since in that case the question for the cause of these qualities would simply have been removed one step farther, but not solved" (Heisenberg, *Wandlungen in den Grundlagen der Naturwissenschaft*, p. 66). The reason why we become suspicious is that only when "in the course of time" the scientists became aware of this logical necessity did they discover that "matter" had no qualities and therefore could no longer be called matter.

mocks us and frustrates our thirst for knowledge, so that wherever we search for that which we are not, we encounter only the patterns of our own minds.

Cartesian doubt, logically the most plausible and chronologically the most immediate consequence of Galileo's discovery, was assuaged for centuries through the ingenious removal of the Archimedean point into man himself, at least so far as natural science was concerned. But the mathematization of physics, by which the absolute renunciation of the senses for the purpose of knowing was carried through, had in its last stages the unexpected and yet plausible consequence that every question man puts to nature is answered in terms of mathematical patterns to which no model can ever be adequate, since one would have to be shaped after our sense experiences.[51] At this point, the connection between thought and sense experience, inherent in the human condition, seems to take its revenge: while technology demonstrates the "truth" of modern science's most abstract concepts, it demonstrates no more than that man can always apply the results of his mind, that no matter which system he uses for the explanation of natural phenomena he will always be able to adopt it as a guiding principle for making and acting. This possibility was latent even in the beginnings of modern mathematics, when it turned out that numerical truths can be fully translated into spatial relationships. If, therefore, present-day science in its perplexity points to technical achievements to "prove" that we deal with an "authentic order" given in nature,[52] it seems it has fallen into a vicious circle, which can be formulated as follows: scientists formulate their hypotheses to arrange their experiments and then use these experiments to verify their hypotheses; during this whole enterprise, they obviously deal with a hypothetical nature.[53]

51. In the words of Erwin Schrödinger: "As our mental eye penetrates into smaller and smaller distances and shorter and shorter times, we find nature behaving so entirely differently from what we observe in visible and palpable bodies of our surrounding that *no* model shaped after our large-scale experiences can ever be 'true' " (*Science and Humanism* [1952], p. 25).

52. Heisenberg, *Wandlungen in den Grundlagen*, p. 64.

53. This point is best illustrated by a statement of Planck, quoted in a very illuminating article by Simone Weil (published under the pseudonym "Emil Novis" and entitled "Réflexions à propos de la théorie des quanta," in *Cahiers du*

In other words, the world of the experiment seems always capable of becoming a man-made reality, and this, while it may increase man's power of making and acting, even of creating a world, far beyond what any previous age dared to imagine in dream and phantasy, unfortunately puts man back once more—and now even more forcefully—into the prison of his own mind, into the limitations of patterns he himself created. The moment he wants what all ages before him were capable of achieving, that is, to experience the reality of what he himself is not, he will find that nature and the universe "escape him" and that a universe construed according to the behavior of nature in the experiment and in accordance with the very principles which man can translate technically into a working reality lacks all possible representation. What is new here is not that things exist of which we cannot form an image—such "things" were always known and among them, for instance, belonged the "soul"—but that the material things we see and represent and against which we had measured immaterial things for which we can form no images should likewise be "unimaginable." With the disappearance of the sensually given world, the transcendent world disappears as well, and with it the possibility of transcending the material world in concept and thought. It is therefore not surprising that the new universe is not only "practically inaccessible but not even thinkable," for "however we think it, it is wrong; not perhaps quite as meaningless as a 'triangular circle,' but much more so than a 'winged lion.' "[54]

Cartesian universal doubt has now reached the heart of physical

---

Sud [December, 1942]), which in the French translation runs as follows: "Le créateur d'une hypothèse dispose de possibilités pratiquement illimitées, il est aussi peu lié par le fonctionnement des organes de ses sens qu'il ne l'est par celui des instruments dont il se sert. . . . On peut même dire qu'il se crée une géométrie à sa fantasie. . . . C'est pourquoi aussi jamais des mesures ne pourront confirmer ni infirmer directement une hypothèse; elles pourront seulement en faire ressortir la convenance plus ou moins grande." Simone Weil points out at length how something "infiniment plus précieux" than science is compromised in this crisis, namely, the notion of truth; she fails, however, to see that the greatest perplexity in this state of affairs arises from the undeniable fact that these hypotheses "work." (I owe the reference to this little known article to Miss Beverly Woodward, a former student of mine.)

54. Schrödinger, *op. cit.*, p. 26.

science itself; for the escape into the mind of man himself is closed if it turns out that the modern physical universe is not only beyond presentation, which is a matter of course under the assumption that nature and Being do not reveal themselves to the senses, but is inconceivable, unthinkable in terms of pure reasoning as well.

41

THE REVERSAL OF CONTEMPLATION AND ACTION

Perhaps the most momentous of the spiritual consequences of the discoveries of the modern age and, at the same time, the only one that could not have been avoided, since it followed closely upon the discovery of the Archimedean point and the concomitant rise of Cartesian doubt, has been the reversal of the hierarchical order between the *vita contemplativa* and the *vita activa*.

In order to understand how compelling the motives for this reversal were, it is first of all necessary to rid ourselves of the current prejudice which ascribes the development of modern science, because of its applicability, to a pragmatic desire to improve conditions and better human life on earth. It is a matter of historical record that modern technology has its origins not in the evolution of those tools man had always devised for the twofold purpose of easing his labors and erecting the human artifice, but exclusively in an altogether non-practical search for useless knowledge. Thus, the watch, one of the first modern instruments, was not invented for purposes of practical life, but exclusively for the highly "theoretical" purpose of conducting certain experiments with nature. This invention, to be sure, once its practical usefulness became apparent, changed the whole rhythm and the very physiognomy of human life; but from the standpoint of the inventors, this was a mere incident. If we had to rely only on men's so-called practical instincts, there would never have been any technology to speak of, and although today the already existing technical inventions carry a certain momentum which will probably generate improvements up to a certain point, it is not likely that our technically conditioned world could survive, let alone develop further, if we ever succeeded in convincing ourselves that man is primarily a practical being.

However that may be, the fundamental experience behind the reversal of contemplation and action was precisely that man's thirst for knowledge could be assuaged only after he had put his trust into the ingenuity of his hands. The point was not that truth and knowledge were no longer important, but that they could be won only by "action" and not by contemplation. It was an instrument, the telescope, a work of man's hands, which finally forced nature, or rather the universe, to yield its secrets. The reasons for trusting *doing* and for distrusting *contemplation* or *observation* became even more cogent after the results of the first active inquiries. After being and appearance had parted company and truth was no longer supposed to appear, to reveal and disclose itself to the mental eye of a beholder, there arose a veritable necessity to hunt for truth behind deceptive appearances. Nothing indeed could be less trustworthy for acquiring knowledge and approaching truth than passive observation or mere contemplation. In order to be certain one had to *make sure*, and in order to know one had to do. Certainty of knowledge could be reached only under a twofold condition: first, that knowledge concerned only what one had done himself—so that its ideal became mathematical knowledge, where we deal only with self-made entities of the mind—and second, that knowledge was of such a nature that it could be tested only through more doing.

Since then, scientific and philosophic truth have parted company; scientific truth not only need not be eternal, it need not even be comprehensible or adequate to human reason. It took many generations of scientists before the human mind grew bold enough to fully face this implication of modernity. If nature and the universe are products of a divine maker, and if the human mind is incapable of understanding what man has not made himself, then man cannot possibly expect to learn anything about nature that he can understand. He may be able, through ingenuity, to find out and even to imitate the devices of natural processes, but that does not mean these devices will ever make sense to him—they do not have to be intelligible. As a matter of fact, no supposedly suprarational divine revelation and no supposedly abstruse philosophic truth has ever offended human reason so glaringly as certain results of modern science. One can indeed say with Whitehead: "Heaven knows what seeming nonsense may not to-morrow be demonstrated truth."[55]

Actually, the change that took place in the seventeenth century was more radical than what a simple reversal of the established traditional order between contemplation and doing is apt to indicate. The reversal, strictly speaking, concerned only the relationship between thinking and doing, whereas contemplation, in the original sense of beholding the truth, was altogether eliminated. For thought and contemplation are not the same. Traditionally, thought was conceived as the most direct and important way to lead to the contemplation of truth. Since Plato, and probably since Socrates, thinking was understood as the inner dialogue in which one speaks with himself (*eme emautō*, to recall the idiom current in Plato's dialogues); and although this dialogue lacks all outward manifestation and even requires a more or less complete cessation of all other activities, it constitutes in itself a highly active state. Its outward inactivity is clearly separated from the passivity, the complete stillness, in which truth is finally revealed to man. If medieval scholasticism looked upon philosophy as the handmaiden of theology, it could very well have appealed to Plato and Aristotle themselves; both, albeit in a very different context, considered this dialogical thought process to be the way to prepare the soul and lead the mind to a beholding of truth beyond thought and beyond speech—a truth that is *arrhēton*, incapable of being communicated through words, as Plato put it,[56] or beyond speech, as in Aristotle.[57]

The reversal of the modern age consisted then not in raising doing to the rank of contemplating as the highest state of which human beings are capable, as though henceforth doing was the ultimate meaning for the sake of which contemplation was to be performed, just as, up to that time, all activities of the *vita activa* had been judged and justified to the extent that they made the *vita con-*

55. *Science and the Modern World*, p. 116.

56. In the *Seventh Letter* 341C: *rhēton gar oudamōs estin hōs alla mathēmata* ("for it is never to be expressed by words like other things we learn").

57. See esp. *Nicomachean Ethics* 1142a25 ff. and 1143a36 ff. The current English translation distorts the meaning because it renders *logos* as "reason" or "argument."

templativa possible. The reversal concerned only thinking, which from then on was the handmaiden of doing as it had been the *ancilla theologiae*, the handmaiden of contemplating divine truth in medieval philosophy and the handmaiden of contemplating the truth of Being in ancient philosophy. Contemplation itself became altogether meaningless.

The radicality of this reversal is somehow obscured by another kind of reversal, with which it is frequently identified and which, since Plato, has dominated the history of Western thought. Whoever reads the Cave allegory in Plato's *Republic* in the light of Greek history will soon be aware that the *periagōgē*, the turning-about that Plato demands of the philosopher, actually amounts to a reversal of the Homeric world order. Not life after death, as in the Homeric Hades, but ordinary life on earth, is located in a "cave," in an underworld; the soul is not the shadow of the body, but the body the shadow of the soul; and the senseless, ghostlike motion ascribed by Homer to the lifeless existence of the soul after death in Hades is now ascribed to the senseless doings of men who do not leave the cave of human existence to behold the eternal ideas visible in the sky.[58]

In this context, I am concerned only with the fact that the Platonic tradition of philosophical as well as political thought started with a reversal, and that this original reversal determined to a large extent the thought patterns into which Western philosophy almost automatically fell wherever it was not animated by a great and original philosophical impetus. Academic philosophy, as a matter of fact, has ever since been dominated by the never-ending reversals of idealism and materialism, of transcendentalism and immanentism, of realism and nominalism, of hedonism and asceticism, and so on. What matters here is the reversibility of all these systems, that they can be turned "upside down" or "downside up" at any moment in history without requiring for such reversal either historical events or changes in the structural elements involved. The concepts themselves remain the same no matter where they

58. It is particularly Plato's use of the words *eidōlon* and *skia* in the story of the Cave which makes the whole account read like a reversal of and a reply to Homer; for these are the key words in Homer's description of Hades in the *Odyssey*.

are placed in the various systematic orders. Once Plato had succeeded in making these structural elements and concepts reversible,
reversals within the course of intellectual history no longer needed
more than purely intellectual experience, an experience within the
framework of conceptual thinking itself. These reversals already
began with the philosophical schools in late antiquity and have remained part of the Western tradition. It is still the same tradition,
the same intellectual game with paired antitheses that rules, to an
extent, the famous modern reversals of spiritual hierarchies, such
as Marx's turning Hegelian dialectic upside down or Nietzsche's
revaluation of the sensual and natural as against the supersensual
and supernatural.

The reversal we deal with here, the spiritual consequence of
Galileo's discoveries, although it has frequently been interpreted
in terms of the traditional reversals and hence as integral to the
Western history of ideas, is of an altogether different nature. The
conviction that objective truth is not given to man but that he can
know only what he makes himself is not the result of skepticism
but of a demonstrable discovery, and therefore does not lead to
resignation but either to redoubled activity or to despair. The
world loss of modern philosophy, whose introspection discovered
consciousness as the inner sense with which one senses his senses
and found it to be the only guaranty of reality, is different not only
in degree from the age-old suspicion of the philosophers toward the
world and toward the others with whom they shared the world;
the philosopher no longer turns from the world of deceptive perishability to another world of eternal truth, but turns away from both
and withdraws into himself. What he discovers in the region of the
inner self is, again, not an image whose permanence can be beheld
and contemplated, but, on the contrary, the constant movement of
sensual perceptions and the no less constantly moving activity of
the mind. Since the seventeenth century, philosophy has produced
the best and least disputed results when it has investigated, through
a supreme effort of self-inspection, the processes of the senses and
of the mind. In this aspect, most of modern philosophy is indeed
theory of cognition and psychology, and in the few instances where
the potentialities of the Cartesian method of introspection were
fully realized by men like Pascal, Kierkegaard, and Nietzsche, one

is tempted to say that philosophers have experimented with their own selves no less radically and perhaps even more fearlessly than the scientists experimented with nature.

Much as we may admire the courage and respect the extraordinary ingenuity of philosophers throughout the modern age, it can hardly be denied that their influence and importance decreased as never before. It was not in the Middle Ages but in modern thinking that philosophy came to play second and even third fiddle. After Descartes based his own philosophy upon the discoveries of Galileo, philosophy has seemed condemned to be always one step behind the scientists and their ever more amazing discoveries, whose principles it has strived arduously to discover *ex post facto* and to fit into some over-all interpretation of the nature of human knowledge. As such, however, philosophy was not needed by the scientists, who—up to our time, at least—believed that they had no use for a handmaiden, let alone one who would "carry the torch in front of her gracious lady" (Kant). The philosophers became either epistemologists, worrying about an over-all theory of science which the scientists did not need, or they became, indeed, what Hegel wanted them to be, the organs of the *Zeitgeist*, the mouthpieces in which the general mood of the time was expressed with conceptual clarity. In both instances, whether they looked upon nature or upon history, they tried to understand and come to terms with what happened without them. Obviously, philosophy suffered more from modernity than any other field of human endeavor; and it is difficult to say whether it suffered more from the almost automatic rise of activity to an altogether unexpected and unprecedented dignity or from the loss of traditional truth, that is, of the concept of truth underlying our whole tradition.

42

THE REVERSAL WITHIN THE *Vita Activa* AND THE VICTORY OF *Homo Faber*

First among the activities within the *vita activa* to rise to the position formerly occupied by contemplation were the activities of making and fabricating—the prerogatives of *homo faber*. This was natural enough, since it had been an instrument and therefore man in so far as he is a toolmaker that led to the modern revolution. From then on, all scientific progress has been most intimately tied up with the ever more refined development in the manufacture of new tools and instruments. While, for instance, Galileo's experiments with the fall of heavy bodies could have been made at any time in history if men had been inclined to seek truth through experiments, Michelson's experiment with the interferometer at the end of the nineteenth century relied not merely on his "experimental genius" but "required the general advance in technology," and therefore "could not have been made earlier than it was."[59]

It is not only the paraphernalia of instruments and hence the help man had to enlist from *homo faber* to acquire knowledge that caused these activities to rise from their former humble place in the hierarchy of human capacities. Even more decisive was the element of making and fabricating present in the experiment itself, which produces its own phenomena of observation and therefore depends from the very outset upon man's productive capacities. The use of the experiment for the purpose of knowledge was already the consequence of the conviction that one can know only what he has made himself, for this conviction meant that one might learn about those things man did not make by figuring out and imitating the processes through which they had come into being. The much discussed shift of emphasis in the history of science from the old questions of "what" or "why" something is to the new question of "how" it came into being is a direct consequence of this conviction, and its answer can only be found in the experiment. The experiment repeats the natural process as though man himself were about to make nature's objects, and although in the early stages of the modern age no responsible scientist would have dreamt of the extent to which man actually is capable of "making" nature, he nevertheless from the onset approached it from the standpoint of the One who made it, and this not for practical reasons of technical applicability but exclusively for the "theoretical" reason that certainty in knowledge could not be gained otherwise: "Give me matter and I will build a world from it, that is, give me matter and

59. Whitehead, *Science and the Modern World*, pp. 116–17.

I will show you how a world developed from it."[60] These words of Kant show in a nutshell the modern blending of making and knowing, whereby it is as though a few centuries of knowing in the mode of making were needed as the apprenticeship to prepare modern man for making what he wanted to know.

Productivity and creativity, which were to become the highest ideals and even the idols of the modern age in its initial stages, are inherent standards of *homo faber*, of man as a builder and fabricator. However, there is another and perhaps even more significant element noticeable in the modern version of these faculties. The shift from the "why" and "what" to the "how" implies that the actual objects of knowledge can no longer be things or eternal motions but must be processes, and that the object of science therefore is no longer nature or the universe but the history, the story of the coming into being, of nature or life or the universe. Long before the modern age developed its unprecedented historical consciousness and the concept of history became dominant in modern philosophy, the natural sciences had developed into historical disciplines, until in the nineteenth century they added to the older disciplines of physics and chemistry, of zoology and botany, the new natural sciences of geology or history of the earth, biology or the history of life, anthropology or the history of human life, and, generally, natural history. In all these instances, development, the key concept of the historical sciences, became the central concept of the physical sciences as well. Nature, because it could be known only in processes which human ingenuity, the ingeniousness of *homo faber*, could repeat and remake in the experiment, became a process,[61] and all particular natural things derived their significance and meaning solely from their functions in the over-all process. In the place of the concept of Being we now find the concept of Process. And whereas it is in the nature of Being to appear and thus disclose

60. "Gebet mir Materie, ich will eine Welt daraus bauen! das ist, gebet mir Materie, ich will euch zeigen, wie eine Welt daraus entstehen soll" (see Kant's Preface to his *Allgemeine Naturgeschichte und Theorie des Himmels*).

61. That "nature is a process," that therefore "the ultimate fact for sense-awareness is an event," that natural science deals only with occurrences, happenings, or events, but not with things and that "apart from happenings there is nothing" (see Whitehead, *The Concept of Nature*, pp. 53, 15, 66), belongs among the axioms of modern natural science in all its branches.

itself, it is in the nature of Process to remain invisible, to be something whose existence can only be inferred from the presence of certain phenomena. This process was originally the fabrication process which "disappears in the product," and it was based on the experience of *homo faber*, who knew that a production process necessarily precedes the actual existence of every object.

Yet while this insistence on the process of making or the insistence upon considering every thing as the result of a fabrication process is highly characteristic of *homo faber* and his sphere of experience, the exclusive emphasis the modern age placed on it at the expense of all interest in the things, the products themselves, is quite new. It actually transcends the mentality of man as a toolmaker and fabricator, for whom, on the contrary, the production process was a mere means to an end. Here, from the standpoint of *homo faber*, it was as though the means, the production process or development, was more important than the end, the finished product. The reason for this shift of emphasis is obvious: the scientist made only in order to know, not in order to produce things, and the product was a mere by-product, a side effect. Even today all true scientists will agree that the technical applicability of what they are doing is a mere by-product of their endeavor.

The full significance of this reversal of means and ends remained latent as long as the mechanistic world view, the world view of *homo faber* par excellence, was predominant. This view found its most plausible theory in the famous analogy of the relationship between nature and God with the relationship between the watch and the watchmaker. The point in our context is not so much that the eighteenth-century idea of God was obviously formed in the image of *homo faber* as that in this instance the process character of nature was still limited. Although all particular natural things had already been engulfed in the process from which they had come into being, nature as a whole was not yet a process but the more or less stable end product of a divine maker. The image of watch and watchmaker is so strikingly apposite precisely because it contains both the notion of a process character of nature in the image of the movements of the watch and the notion of its still intact object character in the image of the watch itself and its maker.

It is important at this point to remember that the specifically modern suspicion toward man's truth-receiving capacities, the mistrust of the given, and hence the new confidence in making and introspection which was inspired by the hope that in human consciousness there was a realm where knowing and producing would coincide, did not arise directly from the discovery of the Archimedean point outside the earth in the universe. They were, rather, the necessary consequences of this discovery for the discoverer himself, in so far as he was and remained an earth-bound creature. This close relationship of the modern mentality with philosophical reflection naturally implies that the victory of *homo faber* could not remain restricted to the employment of new methods in the natural sciences, the experiment and the mathematization of scientific inquiry. One of the most plausible consequences to be drawn from Cartesian doubt was to abandon the attempt to understand nature and generally to know about things not produced by man, and to turn instead exclusively to things that owed their existence to man. This kind of argument, in fact, made Vico turn his attention from natural science to history, which he thought to be the only sphere where man could obtain certain knowledge, precisely because he dealt here only with the products of human activity.[62] The modern discovery of history and historical consciousness owed one of its greatest impulses neither to a new enthusiasm for the greatness of man, his doings and sufferings, nor to the belief that the meaning of human existence can be found in the story of mankind, but to the

62. Vico (*op. cit.*, ch. 4) states explicitly why he turned away from natural science. True knowledge of nature is impossible, because not man but God made it; God can know nature with the same certainty man knows geometry: *Geometrica demonstramus quia facimus; si physica demonstrare possemus, faceremus* ("We can prove geometry because we make it; to prove the physical we would have to make it"). This little treatise, written more than fifteen years before the first edition of the *Scienza Nuova* (1725), is interesting in more than one respect. Vico criticizes all existing sciences, but not yet for the sake of his new science of history; what he recommends is the study of moral and political science, which he finds unduly neglected. It must have been much later that the idea occurred to him that history is made by man as nature is made by God. This biographical development, though quite extraordinary in the early eighteenth century, became the rule approximately one hundred years later: each time the modern age had reason to hope for a new political philosophy, it received a philosophy of history instead.

despair of human reason, which seemed adequate only when confronted with man-made objects.

Prior to the modern discovery of history but closely connected with it in its impulses are the seventeenth-century attempts to formulate new political philosophies or, rather, to invent the means and instruments with which to "make an artificial animal . . . called a Commonwealth, or State."[63] With Hobbes as with Descartes "the prime mover was doubt,"[64] and the chosen method to establish the "art of man," by which he would make and rule his own world as "God hath made and governs the world" by the art of nature, is also introspection, "to read in himself," since this reading will show him "the similitude of the thoughts and passions of one man to the thoughts and passions of another." Here, too, the rules and standards by which to build and judge this most human of human "works of art"[65] do not lie outside of men, are not something men have in common in a worldly reality perceived by the senses or by the mind. They are, rather, inclosed in the inwardness of man, open only to introspection, so that their very validity rests on the assumption that "not . . . the objects of the passions" but the passions themselves are the same in every specimen of the species man-kind Here again we find the image of the watch, this time applied to the human body and then used for the movements of the passions. The establishment of the Commonwealth, the human creation of "an artificial man," amounts to the building of an "automaton [an engine] that moves [itself] by springs and wheels as doth a watch."

In other words, the process which, as we saw, invaded the natural sciences through the experiment, through the attempt to imitate under artificial conditions the process of "making" by which a natural thing came into existence, serves as well or even better as the principle for doing in the realm of human affairs. For here the processes of inner life, found in the passions through introspection, can become the standards and rules for the creation of the "auto-

63. Hobbes's Introduction to the *Leviathan*.

64. See Michael Oakeshott's excellent Introduction to the *Leviathan* (Blackwell's Political Texts), p. xiv.

65. *Ibid.*, p. lxiv.

matic" life of that "artificial man" who is "the great Leviathan." The results yielded by introspection, the only method likely to deliver certain knowledge, are in the nature of movements: only the objects of the senses remain as they are and endure, precede and survive, the act of sensation; only the objects of the passions are permanent and fixed to the extent that they are not devoured by the attainment of some passionate desire; only the objects of thoughts, but never thinking itself, are beyond motion and perishability. Processes, therefore, and not ideas, the models and shapes of the things to be, become the guide for the making and fabricating activities of *homo faber* in the modern age.

Hobbes's attempt to introduce the new concepts of making and reckoning into political philosophy—or, rather, his attempt to apply the newly discovered aptitudes of making to the realm of human affairs—was of the greatest importance; modern rationalism as it is currently known, with the assumed antagonism of reason and passion as its stock-in-trade, has never found a clearer and more uncompromising representative. Yet it was precisely the realm of human affairs where the new philosophy was first found wanting, because by its very nature it could not understand or even believe in reality. The idea that only what I am going to make will be real—perfectly true and legitimate in the realm of fabrication—is forever defeated by the actual course of events, where nothing happens more frequently than the totally unexpected. To act in the form of making, to reason in the form of "reckoning with consequences," means to leave out the unexpected, the event itself, since it would be unreasonable or irrational to expect what is no more than an "infinite improbability." Since, however, the event constitutes the very texture of reality within the realm of human affairs, where the "wholly improbable happens regularly," it is highly unrealistic not to reckon with it, that is, not to reckon with something with which nobody can safely reckon. The political philosophy of the modern age, whose greatest representative is still Hobbes, founders on the perplexity that modern rationalism is unreal and modern realism is irrational—which is only another way of saying that reality and human reason have parted company. Hegel's gigantic enterprise to reconcile spirit with reality (*den Geist mit der Wirklichkeit zu versöhnen*), a reconciliation that is the deepest concern of all modern theories of history, rested on the insight that modern reason foundered on the rock of reality.

The fact that modern world alienation was radical enough to extend even to the most worldly of human activities, to work and reification, the making of things and the building of a world, distinguishes modern attitudes and evaluations even more sharply from those of tradition than a mere reversal of contemplation and action, of thinking and doing, would indicate. The break with contemplation was consummated not with the elevation of man the maker to the position formerly held by man the contemplator, but with the introduction of the concept of process into making. Compared with this, the striking new arrangement of hierarchical order within the *vita activa*, where fabrication now came to occupy a rank formerly held by political action, is of minor importance. We saw before that this hierarchy had in fact, though not expressly, already been overruled in the very beginnings of political philosophy by the philosophers' deep-rooted suspicion of politics in general and action in particular.

The matter is somewhat confused because Greek political philosophy still follows the order laid down by the *polis* even when it turns against it; but in their strictly philosophical writings (to which, of course, one must turn if he wants to know their innermost thoughts), Plato as well as Aristotle tends to invert the relationship between work and action in favor of work. Thus Aristotle, in a discussion of the different kinds of cognition in his *Metaphysics*, places *dianoia* and *epistēmē praktikē*, practical insight and political science, at the lowest rank of his order, and puts above them the science of fabrication, *epistēmē poiētikē*, which immediately precedes and leads to *theōria*, the contemplation of truth.[66] And the reason for this predilection in philosophy is by no means the politically inspired suspicion of action which we mentioned before, but the philosophically much more compelling one that contemplation and fabrication (*theōria* and *poiēsis*) have an inner affinity and do not stand in the same unequivocal opposition to each other as contemplation and action. The decisive point of similarity, at least in Greek philosophy, was that contemplation, the beholding of something, was considered to be an inherent ele-

66. *Metaphysics* 1025b25 ff., 1064a17 ff.

ment in fabrication as well, inasmuch as the work of the craftsman was guided by the "idea," the model beheld by him before the fabrication process had started as well as after it had ended, first to tell him what to make and then to enable him to judge the finished product.

Historically, the source of this contemplation, which we find for the first time described in the Socratic school, is at least twofold. On one hand, it stands in obvious and consistent connection with the famous contention of Plato, quoted by Aristotle, that *thaumazein*, the shocked wonder at the miracle of Being, is the beginning of all philosophy.[67] It seems to me highly probable that this Platonic contention is the immediate result of an experience, perhaps the most striking one, that Socrates offered his disciples: the sight of him time and again suddenly overcome by his thoughts and thrown into a state of absorption to the point of perfect motionlessness for many hours. It seems no less plausible that this shocked wonder should be essentially speechless, that is, that its actual content should be untranslatable into words. This, at least, would explain why Plato and Aristotle, who held *thaumazein* to be the beginning of philosophy, should also agree—despite so many and such decisive disagreements—that some state of speechlessness, the essentially speechless state of contemplation, was the end of philosophy. *Theōria*, in fact, is only another word for *thaumazein*; the contemplation of truth at which the philosopher ultimately arrives is the philosophically purified speechless wonder with which he began.

There is, however, another side to this matter, which shows itself most articulately in Plato's doctrine of ideas, in its content as well as in its terminology and exemplifications. These reside in the experiences of the craftsman, who sees before his inner eye the shape of the model according to which he fabricates his object. To

67. For Plato see *Theaetetus* 155: *Mala gar philosophou touto to pathos, to thaumazein; ou gar allē archē philosophias ē hautē* ("For wonder is what the philosopher endures most; for there is no other beginning of philosophy than this"). Aristotle, who at the beginning of the *Metaphysics* (982b12 ff.) seems to repeat Plato almost verbatim—"For it is owing to their wonder that men both now begin and at first began to philosophize"—actually uses this wonder in an altogether different way; to him, the actual impulse to philosophize lies in the desire "to escape ignorance."

Plato, this model, which craftsmanship can only imitate but not create, is no product of the human mind but given to it. As such it possesses a degree of permanence and excellence which is not actualized but on the contrary spoiled in its materialization through the work of human hands. Work makes perishable and spoils the excellence of what remained eternal so long as it was the object of mere contemplation. Therefore, the proper attitude toward the models which guide work and fabrication, that is, toward Platonic ideas, is to leave them as they are and appear to the inner eye of the mind. If man only renounces his capacity for work and does not do anything, he can behold them and thus participate in their eternity. Contemplation, in this respect, is quite unlike the enraptured state of wonder with which man responds to the miracle of Being as a whole. It is and remains part and parcel of a fabrication process even though it has divorced itself from all work and all doing; in it, the beholding of the model, which now no longer is to guide any doing, is prolonged and enjoyed for its own sake.

In the tradition of philosophy, it is this second kind of contemplation that became the predominant one. Therefore the motionlessness which in the state of speechless wonder is no more than an incidental, unintended result of absorption, becomes now the condition and hence the outstanding characteristic of the *vita contemplativa*. It is not wonder that overcomes and throws man into motionlessness, but it is through the conscious cessation of activity, the activity of making, that the contemplative state is reached. If one reads medieval sources on the joys and delights of contemplation, it is as though the philosophers wanted to make sure that *homo faber* would heed the call and let his arms drop, finally realizing that his greatest desire, the desire for permanence and immortality, cannot be fulfilled by his doings, but only when he realizes that the beautiful and eternal cannot be made. In Plato's philosophy, speechless wonder, the beginning and the end of philosophy, together with the philosopher's love for the eternal and the craftsman's desire for permanence and immortality, permeate each other until they are almost indistinguishable. Yet the very fact that the philosophers' speechless wonder seemed to be an experience reserved for the few, while the craftsmen's contemplative glance was known by many, weighed heavily in favor of a contemplation primarily derived from the experiences of *homo faber*. It already weighed heavily with Plato, who drew his examples from the realm of making because they were closer to a more general human experience, and it weighed even more heavily where some kind of contemplation and meditation was required of everybody, as in medieval Christianity.

Thus it was not primarily the philosopher and philosophic speechless wonder that molded the concept and practice of contemplation and the *vita contemplativa*, but rather *homo faber* in disguise; it was man the maker and fabricator, whose job it is to do violence to nature in order to build a permanent home for himself, and who now was persuaded to renounce violence together with all activity, to leave things as they are, and to find his home in the contemplative dwelling in the neighborhood of the imperishable and eternal. *Homo faber* could be persuaded to this change of attitude because he knew contemplation and some of its delights from his own experience; he did not need a complete change of heart, a true *periagōgē*, a radical turnabout. All he had to do was let his arms drop and prolong indefinitely the act of beholding the *eidos*, the eternal shape and model he had formerly wanted to imitate and whose excellence and beauty he now knew he could only spoil through any attempt at reification.

If, therefore, the modern challenge to the priority of contemplation over every kind of activity had done no more than turn upside down the established order between making and beholding, it would still have remained in the traditional framework. This framework was forced wide open, however, when in the understanding of fabrication itself the emphasis shifted entirely away from the product and from the permanent, guiding model to the fabrication process, away from the question of what a thing is and what kind of thing was to be produced to the question of how and through which means and processes it had come into being and could be reproduced. For this implied both that contemplation was no longer believed to yield truth and that it had lost its position in the *vita activa* itself and hence within the range of ordinary human experience.

43

THE DEFEAT OF *Homo Faber* AND THE PRINCIPLE OF HAPPINESS

If one considers only the events that led into the modern age and reflects solely upon the immediate consequences of Galileo's discovery, which must have struck the great minds of the seventeenth century with the compelling force of self-evident truth, the reversal of contemplation and fabrication, or rather the elimination of contemplation from the range of meaningful human capacities, is almost a matter of course. It seems equally plausible that this reversal should have elevated *homo faber*, the maker and fabricator, rather than man the actor or man as *animal laborans*, to the highest range of human possibilities.

And, indeed, among the outstanding characteristics of the modern age from its beginning to our own time we find the typical attitudes of *homo faber*: his instrumentalization of the world, his confidence in tools and in the productivity of the maker of artificial objects; his trust in the all-comprehensive range of the means-end category, his conviction that every issue can be solved and every human motivation reduced to the principle of utility; his sovereignty, which regards everything given as material and thinks of the whole of nature as of "an immense fabric from which we can cut out whatever we want to resew it however we like";[68] his equation of intelligence with ingenuity, that is, his contempt for all

68. Henri Bergson, *Évolution créatrice* (1948), p. 157. An analysis of Bergson's position in modern philosophy would lead us too far afield. But his insistence on the priority of *homo faber* over *homo sapiens* and on fabrication as the source of human intelligence, as well as his emphatic opposition of life to intelligence, is very suggestive. Bergson's philosophy could easily be read like a case study of how the modern age's earlier conviction of the relative superiority of making over thinking was then superseded and annihilated by its more recent conviction of an absolute superiority of life over everything else. It is because Bergson himself still united both of these elements that he could exert such a decisive influence on the beginnings of labor theories in France. Not only the earlier works of Édouard Berth and Georges Sorel, but also Adriano Tilgher's *Homo faber* (1929), owe their terminology chiefly to Bergson; this is still true of Jules Vuillemin's *L'Être et le travail* (1949), although Vuillemin, like almost every present-day French writer, thinks primarily in Hegelian terms.

thought which cannot be considered to be "the first step . . . for the fabrication of artificial objects, particularly of tools to make tools, and to vary their fabrication indefinitely";[69] finally, his matter-of-course identification of fabrication with action.

It would lead us too far afield to follow the ramifications of this mentality, and it is not necessary, for they are easily detected in the natural sciences, where the purely theoretical effort is understood to spring from the desire to create order out of "mere disorder," the "wild variety of nature,"[70] and where therefore *homo faber*'s predilection for patterns for things to be produced replaces the older notions of harmony and simplicity. It can be found in classical economics, whose highest standard is productivity and whose prejudice against non-productive activities is so strong that even Marx could justify his plea for justice for laborers only by misrepresenting the laboring, non-productive activity in terms of work and fabrication. It is most articulate, of course, in the pragmatic trends of modern philosophy, which are not only characterized by Cartesian world alienation but also by the unanimity with which English philosophy from the seventeenth century onward and French philosophy in the eighteenth century adopted the principle of utility as the key which would open all doors to the explanation of human motivation and behavior. Generally speaking, the oldest conviction of *homo faber*—that "man is the measure of all things"—advanced to the rank of a universally accepted commonplace.

What needs explanation is not the modern esteem of *homo faber* but the fact that this esteem was so quickly followed by the elevation of laboring to the highest position in the hierarchical order of the *vita activa*. This second reversal of hierarchy within the *vita activa* came about more gradually and less dramatically than either the reversal of contemplation and action in general or the reversal of action and fabrication in particular. The elevation of laboring was preceded by certain deviations and variations from the traditional mentality of *homo faber* which were highly characteristic of the modern age and which, indeed, arose almost automatically from the very nature of the events that ushered it in. What changed

69. Bergson, *op. cit.*, p. 140.
70. Bronowski, *op. cit.*

the mentality of *homo faber* was the central position of the concept of process in modernity. As far as *homo faber* was concerned, the modern shift of emphasis from the "what" to the "how," from the thing itself to its fabrication process, was by no means an unmixed blessing. It deprived man as maker and builder of those fixed and permanent standards and measurements which, prior to the modern age, have always served him as guides for his doing and criteria for his judgment. It is not only and perhaps not even primarily the development of commercial society that, with the triumphal victory of exchange value over use value, first introduced the principle of interchangeability, then the relativization, and finally the devaluation, of all values. For the mentality of modern man, as it was determined by the development of modern science and the concomitant unfolding of modern philosophy, it was at least as decisive that man began to consider himself part and parcel of the two superhuman, all-encompassing processes of nature and history, both of which seemed doomed to an infinite progress without ever reaching any inherent *telos* or approaching any preordained idea.

Homo faber, in other words, as he arose from the great revolution of modernity, though he was to acquire an undreamed-of ingenuity in devising instruments to measure the infinitely large and the infinitely small, was deprived of those permanent measures that precede and outlast the fabrication process and form an authentic and reliable absolute with respect to the fabricating activity. Certainly, none of the activities of the *vita activa* stood to lose as much through the elimination of contemplation from the range of meaningful human capacities as fabrication. For unlike action, which partly consists in the unchaining of processes, and unlike laboring, which follows closely the metabolic process of biological life, fabrication experiences processes, if it is aware of them at all, as mere means toward an end, that is, as something secondary and derivative. No other capacity, moreover, stood to lose as much through modern world alienation and the elevation of introspection into an omnipotent device to conquer nature as those faculties which are primarily directed toward the building of the world and the production of worldly things.

Nothing perhaps indicates clearer the ultimate failure of *homo faber* to assert himself than the rapidity with which the principle of utility, the very quintessence of his world view, was found wanting and was superseded by the principle of "the greatest happiness of the greatest number."[71] When this happened it was manifest that the conviction of the age that man can know only what he makes himself—which seemingly was so eminently propitious to a full victory of *homo faber*—would be overruled and eventually destroyed by the even more modern principle of process, whose concepts and categories are altogether alien to the needs and ideals of *homo faber*. For the principle of utility, though its point of reference is clearly man, who uses matter to produce things, still presupposes a world of use objects by which man is surrounded and in which he moves. If this relationship between man and world is no longer secure, if worldly things are no longer primarily considered in their usefulness but as more or less incidental results of the production process which brought them into being, so that the end product of the production process is no longer a true end and the produced thing is valued not for the sake of its predetermined usage but "for its production of something else," then, obviously, the objection can be "raised that . . . its value is secondary only, and a world that contains no primary values can contain no secondary ones either."[72] This radical loss of values within the restricted

71. Jeremy Bentham's formula in *An Introduction to the Principles of Morals and Legislation* (1789) was "suggested to him by Joseph Priestley and closely resembled Beccaria's *la massima felicità divisa nel maggior numero*" (Introduction to the Hafner edition by Laurence J. Lafleur). According to Élie Halévy (*The Growth of Philosophic Radicalism* [Beacon Press, 1955]), both Beccaria and Bentham were indebted to Helvétius' *De l'esprit*.

72. Lafleur, *op. cit.*, p. xi. Bentham himself expresses his dissatisfaction with a merely utilitarian philosophy in the note added to a late edition of his work (Hafner ed., p. 1): "The word *utility* does not so clearly point to the ideas of *pleasure* and *pain* as the words *happiness* and *felicity* do." His chief objection is that utility is not measurable and therefore does not "lead us to the consideration of the *number*," without which a "formation of the standard of right and wrong" would not be possible. Bentham derives his happiness principle from the utility principle by divorcing the concept of utility from the notion of usage (see ch. 1, par. 3). This separation marks a turning point in the history of utilitarianism. For while it is true that the utility principle had been related primarily to the ego prior to Bentham, it is only Bentham who radically emptied the idea of utility of all reference to an independent world of use things and thus transformed utilitarianism into a truly "universalized egoism" (Halévy).

frame of reference of *homo faber* himself occurs almost automatically as soon as he defines himself not as the maker of objects and the builder of the human artifice who incidentally invents tools, but considers himself primarily a toolmaker and "particularly [a maker] of tools to make tools" who only incidentally also produces things. If one applies the principle of utility in this context at all, then it refers primarily not to use objects and not to usage but to the production process. Now what helps stimulate productivity and lessens pain and effort is useful. In other words, the ultimate standard of measurement is not utility and usage at all, but "happiness," that is, the amount of pain and pleasure experienced in the production or in the consumption of things.

Bentham's invention of the "pain and pleasure calculus" combined the advantage of seemingly introducing the mathematical method into the moral sciences with the even greater attraction of having found a principle which resided entirely on introspection. His "happiness," the sum total of pleasures minus pains, is as much an inner sense which senses sensations and remains unrelated to worldly objects as the Cartesian consciousness that is conscious of its own activity. Moreover, Bentham's basic assumption that what all men have in common is not the world but the sameness of their own nature, which manifests itself in the sameness of calculation and the sameness of being affected by pain and pleasure, is directly derived from the earlier philosophers of the modern age. For this philosophy, "hedonism" is even more of a misnomer than for the epicureanism of late antiquity, to which modern hedonism is only superficially related. The principle of all hedonism, as we saw before, is not pleasure but avoidance of pain, and Hume, who in contradistinction to Bentham was still a philosopher, knew quite well that he who wants to make pleasure the ultimate end of all human action is driven to admit that not pleasure but pain, not desire but fear, are his true guides. "If you . . . inquire, why [somebody] desires health, he will readily reply, because sickness is painful. If you push your inquiries further and desire a reason why he hates pain, it is impossible he can ever give any. This is an ultimate end, and is never referred to by any other object."[73] The reason for this impossibility is that only pain is completely inde-

73. Quoted from Halévy, *op. cit.*, p. 13.

pendent of any object, that only one who is in pain really senses nothing but himself; pleasure does not enjoy itself but something besides itself. Pain is the only inner sense found by introspection which can rival in independence from experienced objects the self-evident certainty of logical and arithmetical reasoning.

While this ultimate foundation of hedonism in the experience of pain is true for both its ancient and modern varieties, in the modern age it acquires an altogether different and much stronger emphasis. For here it is by no means the world, as in antiquity, that drives man into himself to escape the pains it may inflict, under which circumstance both pain and pleasure still retain a good deal of their worldly significance. Ancient world alienation in all its varieties—from stoicism to epicureanism down to hedonism and cynicism—had been inspired by a deep mistrust of the world and moved by a vehement impulse to withdraw from worldly involvement, from the trouble and pain it inflicts, into the security of an inward realm in which the self is exposed to nothing but itself. Their moderr counterparts—puritanism, sensualism, and Bentham's hedonism—on the contrary, were inspired by an equally deep mistrust of man as such; they were moved by doubt of the adequacy of the human senses to receive reality, the adequacy of human reason to receive truth, and hence by the conviction of the deficiency or even depravity of human nature.

This depravity is not Christian or biblical either in origin or in content, although it was of course interpreted in terms of original sin, and it is difficult to say whether it is more harmful and repulsive when puritans denounce man's corruptness or when Bentham-ites brazenly hail as virtues what men always have known to be vices. While the ancients had relied upon imagination and memory, the imagination of pains from which they were free or the memory of past pleasures in situations of acute painfulness, to convince themselves of their happiness, the moderns needed the calculus of pleasure or the puritan moral bookkeeping of merits and transgressions to arrive at some illusory mathematical certainty of happiness or salvation. (These moral arithmetics are, of course, quite alien to the spirit pervading the philosophic schools of late antiquity. Moreover, one need only reflect on the rigidity of self-imposed discipline and the concomitant nobility of character, so manifest in those who had been formed by ancient stoicism or epicureanism, to become aware of the gulf by which these versions of hedonism are separated from modern puritanism, sensualism, and hedonism. For this difference, it is almost irrelevant whether the modern character is still formed by the older narrow-minded, fanatic self-righteousness or has yielded to the more recent self-centered and self-indulgent egotism with its infinite variety of futile miseries.) It seems more than doubtful that the "greatest happiness principle" would have achieved its intellectual triumphs in the English-speaking world if no more had been involved than the questionable discovery that "nature has placed mankind under the governance of two sovereign masters, pain and pleasure,"[74] or the absurd idea of establishing morals as an exact science by isolating "in the human soul that feeling which seems to be the most easily measurable."[75]

Hidden behind this as behind other, less interesting variations of the sacredness of egoism and the all-pervasive power of self-interest, which were current to the point of being commonplace in the eighteenth and early nineteenth centuries, we find another point of reference which indeed forms a much more potent principle than any pain-pleasure calculus could ever offer, and that is the principle of life itself. What pain and pleasure, fear and desire, are actually supposed to achieve in all these systems is not happiness at all but the promotion of individual life or a guaranty of the survival of mankind. If modern egoism were the ruthless search for pleasure (called happiness) it pretends to be, it would not lack what in all truly hedonistic systems is an indispensable element of argumentation—a radical justification of suicide. This lack alone indicates that in fact we deal here with life philosophy in its most vulgar and least critical form. In the last resort, it is always life itself which is the supreme standard to which everything else is referred, and the

74. This, of course, is the first sentence of the *Principles of Morals and Legislation*. The famous sentence is "copied almost word for word from Helvétius" (Halévy, *op. cit.*, p. 26). Halévy rightly remarks that "it was natural that a current idea should on all sides rather tend to find expression in the same formulae" (p. 22). This fact, incidentally, clearly shows that the authors we deal with here are not philosophers; for no matter how current certain ideas might be during a given period, there never are two philosophers who could arrive at identical formulations without copying from each other.

75. *Ibid.*, p. 15.

interests of the individual as well as the interests of mankind are always equated with individual life or the life of the species as though it were a matter of course that life is the highest good.

The curious failure of *homo faber* to assert himself under conditions seemingly so extraordinarily propitious could also have been illustrated by another, philosophically even more relevant, revision of basic traditional beliefs. Hume's radical criticism of the causality principle, which prepared the way for the later adoption of the principle of evolution, has often been considered one of the origins of modern philosophy. The causality principle with its twofold central axiom—that everything that is must have a cause (*nihil sine causa*) and that the cause must be more perfect than its most perfect effect—obviously relies entirely on experiences in the realm of fabrication, where the maker is superior to his products. Seen in this context, the turning point in the intellectual history of the modern age came when the image of organic life development—where the evolution of a lower being, for instance the ape, can cause the appearance of a higher being, for instance man—appeared in the place of the image of the watchmaker who must be superior to all watches whose cause he is.

Much more is implied in this change than the mere denial of the lifeless rigidity of a mechanistic world view. It is as though in the latent seventeenth-century conflict between the two possible methods to be derived from the Galilean discovery, the method of the experiment and of making on one hand and the method of introspection on the other, the latter was to achieve a somewhat belated victory. For the only tangible object introspection yields, if it is to yield more than an entirely empty consciousness of itself, is indeed the biological process. And since this biological life, accessible in self-observation, is at the same time a metabolic process between man and nature, it is as though introspection no longer needs to get lost in the ramifications of a consciousness without reality, but has found within man—not in his mind but in his bodily processes— enough outside matter to connect him again with the outer world. The split between subject and object, inherent in human consciousness and irremediable in the Cartesian opposition of man as a *res cogitans* to a surrounding world of *res extensae*, disappears altogether in the case of a living organism, whose very survival depends upon the incorporation, the consumption, of outside matter. Naturalism, the nineteenth-century version of materialism, seemed to find in life the way to solve the problems of Cartesian philosophy and at the same time to bridge the ever-widening chasm between philosophy and science.[76]

44

LIFE AS THE HIGHEST GOOD

Tempting as it may be for the sake of sheer consistency to derive the modern life concept from the self-inflicted perplexities of modern philosophy, it would be a delusion and a grave injustice to the seriousness of the problems of the modern age if one looked upon them merely from the viewpoint of the development of ideas. The defeat of *homo faber* may be explainable in terms of the initial transformation of physics into astrophysics, of natural sciences into a "universal" science. What still remains to be explained is why this defeat ended with a victory of the *animal laborans;* why, with the rise of the *vita activa*, it was precisely the laboring activity that was to be elevated to the highest rank of man's capacities or, to put it another way, why within the diversity of the human condition with its various human capacities it was precisely life that overruled all other considerations.

The reason why life asserted itself as the ultimate point of reference in the modern age and has remained the highest good of mod-

76. The greatest representatives of modern life philosophy are Marx, Nietzsche, and Bergson, inasmuch as all three equate Life and Being. For this equation, they rely on introspection, and life is indeed the only "being" man can possibly be aware of by looking merely into himself. The difference between these and the earlier philosophers of the modern age is that life appears to be more active and more productive than consciousness, which seems to be still too closely related to contemplation and the old ideal of truth. This last stage of modern philosophy is perhaps best described as the rebellion of the philosophers against philosophy, a rebellion which, beginning with Kierkegaard and ending in existentialism, appears at first glance to emphasize action as against contemplation. Upon closer inspection, however, none of these philosophers is actually concerned with action as such. We may leave aside here Kierkegaard and his non-worldly, inward-directed acting. Nietzsche and Bergson describe action in terms of fabrication—*homo faber* instead of *homo sapiens*—just as Marx thinks of acting in terms of making and describes labor in terms of work. But their ultimate point of reference is not work and worldliness any more than action; it is life and life's fertility.

ern society is that the modern reversal operated within the fabric of a Christian society whose fundamental belief in the sacredness of life has survived, and has even remained completely unshaken by, secularization and the general decline of the Christian faith. In other words, the modern reversal followed and left unchallenged the most important reversal with which Christianity had broken into the ancient world, a reversal that was politically even more far-reaching and, historically at any rate, more enduring than any specific dogmatic content or belief. For the Christian "glad tidings" of the immortality of individual human life had reversed the ancient relationship between man and world and promoted the most mortal thing, human life, to the position of immortality, which up to then the cosmos had held.

Historically, it is more than probable that the victory of the Christian faith in the ancient world was largely due to this reversal, which brought hope to those who knew that their world was doomed, indeed a hope beyond hope, since the new message promised an immortality they never had dared to hope for. This reversal could not but be disastrous for the esteem and the dignity of politics. Political activity, which up to then had derived its greatest inspiration from the aspiration toward worldly immortality, now sank to the low level of an activity subject to necessity, destined to remedy the consequences of human sinfulness on one hand and to cater to the legitimate wants and interests of earthly life on the other. Aspiration toward immortality could now only be equated with vainglory; such fame as the world could bestow upon man was an illusion, since the world was even more perishable than man, and a striving for worldly immortality was meaningless, since life itself was immortal.

It is precisely individual life which now came to occupy the position once held by the "life" of the body politic, and Paul's statement that "death is the wages of sin," since life is meant to last forever, echoes Cicero's statement that death is the reward of sins committed by political communities which were built to last for eternity.[77] It is as though the early Christians—at least Paul,

77. Cicero's remark: *Civitatibus autem mors ipsa poena est ... debet enim constituta sic esse civitas ut aeterna sit* (*De re publica* iii. 23). For the conviction in antiquity that a well-founded body politic should be immortal, see also Plato, *Laws* 713,

who after all was a Roman citizen—consciously shaped their concept of immortality after the Roman model, substituting individual life for the political life of the body politic. Just as the body politic possesses only a potential immortality which can be forfeited by political transgressions, individual life had once forfeited its guaranteed immortality in Adam's fall and now, through Christ, had regained a new, potentially everlasting life which, however, could again be lost in a second death through individual sin.

Certainly, Christian emphasis on the sacredness of life is part and parcel of the Hebrew heritage, which already presented a striking contrast to the attitudes of antiquity: the pagan contempt for the hardships which life imposes upon man in labor and giving birth, the envious picture of the "easy life" of the gods, the custom of exposing unwanted offspring, the conviction that life without health is not worth living (so that the physician, for instance, is held to have misunderstood his calling when he prolongs life where he cannot restore health)[78] and that suicide is a noble gesture to escape a life that has become burdensome. Still, one need only remember how the Decalogue enumerates the offense of murder, without any special emphasis, among a number of other transgressions—which to our way of thinking can hardly compete in gravity with this supreme crime—to realize that not even the Hebrew legal code, though much closer to our own than any pagan scale of offenses, made the preservation of life the cornerstone of the legal system of the Jewish people. This intermediary position which the Hebrew legal code occupies between pagan antiquity and all Christian or post-Christian legal systems may be explicable by the Hebrew creed which stresses the potential immortality of the people, as distinguished from the pagan immortality of the world on one side and the Christian immortality of individual life on the other. At any event, this Christian immortality that is bestowed upon the person, who in his uniqueness begins life by birth on earth, resulted not only in the more obvious increase of otherworldliness, but also in an enormously increased importance of life

---

where the founders of a new *polis* are told to imitate the immortal part in man (*hoson en hēmin athanasias enest*).

78. See Plato *Republic* 405C.

on earth. The point is that Christianity—except for heretical and gnostic speculations—always insisted that life, though it had no longer a final end, still has a definite beginning. Life on earth may be only the first and the most miserable stage of eternal life; it still is life, and without this life that will be terminated in death, there cannot be eternal life. This may be the reason for the undisputable fact that only when the immortality of individual life became the central creed of Western mankind, that is, only with the rise of Christianity, did life on earth also become the highest good of man.

Christian emphasis on the sacredness of life tended to level out the ancient distinctions and articulations within the *vita activa;* it tended to view labor, work, and action as equally subject to the necessity of present life. At the same time it helped to free the laboring activity, that is, whatever is necessary to sustain the biological process itself, from some of the contempt in which antiquity had held it. The old contempt toward the slave, who had been despised because he served only life's necessities and submitted to the compulsion of his master because he wanted to stay alive at all costs, could not possibly survive in the Christian era. One could no longer with Plato despise the slave for not having committed suicide rather than submit to a master, for to stay alive under all circumstances had become a holy duty, and suicide was regarded as worse than murder. Not the murderer, but he who had put an end to his own life was refused a Christian burial.

Yet contrary to what some modern interpreters have tried to read into Christian sources, there are no indications of the modern glorification of laboring in the New Testament or in other premodern Christian writers. Paul, who has been called "the apostle of labor,"[79] was nothing of the sort, and the few passages on which

79. By the Dominican Bernard Allo, *Le travail d'après St. Paul* (1914). Among the defenders of the Christian origin of modern glorification of labor are: in France, Étienne Borne and François Henry, *Le travail et l'homme* (1937); in Germany, Karl Müller, *Die Arbeit: Nach moral-philosophischen Grundsätzen des heiligen Thomas von Aquino* (1912). More recently, Jacques Leclercq from Louvain, who has contributed one of the most valuable and interesting works to the philosophy of labor in the fourth book of his *Leçons de droit naturel*, entitled *Travail, propriété* (1946), has rectified this misinterpretation of the Christian sources: "Le christianisme n'a pas changé grand'chose à l'estime du travail";

this claim is based either are addressed to those who out of laziness "ate other men's bread" or they recommend labor as a good means to keep out of trouble, that is, they reinforce the general prescription of a strictly private life and warn of political activities.[80] It is even more relevant that in later Christian philosophy, and particularly in Thomas Aquinas, labor had become a duty for those who had no other means to keep alive, the duty consisting in keeping one's self alive and not in laboring; if one could provide for himself through beggary, so much the better. Whoever reads the sources without modern prolabor prejudices will be surprised at how little the church fathers availed themselves even of the obvious opportunity to justify labor as punishment for original sin. Thus Thomas does not hesitate to follow Aristotle rather than the Bible in this question and to assert that "only the necessity to keep alive compels to do manual labor."[81] Labor to him is nature's way of keeping the human species alive, and from this he concludes that it is by no means necessary that all men earn their bread by the sweat of their brows, but that this is rather a kind of last and desperate resort to solve the problem or fulfil the duty.[82] Not even the use of labor as a means with which to ward off the dangers of otiosity is a new Christian discovery, but was already a commonplace of Roman morality. In complete agreement with ancient convictions about the character of the laboring activity, finally, is the frequent Christian use for the mortification of the flesh, where labor, especially in the monasteries, sometimes played the same role as other painful exercises and forms of self-torture.[83]

and in Aquinas' work "la notion du travail n'apparaît que fort accidentellement" (pp. 61–62).

80. See I Thess. 4:9–12 and II Thess. 3:8–12.

81. *Summa contra Gentiles* iii. 135: *Sola enim necessitas victus cogit manibus operari.*

82. *Summa theologica* ii. 2. 187. 3, 5.

83. In the monastic rules, particularly in the *ora et labora* of Benedict, labor is recommended against the temptations of an idle body (see ch. 48 of the rule). In the so-called rule of Augustine (*Epistolae* 211), labor is considered to be a law of nature, not a punishment for sin. Augustine recommends manual labor—he uses the words *opera* and *labor* synonymously as the opposite of *otium*—for three

The Human Condition

The reason why Christianity, its insistence on the sacredness of life and on the duty to stay alive notwithstanding, never developed a positive labor philosophy lies in the unquestioned priority given to the *vita contemplativa* over all kinds of human activities. *Vita contemplativa simpliciter melior est quam vita activa* ("the life of contemplation is simply better than the life of action"), and whatever the merits of an active life might be, those of a life devoted to contemplation are "more effective and more powerful."[84] This conviction, it is true, can hardly be found in the preachings of Jesus of Nazareth, and it is certainly due to the influence of Greek philosophy; yet even if medieval philosophy had kept closer to the spirit of the Gospels, it could hardly have found there any reason for a glorification of laboring.[85] The only activity Jesus of Nazareth recommends in his preachings is action, and the only human capacity he stresses is the capacity "to perform miracles."

However that may be, the modern age continued to operate under the assumption that life, and not the world, is the highest good of man; in its boldest and most radical revisions and criticisms of traditional beliefs and concepts, it never even thought of challenging this fundamental reversal which Christianity had brought into

reasons: it helps to fight the temptations of otiosity; it helps the monasteries to fulfil their duty of charity toward the poor; and it is favorable to contemplation because it does not engage the mind unduly like other occupations, for instance, the buying and selling of goods. For the role of labor in the monasteries, compare Étienne Delaruelle, "Le travail dans les règles monastiques occidentales du 4e au 9e siècle," *Journal de psychologie normale et pathologique*, Vol. XLI, No. 1 (1948). Apart from these formal considerations, it is quite characteristic that the Solitaires de Port-Royal, looking for some instrument of really effective punishment, thought immediately of labor (see Lucien Fèbre, "Travail: Évolution d'un mot et d'une idée," *Journal de psychologie normale et pathologique*, Vol. XLI, No. 1 [1948]).

84. Aquinas *Summa theologica* ii. 2. 182. 1, 2. In his insistence on the absolute superiority of the *vita contemplativa*, Thomas shows a characteristic difference from Augustine, who recommends the *inquisitio, aut inventio veritatis: ut in ea quisque proficiat*—"inquisition or discovery of truth so that somebody may profit from it" (*De civitate Dei* xix. 19). But this difference is hardly more than the difference between a Christian thinker formed by Greek, and another by Roman, philosophy.

85. The Gospels are concerned with the evil of earthly possessions, not with the praise of labor or laborers (see esp. Matt. 6:19–32, 19:21–24; Mark 4:19; Luke 6:20–34, 18:22–25; Acts 4:32–35).

the dying ancient world. No matter how articulate and how conscious the thinkers of modernity were in their attacks on tradition, the priority of life over everything else had acquired for them the status of a "self-evident truth," and as such it has survived even in our present world, which has begun already to leave the whole modern age behind and to substitute for a laboring society the society of jobholders. But while it is quite conceivable that the development following upon the discovery of the Archimedean point would have taken an altogether different direction if it had taken place seventeen hundred years earlier, when not life but the world was still the highest good of man, it by no means follows that we still live in a Christian world. For what matters today is not the immortality of life, but that life is the highest good. And while this assumption certainly is Christian in origin, it constitutes no more than an important attending circumstance for the Christian faith. Moreover, even if we disregard the details of Christian dogma and consider only the general mood of Christianity, which resides in the importance of faith, it is obvious that nothing could be more detrimental to this spirit than the spirit of distrust and suspicion of the modern age. Surely, Cartesian doubt has proved its efficiency nowhere more disastrously and irretrievably than in the realm of religious belief, where it was introduced by Pascal and Kierkegaard, the two greatest religious thinkers of modernity. (For what undermined the Christian faith was not the atheism of the eighteenth century or the materialism of the nineteenth—their arguments are frequently vulgar and, for the most part, easily refutable by traditional theology—but rather the doubting concern with salvation of genuinely religious men, in whose eyes the traditional Christian content and promise had become "absurd.")

Just as we do not know what would have happened if the Archimedean point had been discovered before the rise of Christianity, we are in no position to ascertain what the destiny of Christianity would have been if the great awakening of the Renaissance had not been interrupted by this event. Before Galileo, all paths still seemed to be open. If we think back to Leonardo, we may well imagine that a technical revolution would have overtaken the development of humanity in any case. This might well have led to flight, the realization of one of the oldest and most persistent dreams of man, but it hardly would have led into the universe; it might well have brought about the unification of the earth, but it hardly would have brought about the transformation of matter into energy and the adventure into the microscopic universe. The only thing we can be sure of is that the coincidence of the reversal of doing and contemplating with the earlier reversal of life and world became the point of departure for the whole modern development. Only when the *vita activa* had lost its point of reference in the *vita contemplativa* could it become active life in the full sense of the word; and only because this active life remained bound to life as its only point of reference could life as such, the laboring metabolism of man with nature, become active and unfold its entire fertility.

45

THE VICTORY OF THE *Animal Laborans*

The victory of the *animal laborans* would never have been complete had not the process of secularization, the modern loss of faith inevitably arising from Cartesian doubt, deprived individual life of its immortality, or at least of the certainty of immortality. Individual life again became mortal, as mortal as it had been in antiquity, and the world was even less stable, less permanent, and hence less to be relied upon than it had been during the Christian era. Modern man, when he lost the certainty of a world to come, was thrown back upon himself and not upon this world; far from believing that the world might be potentially immortal, he was not even sure that it was real. And in so far as he was to assume that it was real in the uncritical and apparently unbothered optimism of a steadily progressing science, he had removed himself from the earth to a much more distant point than any Christian otherworldliness had ever removed him. Whatever the word "secular" is meant to signify in current usage, historically it cannot possibly be equated with worldliness; modern man at any rate did not gain this world when he lost the other world, and he did not gain life, strictly speaking, either; he was thrust back upon it, thrown into the closed inwardness of introspection, where the highest he could experience were the empty processes of reckoning of the mind, its play with itself. The only contents left were appetites and desires, the senseless urges of his body which he mistook for passion and which he deemed to be "unreasonable" because he found he could not "reason," that is, not reckon with them. The only thing that could now be potentially immortal, as immortal as the body politic in antiquity and as individual life during the Middle Ages, was life itself, that is, the possibly everlasting life process of the species mankind.

We saw before that in the rise of society it was ultimately the life of the species which asserted itself. Theoretically, the turning point from the earlier modern age's insistence on the "egoistic" life of the individual to its later emphasis on "social" life and "socialized man" (Marx) came when Marx transformed the cruder notion of classical economy—that all men, in so far as they act at all, act for reasons of self-interest—into forces of interest which inform, move, and direct the classes of society, and through their conflicts direct society as a whole. Socialized mankind is that state of society where only one interest rules, and the subject of this interest is either classes or man-kind, but neither man nor men. The point is that now even the last trace of action in what men were doing, the motive implied in self-interest, disappeared. What was left was a "natural force," the force of the life process itself, to which all men and all human activities were equally submitted ("the thought process itself is a natural process")[86] and whose only aim, if it had an aim at all, was survival of the animal species man. None of the higher capacities of man was any longer necessary to connect individual life with the life of the species; individual life became part of the life process, and to labor, to assure the continuity of one's own life and the life of his family, was all that was needed. What was not needed, not necessitated by life's metabolism with nature, was either superfluous or could be justified only in terms of a peculiarity of human as distinguished from other animal life—so that Milton was considered to have written his *Paradise Lost* for the same reasons and out of similar urges that compel the silkworm to produce silk.

If we compare the modern world with that of the past, the loss of human experience involved in this development is extraordinarily striking. It is not only and not even primarily contemplation which

86. In a letter Marx wrote to Kugelmann in July, 1868.

has become an entirely meaningless experience. Thought itself, when it became "reckoning with consequences," became a function of the brain, with the result that electronic instruments are found to fulfil these functions much better than we ever could. Action was soon and still is almost exclusively understood in terms of making and fabricating, only that making, because of its worldliness and inherent indifference to life, was now regarded as but another form of laboring, a more complicated but not a more mysterious function of the life process.

Meanwhile, we have proved ingenious enough to find ways to ease the toil and trouble of living to the point where an elimination of laboring from the range of human activities can no longer be regarded as utopian. For even now, laboring is too lofty, too ambitious a word for what we are doing, or think we are doing, in the world we have come to live in. The last stage of the laboring society, the society of jobholders, demands of its members a sheer automatic functioning, as though individual life had actually been submerged in the over-all life process of the species and the only active decision still required of the individual were to let go, so to speak, to abandon his individuality, the still individually sensed pain and trouble of living, and acquiesce in a dazed, "tranquilized," functional type of behavior. The trouble with modern theories of behaviorism is not that they are wrong but that they could become true, that they actually are the best possible conceptualization of certain obvious trends in modern society. It is quite conceivable that the modern age—which began with such an unprecedented and promising outburst of human activity—may end in the deadliest, most sterile passivity history has ever known.

But there are other more serious danger signs that man may be willing and, indeed, is on the point of developing into that animal species from which, since Darwin, he imagines he has come. If, in concluding, we return once more to the discovery of the Archimedean point and apply it, as Kafka warned us not to do, to man himself and to what he is doing on this earth, it at once becomes manifest that all his activities, watched from a sufficiently removed vantage point in the universe, would appear not as activities of any kind but as processes, so that, as a scientist recently put it, modern motorization would appear like a process of biological mutation in which human bodies gradually begin to be covered by shells of steel. For the watcher from the universe, this mutation would be no more or less mysterious than the mutation which now goes on before our eyes in those small living organisms which we fought with antibiotics and which mysteriously have developed new strains to resist us. How deep-rooted this usage of the Archimedean point against ourselves is can be seen in the very metaphors which dominate scientific thought today. The reason why scientists can tell us about the "life" in the atom—where apparently every particle is "free" to behave as it wants and the laws ruling these movements are the same statistical laws which, according to the social scientists, rule human behavior and make the multitude behave as it must, no matter how "free" the individual particle may appear to be in its choices—the reason, in other words, why the behavior of the infinitely small particle is not only similar in pattern to the planetary system as it appears to us but resembles the life and behavior patterns in human society is, of course, that we look and live in this society as though we were as far removed from our own human existence as we are from the infinitely small and the immensely large which, even if they could be perceived by the finest instruments, are too far away from us to be experienced.

Needless to say, this does not mean that modern man has lost his capacities or is on the point of losing them. No matter what sociology, psychology, and anthropology will tell us about the "social animal," men persist in making, fabricating, and building, although these faculties are more and more restricted to the abilities of the artist, so that the concomitant experiences of worldliness escape more and more the range of ordinary human experience.[87]

Similarly, the capacity for action, at least in the sense of the releasing of processes, is still with us, although it has become the exclusive prerogative of the scientists, who have enlarged the

87. This inherent worldliness of the artist is of course not changed if a "non-objective art" replaces the representation of things; to mistake this "non-objectivity" for subjectivity, where the artist feels called upon to "express himself," his subjective feelings, is the mark of charlatans, not of artists. The artist, whether painter or sculptor or poet or musician, produces worldly objects, and his reification has nothing in common with the highly questionable and, at any rate, wholly unartistic practice of expression. Expressionist art, but not abstract art, is a contradiction in terms.

realm of human affairs to the point of extinguishing the time-honored protective dividing line between nature and the human world. In view of such achievements, performed for centuries in the unseen quiet of the laboratories, it seems only proper that their deeds should eventually have turned out to have greater news value, to be of greater political significance, than the administrative and diplomatic doings of most so-called statesmen. It certainly is not without irony that those whom public opinion has persistently held to be the least practical and the least political members of society should have turned out to be the only ones left who still know how to act and how to act in concert. For their early organizations, which they founded in the seventeenth century for the conquest of nature and in which they developed their own moral standards and their own code of honor, have not only survived all vicissitudes of the modern age, but they have become one of the most potent power-generating groups in all history. But the action of the scientists, since it acts into nature from the standpoint of the universe and not into the web of human relationships, lacks the revelatory character of action as well as the ability to produce stories and become historical, which together form the very source from which meaningfulness springs into and illuminates human existence. In this existentially most important aspect, action, too, has become an experience for the privileged few, and these few who still know what it means to act may well be even fewer than the artists, their experience even rarer than the genuine experience of and love for the world.

Thought, finally—which we, following the premodern as well as the modern tradition, omitted from our reconsideration of the *vita activa*—is still possible, and no doubt actual, wherever men live under the conditions of political freedom. Unfortunately, and contrary to what is currently assumed about the proverbial ivory-tower independence of thinkers, no other human capacity is so vulnerable, and it is in fact far easier to act under conditions of tyranny than it is to think. As a living experience, thought has always been assumed, perhaps wrongly, to be known only to the few. It may not be presumptuous to believe that these few have not become fewer in our time. This may be irrelevant, or of restricted relevance, for the future of the world; it is not irrelevant for the future of man. For if no other test but the experience of being active, no other measure but the extent of sheer activity were to be applied to the various activities within the *vita activa*, it might well be that thinking as such would surpass them all. Whoever has any experience in this matter will know how right Cato was when he said: *Numquam se plus agere quam nihil cum ageret, numquam minus solum esse quam cum solus esset*—"Never is he more active than when he does nothing, never is he less alone than when he is by himself."

Acknowledgments

The present study owes its origin to a series of lectures, delivered under the auspices of the Charles R. Walgreen Foundation in April, 1956, at the University of Chicago, under the title, "Vita Activa." In the initial stages of this work, which goes back to the early fifties, I was given a grant by the Simon Guggenheim Memorial Foundation, and in the last stage I was greatly helped by a grant from the Rockefeller Foundation. In the fall of 1953, the Christian Gauss Seminar in Criticism of Princeton University offered me the opportunity to present some of my ideas in a series of lectures under the title "Karl Marx and the Tradition of Political Thought." I am still grateful for the patience and encouragement with which these first attempts were received and for the lively exchange of ideas with writers from here and abroad for which the Seminar, unique in this respect, provides a sounding board.

Rose Feitelson, who has helped me ever since I began to publish in this country, was again of great assistance in the preparation of manuscript and index. If I had to be grateful for what she has done over a period of twelve years, I would be altogether helpless.

HANNAH ARENDT

Publisher's Note:

In preparation of this second edition, we have corrected several minor typographical errors. The observant reader may notice, however, that the Greek letter *chi* has been transliterated as *kh* in some instances and as the more standard *ch* in others. We have chosen to let stand instances of the former, probably a Germanism.

This second edition includes a completely new and expanded index, which replaces that of the first edition.

Index

Abraham, 243–44
absolutes, 270
abstract art, 323n
Achilles, 25, 193–94
action, 175–247; the agent as disclosed in, 175–81; *archein*, 177, 189, 222–23, 224; in Aristotle's *bios politikos*, 12–13, 25; behavior as replacing, 41, 45; and being together, 23; and birth, 178; as boundless, 190–91; capacity for as still with us, 323–24; contemplation opposed to in traditional thought, 14–17, 85; and contemplation reversed in modern age, 289–94; courage as required for, 186; as creating its own remembrance, 207–8; defined, 7; doing and suffering as two sides of same coin, 190; as exclusive prerogative of man, 22–23; fabrication distinguished from, 188, 192; futility of, 173, 197; greatness as criterion of, 205; Greek and Latin verbs for "act," 189; history as a story of, 185; in *homo faber's* redemption, 236; as idleness for *animal laborans* and *homo faber*, 208; interests as concern of, 182; irreversibility of, 233, 236–43; in Jesus' preaching, 318; in life philosophies, 313n; location of human activities, 73–78; as miracle working, 246–47, 247n; natality and mortality as connected with, 8–9; people distinguishing themselves by, 176; Plato as separating from thought, 223–27; plurality as condition of, 7, 8; plurality as source of calamities of, 220; the *polis* as giving permanence to, 198; political realm arising out of acting together, 198; *prattein*, 189, 222–23; process character of, 230–36; products of, 95; and reac-

tion, 190; reification of, 95, 187; as revealing itself fully only to the storyteller, 191–92; revelatory character of, 178–80, 187; society as excluding, 40–41; and speech, 26, 177n.1, 178–81; stories resulting from, 97; strength of, 188–89, 233; as superstructure, 33; as taking initiative, 177; threefold frustration of, 220; traditional substitution of making for, 220–30; understood as fabricating, 322; unpredictability of, 144, 191–92, 233, 237, 243–47; in *vita activa*, 7, 205, 301; in web of relationships, 184; and work in Greek political philosophy, 301–2. *See also* deeds; *vita activa*
admiration, public, 56–57
Agamemnon, 190
agent, the: as disclosed in speech and action, 175–81; stories revealing, 184
agere, 189
aging, 51n.43
agora, the, 160
agriculture: Hesiod on, 83n.8; as liberal art, 91, 91n.24; tilling of the soil, 138
alienation: earth alienation, 264–65; Marxian self-alienation, 89n.21, 162, 210, 254; world alienation, 6, 209, 248–57, 301, 307, 310
alteritas, 176
American Revolution, 228
analytical geometry, 266–67
Anders, Günther, 150n
animal laborans: abundance as ideal of, 126, 134; on action and speech as idleness, 208; as animal, 84; *homo faber* contrasted with, 136; instruments of work and, 144–53; Marx on, 86n.14, 87, 99n.36, 102, 105, 117; in modern age, 85; natural fer-

[*329*]

Index

Index

atom bomb, 149, 150n
atomic revolution, 121, 149
atomic theory of matter, 259n.10, 323
Augean stables, 101
Augustine, St.: on beginnings, 177, 177n.3; on burden of active life, 16n.14; on charity as basis for human relationships, 53; on citizenship in the *polis*, 14; on government, 229n; on human and animal life, 8n; on human nature, 10, 10n; on minding one's own business, 55n; Tertullian contrasted with, 74n.83; *uti* and *frui* distinguished by, 252n; on *vita negotiosa* or *actuosa*, 12
Augustine, rule of, 317n.83
automation: as culmination of modern technological development, 149, 149n.12; danger of, 132; in emancipation from consumption, 131; as liberating mankind from labor, 4; mechanism contrasted with, 151

Babylon, 195n.22
banausoi, 81–82, 82n
Barrow, R. H.: on epitaphs for slaves, 55n; on free labor in Roman Empire, 66n.70; on industrial development in ancient world, 65n.69, 148n.9; on paternal power in Athens, 29n.16; on property owning by slaves, 62n.58; on slaves as shadowy types, 50n; on torture of slaves, 129n.78
beautiful, the, 13, 152, 172–73, 226, 226n.65
beginners, 177, 189–90
beginning: action as, 177–78; *archein*, 177, 189, 222–23, 224; Augustine on, 177, 177n.3; mortality as interrupted by, 246–47
behavioral sciences, 45
behaviorism, 43, 322
being, and appearance, 275, 276–77
Bell, Daniel, 149n.12
Bellarmine, Cardinal, 260, 263
Benedict, rule of, 54n.48, 317n.83
benefactors, 196
Bentham, Jeremy: hedonism of, 309–11; on utility principle, 308nn. 71, 72
Bergson, Henri, 117, 117n, 136n, 172, 305n, 313n
Berth, Édouard, 117n, 305n
biography, 97
birth: and action, 178; in Eleusinian

Mysteries, 63n.61; household as realm of, 62–63; labor, work, and action as connected with, 8–9; and nature, 96–97; pain in, 115, 121; woman's labor in giving, 30
Bodin, Jean, 68, 183n
"Book of Customs," 65n.67
Borne, Étienne, 316n
boundaries, 30, 190–91
Braun, Wernher von, 231n
Brizon, Pierre, 67n.70, 73n.81
Brochard, V., 113nn. 61, 62
Bronowski, J., 264n.19
Bruno, Giordano, 258, 259, 260
Bücher, Karl, 145n
Burckhardt, Jacob, 82n
bureaucracy, 40, 45, 93n.28
Buridan, Jean, 166n
business companies, 35

Caligula, 130n.81
Calvin, John, 251
canonist doctrine, 42n
capital: accumulation of, 255–56; etymology of term, 68n.74; wealth becoming, 68, 255
capitalism: commercial society, 162, 163, 210, 307; conditions for rise of, 255; conspicuous production of manufacturing, 162; innerworldly asceticism of, 251, 252n, 254; labor power exploited in, 88; Marx's criticism of, 89n.22; use value changing to exchange value in, 165; working hours in early, 132
Cartesian doubt, 273–80; abandoning attempt to understand nature as consequence of, 298; *Dieu trompeur*, 277, 281; in natural sciences, 287–89; religion affected by, 319, 320; as response to Galileo, 260–61, 274, 287; universality of, 275
Cassirer, Ernst, 264n.18, 285n.47
Catholic Church: beatifying saints only after death, 192; as corrupting influence for Machiavelli, 77–78; Counter Reformation, 253; and Galileo, 260; on labor, 127n; property expropriated, 66–67, 252; as standing between modernity and the believers, 277; as substitute for citizenship, 34
Cato, 325
causality principle, 312
Cave parable, 20, 75, 226, 226n.66, 292

Index

certainty: doubt as certain, 279–80; introspection as yielding, 280, 300; loss of, 277–78; twofold condition for, 290

certitudo salutis, 277, 320

character, 181

charity: Augustine on burden of, 16n.14; Augustine on founding human relationships on, 53; in monastic orders, 54

Chenu, M. D., 141n

Christianity: abstention from the worldly, 54–55; on action and contemplation, 14–16, 85; as a body, 53, 53n.46; Cartesian doubt affecting, 319; on the common good, 35, 55; on earthly immortality, 21; eschatological expectations of early, 120; the family as model for, 53–54; goodness as arising with, 73–74; the Gospels, 318; on labor, 316–18, 316n; on life as sacred, 314, 315, 316, 318; on manual labor, 146n; on minding one's own business, 60, 60n; monastic orders, 54, 66, 317, 317n.83; on the public realm, 74; the Reformation, 248, 249, 251–52; on *res publica*, 74; on *vita contemplativa*, 318. *See also* Catholic Church; Jesus of Nazareth

Cicero: on death as reward of sin, 314, 314n; on Epicurus's view of pleasure, 113n.61; on liberal and servile arts, 91n.23; on *mediocris utilitas*, 91n.25; on property acquisition, 110n.56; on *utilitas*, 91n.23, 183n

citizenship: Catholic Church as substitute for, 34; legislators as not necessarily citizens, 194; sons not citizens in father's lifetime in antiquity, 62n.59; as time-consuming in city-states, 14n.10, 82n; tyrannies banning citizens from public realm, 221–22, 224; working for a living precluding in antiquity, 65n.18

city-state. See *polis*, the

class: class membership versus family membership, 256, 257; Marx's classless society, 131n.82; as not existing in the modern age, 5. *See also* working class

classical economics: communistic fiction in, 43–44; on happiness for the greatest number, 133; productivity

as standard of, 306; on self-interest, 42n; on utility, 172; on value, 165

coal, 148n.9

cogito ergo sum, 270, 280, 280n.40

cognition: modern philosophy as theory of, 293; thought distinguished from, 170–71. *See also* thought

collective ownership, 256, 257

comfort, 154

commands, 189

commercial society, 162, 163, 210, 307

common, the, 50–58

common good, 35, 55

common sense: Cartesian doubt of, 277; as fitting into reality as a whole, 208–9; introspection and loss of, 280–84; Whitehead on retreat of, 283

communication, speech in, 179

communism, 72, 118n.65, 131n.82

communistic fiction, 43–44, 44n, 46, 256

computers, 172

conformism, 40, 41–42, 46, 58

conspicuous consumption, 160, 162

Constant, Benjamin, 79

consumer goods: tools providing abundance of, 122; use objects contrasted with, 94, 137–38; use objects transformed into, 124, 125–26, 230

consumptibiles, 69

consumption: in biological cycle, 99; city-state as center of, 66n.69, 119; conspicuous consumption, 160, 162; consumers' society, 126–35; and labor, 99, 100, 102, 126; Marx on emancipation from, 131. *See also* consumer goods

contemplation: action opposed to in traditional thought, 14–17, 85; and action reversed in modern age, 289–94; eternity as object of, 20–21; in Greek political philosophy, 301–2; manual labor as not interfering with, 146n; *nous*, 27; speechless wonder in, 303–4; thought distinguished from, 16. See also *theōria*; *vita contemplativa*

contract theories, 244

Copernican system, 258–60, 260n.12, 285

Copernicus, 258, 259, 264, 274, 274n

Cornford, Francis M., 17n, 142n

Counter Reformation, 253

courage, 36, 186–87

Index

covenants, 243–44
craftsmen: the *agora* as meeting place
    for Greek, 160; Aristotle as ignor-
    ing, 12n.4; and automation, 149n.12;
    dēmiourgoi, 81, 81n.6, 159; freemen
    distinguished from, 65n.67; genius
    contrasted with craftsmanship, 210;
    Greek mistrust of, 82; Greeks con-
    trasting with slaves, 80; and happi-
    ness, 134; Homer on craftsmanship,
    83n.7; lawmakers as for the Greeks,
    194–95; mastership, 161, 161n.28;
    medieval craft guilds, 123, 159n.26;
    philosopher-king compared with,
    227; and Plato's doctrine of ideas,
    142n, 302–3; poets as, 170n; songs
    of, 145n
creation, in Genesis, 8, 8n, 107n
cultivated land, 138–39
cynicism, 310

daimōn, 179–80, 182n.7, 193, 193n.18
Dante, 175, 208n.41
de (in names). *See under* substantive
    part of name
death: in Eleusinian Mysteries, 63n.61;
    eudaimonia obtaining only after, 192;
    experience of the eternal as kind of,
    20; household as realm of, 62–63; la-
    bor, work, and action as connected
    with, 8; and nature, 96–97; Old Tes-
    tament view of, 107, 107n; pain as
    borderline between life and, 51,
    51n.43; and sin, 314; suicide, 315,
    316; underworld deities, 30n.19
Decalogue, 315
decay, 97–98
deeds: endurance capacity of, 233; futil-
    ity of, 173; good deeds, 76, 240; he-
    roic deeds, 101; history as a story of,
    185; immortality of, 19, 19n.19, 197,
    198; inserting ourselves in the world
    with, 176
dēmiourgoi, 81, 81n.6, 159
democracy, 220, 222
Democritus, 170n, 206, 259n.10, 275,
    275n.31
Demosthenes, 26n.8, 64n.66
Descartes, René: analytical geometry
    of, 266–67; on Archimedean point,
    284, 284n.45, 287; *cogito* argument,
    279, 280, 280n.40; doubts about
    God's goodness, 281–82; and Gali-
    leo, 273; introspection in, 280, 282;

Jaspers on science of, 272n.27; in
    modern philosophy, 271–72, 273; on
    philosophers before himself, 249n;
    reason for, 283–84; on secondary
    qualities, 114n.63; subjectivism of,
    272; subject/object split in, 312. *See
    also* Cartesian doubt
design, 152
despotism: the despot's life as unfree,
    13, 13n.7, 32n.22; head of house-
    hold's power as despotic, 27–28, 32
dialectic, 26n.9
Diebold, John, 148n.10, 149n.12
Dinesen, Isak, 113n.61, 175, 211n
distinction, 175–76, 197
division of labor, 123–26; in classical
    economics, 88, 88n.18; in labor pro-
    ductivity's growth, 47; sexual divi-
    sion of labor, 48n.38; and skill, 90;
    Smith on, 88n.18, 161n.29; special-
    ization distinguished from, 47n, 123,
    214n.48; teamwork as, 161–62
Dolléans, Édouard, 133n
dominus, 27, 28n, 130n.81
doubt, Cartesian. *See* Cartesian doubt
drama, 187, 187n.11, 205n.33
dreams, 199, 281
drug addiction, 113n.61
Dunkmann, Karl, 101n
dynamis: Pericles' faith in, 205; poten-
    tial character of, 200. *See also* power

Ecclesiastes, 204
economics: conformism at root of, 42;
    as housekeeping, 28–29; "invisible
    hand" doctrine, 42n, 44, 48n.38,
    185; Marx on laws of, 209; physio-
    crats, 87n.16; political economy, 29,
    33n, 42n; statistics as tool of, 42. *See
    also* classical economics
Eddington, Arthur S., 261
egoism, 311
Einstein's theory of relativity, 263–64
élan vital, 117n
electricity, 148–49
Eleusinian Mysteries, 63n.61
Else, Gerard F., 142n
empiricism, English, 272
end in itself, 154–55, 156, 206
ends and means: as characteristic of
    homo faber, 145, 157, 305; the end as
    justifying the means, 229; Kant on,
    155–56; men and machines as, 145,
    152; and product of fabrication, 153;

Index

ends and means (*continued*)
in utilitarianism, 154–57; work of man as beyond, 207
energeia, 206, 206n.35
Engels, Friedrich, 86n.14, 88, 116, 131n.82
English empiricism, 272
entelecheia, 206
Epicureanism, 112, 235, 309, 310, 311
Epicurus, 113, 113nn. 61, 62
equality: before God, 215; of men and women, 48n.38; and plurality, 175; in the *polis*, 32; in public realm, 215; in society, 39, 40, 41
ergazesthai, 80. *See also* work
ergon, 83n.8. *See also* work
eternal recurrence, 97, 232
eternity: as center of thought for Socrates and Plato, 20; contemplation for experiencing, 20–21; experience of as kind of death, 20; versus immortality, 17–21
eudaimonia: as bought only at price of life, 194; freedom as condition of, 31; meaning of, 192–93; as obtaining only after death, 192
Euripides, 84n.10
Eutherus, 31n
evolution, 312
excellence, 48–49, 49n, 73, 173
exchange market, 159–67; as prior to manufacturing class, 163; as public realm of *homo faber*, 160, 162, 209–10; relativity of, 166; things becoming values in, 163–65; as unknown in Middle Ages, 166n; the work of our hands in, 136
exchange value: triumph over use value, 307; use value changing to in capitalism, 165; use value distinguished from, 163
existentialism, 235n.74, 272, 313n
experiment, 150n, 231, 286, 287–88, 295, 312
expressionist art, 323n
expropriation: in accumulation of wealth, 254–55; of Church property, 66–67, 252; modern age starting with expropriation of the poor, 61, 66; of the peasantry, 251; socialization of man carried through by means of, 72; and theory of private property, 109; and world alienation coinciding, 253

fabrication (making): action distinguished from, 188, 192; action understood as, 322; Bergson on, 305n; in experiments, 295; lawmaking as, 195; means and ends in, 153; Platonic ideas influenced by, 142–43, 142n, 225–26, 302–4; poems as made, 170n; *poiēsis*, 195, 301; as reification, 139–44; reversal within *vita activa* and, 294–304; as taking place in a world, 188; traditional substitution of making for action, 220–30; violence as element of, 139–40, 153; in *vita activa*, 141; world alienation affecting, 307. See also *homo faber*
faith, 247n, 253–54, 271, 319, 320
family, the: Christian community modeled on, 53–54; class membership versus family membership, 256, 257; declining with society's emergence, 40, 256; nation compared with, 256; *paterfamilias*, 27, 28n; society as super-human family, 29, 39. *See also* household
Faulkner, William, 181n
fertility: of *animal laborans*, 112, 122; capital accumulation compared with, 255; labor and, 101–9, 117; of labor power, 118; in life philosophies, 313n; love distinguished from, 242n.82; Marx equating productivity with, 106
feudalism, 29n.13, 34–35, 252
forgiveness: irreversibility and the power to forgive, 236–43; Jesus on, 238–41, 239nn. 76, 77, 240n.80, 247; and love, 242; as personal, 241; punishment contrasted with, 241; as unrealistic, 243; vengeance contrasted with, 240–41
fortune, good, 108, 193
France: annual working days before Revolution, 132n.85; the *sans-culottes*, 218, 218n.54; the "small things" in, 52; utility principle in French philosophy, 306
Franklin, Benjamin, 144, 159
freedom: in Aristotle's *bios politikos*, 12–13, 12n.4; as condition of thought, 324; courage as required for, 187; the despot's life as unfree, 13, 13n.7, 32n.22; as entangling people in web of relationships, 233–34; the house-

Index

Grimm, Jacob and Wilhelm, 80n.3, 81n.5
growing old, 51n.43
growth, 97–98
guilds, 123, 159n.26

Halbwachs, M., 212n.44
Halévy, Élie, 308n.72, 311n.74
happiness: as absence of pain, 112–15, 113n.61; Bentham on, 309; *eudaimonia* contrasted with, 192–93; greatest happiness principle, 133, 308–9, 311; universal demand for in modern age, 134
Hearth, the, 62n.60
Hebrew legal code, 315
hedonism, 51n.43, 112–13, 309–11
Hegel, Georg Wilhelm Friedrich, 86n.14, 254n, 293, 294, 300
Heisenberg, Werner, 153n, 261, 261n.16, 286n.50
heliocentric system, 258–60, 273
Henry, François, 316n
Heraclitus: on human/animal distinction, 19; on *nomos*, 63n.62; on not entering the same river twice, 137; on oracles, 182; on strife as father of all things, 158n; on those dreaming having their own world, 199n.29
Hercules, 101
Herodotus, 18, 18n, 32n.22, 120
heroic deeds, 101
hero of a story, 184–87, 186n, 194
Hesiod: on deeds of gods and men, 23n.1; on founding new cities away from the sea, 132n.84; on labor, 48n.39, 82, 82n, 83n.8; on life of hearth and household, 25n.6
Hestia, 25n.6
highest good, life as the, 313–20
history: action as creating the condition for, 9; fatality as mark of, 246; historian knowing the event better than the actors, 192; law of large numbers applied to, 42; Marx as Darwin of, 116; Marx on violence in, 228; political nature of, 185; science as, 296; as story of events not forces, 252; as story without authors, 184–85; as system of process, 232; Vico on, 298, 298n
Hobbes, Thomas: acquisitive society of, 31; making introduced to political philosophy by, 299, 300; on political philosophy as starting with him-

self, 249n; on rationality, 172, 283; subjectivism of, 272; on vainglory, 56–57; on will to power, 203
hobbies, 118, 118n.65, 128, 128n.76
holidays, 132n.85
Homer: Achilles of, 25; on craftsmanship, 83; Democritus on, 170n; on heroes, 186, 186n; heroes of concerned to be the best, 41n; as immortalizing the Trojan War, 197; on kingship, 221n.57; on leader's role, 189; on necessity in labor, 131n.83; Plato's Cave parable inverting world order of, 292; and *pragmata*, 19n.19; religion of, 25n.6; on slaves losing excellence, 49n
Homeric gods, 18, 23n.1
homo faber: acting and speaking men requiring help of, 173; on action and speech as idleness, 208; *animal laborans* contrasted with, 136; Bergson on, 305n; defeat of, 305–13; etymology of, 136n; exchange market as public realm of, 160, 162, 209–10; excluded from public realm in antiquity, 159; fabrication as reification, 139–44; Greek mistrust of, 82; hands as primordial tools of, 144; ideals of, 126; instruments of work and, 153–59; intellectual worker and, 91; as lord and master, 139, 144; means and ends as characteristic of, 145, 157, 305; mentality of, 305–6; as merchant and trader, 163; modern age seeing man as, 228, 229–30, 305; nature for, 135, 155; and process, 307, 308; redemption of, 236; reversal within *vita activa* and victory of, 294–304; single-minded work orientation of, 151; solitary worker as not, 22; and the space of appearance, 207–12; in telescope's invention, 274; thought as inspiration of, 171; tools made by, 121; utilitarianism as philosophy of, 154–55. *See also* craftsmen
honor, 73
horoi, 30
household: Aquinas contrasting with the *polis*, 27; despotic power of head of, 27–28, 32; distinctive trait of, 30, 45; freedom not existing in, 32; gods of, 30; inequality in, 32; monarchical rule in ancient, 40; the *polis* as opposed to, 24, 24n.6, 28–37; the *polis*

Index

Kierkegaard, Søren, 275, 275n.32, 293, 313n, 319
kingship, monarchy contrasted with, 221n.57
Kronstadt rebellion, 216n

labor, 79–135; admission to public realm, 46–48, 218; in Aristotle's *bios politikos*, 12, 13; automation as liberating mankind from, 4; the blessing of, 106–8; Christianity on, 316–18, 316n; collective nature of, 213, 213n.47; and consumption, 99, 100, 102, 126; contempt for in ancient world, 81–85; defined, 7; elevation of, 306–7, 313; elimination of, 322; emancipation of, 126–35; end of, 98, 143, 144; European words for, 48n.39; and fertility, 101–9, 117; in giving birth, 30; of Hercules, 101; joys of, 140; and life, 96–101, 120; living without, 176; in Locke's theory of property, 70, 101, 105, 110–12, 110n.56, 115–16; loneliness of the laborer, 212–14, 212n.44; manual and intellectual, 85, 90–93, 146n; Marx on, 85–90, 86n.14, 88n.20, 93, 98–99, 99n.34, 101–2, 104–9, 104n.48; modern society as laboring society, 4–5, 46, 85, 126; in monasteries, 317, 317n.83; natality and mortality as connected with, 8–9; and nature's cyclical movement, 98–100, 98n; Old Testament view of, 107, 107n; pain and effort associated with, 48, 48n.39, 80n.3, 81n.5; play opposed to, 127–28, 127n; poverty associated with, 48n.39, 110n.56; as preferable to servitude, 31, 31n.8; productive and unproductive, 85–89, 87n.16; a proper location for, 73; repetition as mark of, 125, 142; rhythm of, 145n, 214; skilled and unskilled, 85, 89–90; slavery as social condition of laboring classes, 119; as slavery to necessity, 83–84; in solitude, 22; as source of wealth for Smith, 101; in *vita activa*, 7; work as now performed in mode of, 230; work distinguished from, 79–93, 80n.3, 81n.5, 83n.8, 94, 103–4, 138. See also *animal laborans*; consumer goods; division of labor; free labor; labor power
labor collectives, 123
laboring class. *See* working class

labor movement, 212–20
labor power: in cycle of biological life, 99, 143; in division of labor, 123; emancipation of, 255; fertility of, 118; laborers as owners of, 162; Marx on, 70, 88, 108, 111; as never being lost, 133; skill contrasted with, 90; as spent in consuming, 131
labor songs, 145n
Lacroix, Jean, 141n
landed wealth, 66
Landshut, Siegfried, 45n
Lares, 62n.60
Last Judgment, 239
laws: Hebrew legal code, 315; legislation, 63, 194–95, 196; as limitations, 191, 191n; making, 188; Marx on economic, 209; Montesquieu on, 190n; *nomos*, 15, 63n.62; of the *polis*, 63–64, 63n.62, 194–95; of science, 263
laziness, 82n
leaders, 189–90
learned societies, 278
Leclercq, Jacques, 107n, 127n, 316n
Leclercq, Jean, 136n
legislation, 63, 194–95, 196
Leibniz, Gottfried Wilhelm, 267–68, 281
leisure, 82n, 131, 132n.84
Leonardo da Vinci, 51n.43, 259, 319
Lessing, Gotthold Ephraim, 154
Levasseur, E., 34n.26, 62n.57, 132n.85, 161n.28
lex, 63n.62
liberal arts, 91–93, 128, 128n.77
life: Aristotle on the good, 36–37; biography, 97; as a burden, 119; Christianity on sacredness of, 314, 315, 316, 318; creating in a test tube, 2, 269; durability used up by, 96; *élan vital*, 117n; as highest good, 313–20; as immortal for Christianity, 314–16; labor and, 96–101, 120; as mortal again in modern age, 320; necessity as driving force of, 70–71; origin of, 189; pain as borderline between death and, 51, 51n.43; as price of *eudaimonia*, 194; subject/object split disappearing in, 312–13; trust in reality of, 120; worldliness as redemption of, 236
life expectancy, 2, 133n
life philosophies, 117, 172, 311–12, 313n

Index

limitations, 190–91
Lipmann, Otto, 127n
literati, the, 211
literature: men of letters, 56; the novel, 39; poetry, 39, 169–70, 170n, 196, 242n.81; science fiction, 2
Livius, 23n.3, 63n.62
Locke, John: on marketable value, 164; on men treated as means, 155; on necessities of life as of short duration, 96, 100; on politics as protecting society, 31; on property, 70, 101, 105, 110–12, 110n.56, 115–16; work and labor distinguished by, 80, 103–4; worth and value distinguished by, 164n.34, 165
Loening, Edgar, 128n.77
logical positivism, 272
logical reasoning, 171–72
loneliness: of the laborer, 212–14, 212n.44; of lover of goodness, 76; as mass phenomenon, 59
love: fertility distinguished from, 242n.82; and forgiveness, 242; as private, 51–52; and romance, 242n.81; as unworldly, 242
luck, 108, 193
Lucretius, 114n.63
Luther, Martin, 139n, 248, 251
lying, 129n.78, 278n.35

Machiavelli, Niccolò, 35, 35n.29, 77–78
machines: design of, 152; instrumentality of, 151; men as servants of, 145, 147; rhythm of, 125, 132, 145n, 146; tools contrasted with, 147
machine tools, 148
Madison, James, 110n.54
making. *See* fabrication
"making a living," 127, 128
Man, Hendrik de, 140n.4, 145n
Manes, 62n.60
"man is the measure of all things," 155, 157–58, 158n, 166, 306
manufacturing: conspicuous production of manufacturing capitalism, 162; as continuous process, 149, 149n.12, 151–52; exchange market as prior to manufacturing class, 163; mass production, 125. *See also* automation
market, exchange. *See* exchange market
marketable value, 164
Marshall, Alfred, 163n.32
Marx, Karl: on *animal laborans*, 86n.14, 87, 99n.36, 102, 105, 117; on being/

appearance distinction in science, 275; on capital accumulation, 255; on capitalism, 89n.22; classical economics developed by, 42n, 43–44; classless and stateless society of, 131n.82; on contradiction in modern conception of government, 69; contradiction in system of, 101–2, 104–5, 104n.48; critics of, 79; as Darwin of history, 116; on dehumanization of commercial society, 210; on division of labor, 214n.48; on economic laws, 209; on emancipation of man from labor, 130–31; on Franklin's definition of man, 159; on freedom supplanting necessity, 104; on growing wealth as natural process, 111; Hegelian dialectic inverted by, 293; on labor, 85–90, 86n.14, 88n.20, 93, 98–99, 99n.34, 101–2, 104–9, 104n.48; on laborers as owners of their labor power, 162; labor misrepresented as work by, 306; on labor power, 70, 88, 90, 99, 108, 111, 133; life philosophy of, 313n; on man as member of a species, 89n.21, 116; materialism of, 183n; on politics as function of society, 31, 33; on private appropriation hindering social productivity, 67; on professions as hobbies under communism, 118, 118n.65, 128, 128n.76; on reckoning, 172; on reification, 102, 102n.41; on self-alienation, 89n.21, 162, 210, 254; on socialized man, 42n, 44, 72, 89, 89n.21, 117–18, 321; and tradition of Western political thought, 12, 17; on values as social, 165; on violence, 228, 228n; on withering away of the state, 45, 60; on world alienation, 254n
mass culture, 134
mass hysteria, 58
mass production, 125
mass society: behavioral sciences associated with, 45; conformism of, 58; as embracing all members of community equally, 41; of laborers, 118; loneliness in, 59; as not having power to gather people together, 52–53; the one-ness of man-kind as root of, 46
mastership, 161, 161n.28
materialism, 183, 183n, 313, 319
mathematics: as leading science of mod-

Index

mathematics (*continued*)
ern age, 265–66, 283; in science, 4, 267, 284, 285–87. *See also* geometry
meaningfulness, utility distinguished from, 154–55
means and ends. *See* ends and means
medicine, 91, 128, 128n.77
memory. *See* remembrance
men of letters, 56
mental images, 141, 161, 173
mental processes, 280–81
Michelson, A. A., 295
Middle Ages: annual working days during, 132n.85; common good concept, 35; community foundation in, 62n.57; conspicuous production of, 209; *corpus rei publicae*, 54n.46; craft guilds, 123, 159n.26; craftsmen and freemen distinguished in, 65n.67; feudalism, 29n.13, 34–35, 252; genius as unknown in, 210; hired labor in, 66n.70; market places of, 161; money in economic theory of, 69n; *opera liberalia*, 92n.27; philosophy as handmaiden of theology, 291; public/private gulf in, 33–34; on value, 164n.34, 166n
Mill, James, 23n
Mill, John Stuart, 23n
Milton, John, 100n.36, 321
mimēsis, 187–88, 187n.12
Minkowski, Hermann, 286n.49
Mirabeau, Marquis de, 87n.16
miracles, action as miracle-working faculty of man, 246–47, 247n
mob rule, 203
modern age: as beginning with expropriation of the poor, 61, 66; cardinal virtues of, 278; contemplation and action reversed in, 289–94; faith as lost in, 253–54; as forcing all men under yoke of necessity, 130; genius as idolized in, 210–11; human material in, 188n.14; as laboring society, 4–5, 46, 85, 126; life as the highest good in, 313–20; man as *homo faber* in, 228, 229–30, 305; mathematics as leading science of, 265–66, 283; modern world distinguished from, 6, 268; on money as begetting money, 105; necessity as triumphed over, 134; political man excluded from its public realm, 159; process as key concept of, 105, 232; property as concern of, 109; secularism of, 253; soci-

ety as victorious in, 45; spirit of distrust and suspicion of, 319; technological development in, 147–50; thought and modern world view, 285–89; three events at threshold of, 248; as turning away from heavenly God, 2; violence as declining in, 129–30, 130n.80; violence in political thought of, 228; *vita activa* and, 248–325; women and working class emancipated in, 73
modern world: equality in, 41; modern age distinguished from, 6, 268; radical subjectivization of, 141
monarchy: in ancient household, 40; as form of rule, 222; kingship contrasted with, 221n.57; Plato's philosopher-king, 221, 224, 226, 227, 229n; as salvation from plurality, 221
monastic orders, 54, 66, 317, 317n.83
money: as begetting money, 105; as equalizing factor, 215; as lacking independent and objective existence, 166; Locke on, 102; in medieval economic theory, 69n; Plato on art of making, 128, 128n.77; public admiration and monetary reward, 56, 57
Montesquieu, Charles, Baron de, 190n, 202, 203n
morality: Cartesian doubt's effect on, 277–78; and forgiving and promising, 245–46; political principles contrasted with, 237–38. *See also* good, the; goodness
mortality: beginning as interrupting, 246–47; immortal gods contrasted with mortal men, 18–19; labor, work, and action as connected with, 8, 9; life as mortal again in modern age, 320. *See also* death
Müller, Karl, 316n
music, 39, 169
Myrdal, Gunnar, 29n.13, 33n, 44n

natality: labor, work, and action as connected with, 8–9; as miracle that saves the world, 247; and the political, 9. *See also* birth
nation, the, 29, 29n.14, 256
nationalism, organic theories of, 256
naturalism, 313
natural science: Archimedean standpoint of, 11, 257–68; astrophysics, 258, 261, 269, 313; Cartesian doubt in, 287–89; circularity confronting,

Index

285–89; crisis in, 3–4; earth alienation in development of, 264; Galileo's laws of falling bodies as beginning of, 258; as historical discipline, 296; as sciences of process, 116, 231–32; telescope's invention, 248, 257–58, 274, 290; universal science versus, 268–73, 313; Vico on, 298, 298n

nature: for *animal laborans*, 134–35; *animal laborans* as servant of, 139; and God as watch and watchmaker, 297; as growth and decay, 97–98; and history, 185; for *homo faber*, 135, 155; labor and cyclical movement of, 98–100, 98n; life within cyclical movement of, 96; natural processes, 150–51; *physis*, 15, 150; as process, 296–97, 296n.61; science as making, 231–32, 295; state of nature, 32; work as destructive for, 100, 139, 153. *See also* human nature

Naville, Pierre, 98n

necessity: as driving force of life, 70–71; emancipation of labor as emancipation from, 131; and freedom, 70, 71; freedom as supplanting for Marx, 104; freedom conceived in terms of, 121; freedom luring people into, 234; household community as born of, 30; labor as slavery to, 83–84; modern age as forcing all men under yoke of, 130; modern age as triumphing over, 134; political activity reduced to, 85, 314; private realm associated with, 73; short duration of necessities of subsistence, 96, 100; torture compared with natural, 129

Nepos, Cornelius, 23n.3

Newton, Isaac, 258, 264, 265, 272

Nicholas of Cusa, 258, 260

Nietzsche, Friedrich: on *cogito* argument, 280n.40; on eternal recurrence, 97; on life as creator of values, 117; life philosophy of, 313n; on man's ignorance of motives and consequences, 233n; on modern philosophy as school of suspicion, 260; philosophy as revaluation, 293; on promising, 245, 245n; on reckoning, 172; and tradition of Western political thought, 17; on will to power, 203, 204n, 245n

Nitti, F., 127n

nomos: etymology of, 63n.62; *physis* contrasted with, 15. *See also* laws

non-Euclidean geometries, 285

non-objective art, 323n

nous, 27

novel, the, 39

Oakeshott, Michael, 299n.64

objectivity: of durable things, 137; and the human condition as supplementary, 9; of interests, 182; money as objective, 57; private realm as deprived of, 58; of Royal Society, 271n.26; thing-character of the world, 9, 93–96

object/subject split, 312–13

ochlocracy, 203

Oedipus, 193n.18

Old Testament: Genesis, 8, 8n, 107n, 139n; on labor and death, 107, 107n; on man's role in creation, 139n

oligarchy, 222

omnipotence, 202

one-man rule. *See* monarchy

oracles, 182, 182n.7

original sin, 310

Osiander of Nuremberg, 260n.12

Otanes, 32n.22

otherness, 176

otherworldliness, 76–77, 320

otiosity, 317, 318n.83

ownership, collective, 256, 257

pain: Bentham's calculus of, 309–11; in birth, 115, 121; happiness as absence of, 112–15, 113n.61; labor associated with, 48, 48n.39, 80n.3, 81n.5; privacy of, 50–51, 51n.43

Paine, Thomas, 110n.54

painting, 82n, 93

Park, M. E., 66n.70

Pascal, Blaise, 276n, 293, 319

passive resistance, 201

paterfamilias, 27, 28n

patriarchs, Old Testament, 107

Paul, St., 8n, 314–15, 316–17

peasants, 83n.9

Peisistratus, 221

penates, 30

people's councils, 216, 216n

Periandros, 221

Pericles, 133, 188n.14, 197, 205–6, 205n

Persians, 18n, 43

personal identity, 179, 193

Index

peuple, le, 219, 219n
phenomenological existentialism, 272
Phidias, 93n.30
philosopher-king, 221, 224, 226, 227, 229n
philosophy: the body as resented in, 16n.15; declining influence of, 294; Descartes in modern, 271–72, 273; Descartes on his predecessors', 249n; dialectic, 26n.9; life philosophies, 117, 172, 311–12, 313n; as love of beauty, 226; as love of wisdom, 75; in the Roman Republic, 59; and science, 272, 290, 294, 313; the self as concern of modern, 254; as series of reversals, 292–93; subjectivism of modern, 272; and theology, 291; tradition opposed by modern, 276; utility principle in modern, 306; wonder as origin of, 273, 302, 302n. *See also* political philosophy; *and philosophers and doctrines by name*
physiocrats, 87n.16
physis: in etymology of "nature," 150; *nomos* contrasted with, 15. *See also* nature
piety, 75n
Pindar, 63n.61
Planck, Max, 287n.53
Plato: on action and work, 301; on the body only as living in the city, 16n.15; Cave parable, 20, 75, 226, 226n.66, 292; on city-states away from the sea, 132n.84; on contemplation's superiority over action, 14; on the demiurge, 22; on durability, 172; on the eternal as center of thought, 20; on God as Platonic idea of man, 11; on human affairs, 19n.19, 25, 185; knowing separated from doing by, 223–27; on laborers, 118; on lawmaking, 195; on man as social animal, 24; on mathematics, 265–66; on money making, 128, 128n.77; on origin of the *polis*, 183n; peasants classed with slaves by, 83n.9; on Pericles, 188n.14; on a philosopher-king, 221, 224, 226, 227, 229n; on philosophy and theology, 291; on pleasure and pain, 113n.61; on the *polis* and the household, 37, 223, 223n.62; politics as fabrication for, 230; on poverty, 110n.56; on private property, 30; on Protagoras, 157–58, 158n, 166; on ruler and ruled, 222,

224, 227n.69, 237–38; on slaves' natural slavishness, 36n.30, 316; on speech and truth, 178n; on thought as inner dialogue, 76, 291; on treatment of slaves, 34n.37; on two modes of action, 222–23; on wonder as beginning of philosophy, 302, 302n. *See also* ideas, Platonic
play, labor opposed to, 127–28, 127n
pleasure: Bentham's calculus of, 309–11; hedonism, 51n.43, 112–13, 309–11
Plinius, 120n
plurality: action's calamities arising from, 220; as condition of action, 7, 8; as condition of politics, 7–8; destruction of, 58; forgiving and promising as dependent on, 237; monarchy as salvation from, 221; otherness as aspect of, 176; space of appearance as depending on, 220; and speech, 178; twofold character of, 175–76; as weakness, 234
Plutarch, 30
poetry, 39, 169–70, 170n, 196, 242n.81
poiēsis: affinity with *theōria*, 301; as making, 195. *See also* fabrication
polis, the: agonal spirit of, 41, 194; Aquinas contrasts with the household, 27; in Aristotle's *bios politikos*, 13; as consumption center, 66n.69, 119; equality in, 32; the eternal contrasted with, 21; freedom in, 30–31; functions of, 196–97; the household as opposed to, 24, 24n.6, 28–37; household compared with by Plato, 37, 223, 223n.62; law of, 63–64, 63n.62, 194–95; as "man writ large" for Plato, 224; occupational classification in, 81–83, 82n; original connotation of, 64n.64; origins of, 183n; as paradigmatic for Western political organization, 201; philosophers discovering higher realm than, 18; Plato on founding new states away from the sea, 132n.84; Plato's reorganization of, 14; prosperity in, 59; small population of, 43; as space of appearance, 198–99; as space of relative permanence, 56; as superseding kinship units, 24; as talkative, 26
political economy, 29, 33n, 42n
political parties, trade unions contrasted with workers', 215–17
political philosophy: action and work in

Index

Index

Ptolemaic system, 258–59, 285
public admiration, 56–57
public realm, 22–78; *animal laborans* in,
134; Christianity as hostile to, 74;
the common, 50–58; equality in,
215; excellence assigned to, 49; exchange market as *homo faber*'s, 160,
162, 209–10; freedom associated
with, 73; goodness as destructive of,
77; labor admitted to, 46–48, 218;
and the political, 28; private property as hiding place from, 71; as reserved for individuality in ancient
world, 41; society's emergence alter-
ing, 38, 257; space of appearance as
preceding, 199; tendency to grow of,
45; two meanings of, 50–52; tyrannies banning citizens from, 221–22,
224; wealth as condition of admission to, 64–65, 65n.67; as the work
of man, 208
punishment, forgiveness compared
with, 241
puritanism, 310, 311
Pythagoreans, 142n, 273

qualities: primary, 115; secondary,
114n.63, 115

rationality. *See* reason
reaction, 190, 241
reality: Cartesian doubt of, 274–79; as
common to us all, 50, 208; introspection confirming, 280; of life, 120;
modern rationalism foundering on
rock of, 300–301; senses' adequacy
to reveal as challenged, 261–62; of
the world, 120
reason: *animal rationale*, 27, 84–85,
171–72, 284; Cartesian, 283–84;
Cartesian doubt of, 274–79; modern
rationalism foundering on rock of reality, 300–301; scientific results offending, 290; thought distinguished
from logical, 171–72
Reformation, 248, 249, 251–52
reification: of action and speech, 95,
187; in art, 95, 168–69; in fabrication, 139–44; Marx on, 102,
102n.41; world alienation affecting,
301
relationships, human. *See* web of relationships
relativism, 263–64, 270
relativity theory, 263–64

religion: Cartesian doubt affecting, 319,
320; Eleusinian Mysteries, 63n.61;
faith, 247n, 253–54, 271, 319, 320;
loss of *certitudo salutis*, 277, 320;
Olympian religion, 25n.6; piety, 75n;
religious values, 235n.74; secularization, 253, 320; Vesta cult, 24n.6. *See
also* Christianity; God; gods
Rembrandt, 51n.43
remembrance: memory of past pleasures, 310; as mother of all arts, 95;
the *polis* as organized, 198; rhythm
in poetry's, 169–70; speech and action creating their own, 207–8;
thought owing its existence to, 76,
90
Renaissance: the arts in, 82n; and Christianity, 319; genius as ideal of, 210;
Jaspers on modern science and philosophy of the, 249n; rebellion
against scholasticism, 264
respect, 243
res publica: Christian antagonism toward, 74; *corpus rei publicae*, 53n.46;
household as *res publica* to slaves, 59;
as space of relative permanence, 56
revelation: action of scientists as not revelatory, 324; doubts about divine,
281; oracles, 182, 182n.7; self-
revelation of love, 242; speech and,
178–80, 187; truth as, 17
revenge, 240–41
revolutions: American Revolution, 228;
Hungarian Revolution of 1956, 215,
217, 219; popular revolt, 200–201;
violence in, 228
rhetoric, 26, 26n.9
rhythm: of labor, 145n, 214; of machines, 125, 132, 145n, 146; in poetry, 169; for work, 145n
Ricardo, David, 165n.37
Riesman, David, 59n.52
Rilke, Ranier Maria, 51n.42, 168n
Roman Catholicism. *See* Catholic
Church
romance, 242n.81
Roman Empire: fall of, 21, 34, 45, 74;
intellectuals in, 92; old freedoms
abolished in, 28n, 130n.81; *servi publici*, 84n.11; slave dress, 218n.53;
slavery in, 36n.30
Roman Republic: being a philosopher
in, 59; slavery in, 36n.30. See also *res
publica*
Romans: agriculture as liberal art for,

Index

91n.24; industrial development as limited, 65n.69; inviolability of agreements, 243; *opus* and *operae* distinguished by, 92n.26; the *plebs*, 62n.59; private and public as coexisting among, 59; slaves of, 59, 59n.54; on sparing the vanquished, 239; territory and law for, 195n.21; Vesta cult, 24n.6; on violence in founding a new body politic, 228. *See also* Roman Empire; Roman Republic

Romanticists, 39

Rousseau, Jean-Jacques, 39, 41, 79

Royal Academies, 278

Royal Society, 271n.26

rule: *archein*, 177, 189, 222–23, 224; Aristotle on, 222; contract contrasted with, 244; democracy, 220, 222; as escape from politics, 222; of feudal lords, 34; Greek and Latin words for, 32n.22; leaders becoming rulers, 189–90; master/slave compared with ruler/ruled, 223–24; ochlocracy, 203; oligarchy, 222; Plato on, 222, 224, 227n.69, 237–38. *See also* monarchy; tyranny

Russell, Bertrand, 267n

sans-culottes, 218, 218n.54

satellites, artificial, 1, 269

saving the appearances, 259n.10, 260, 266, 285

Schachermeyr, M. F., 195n.22

Schelsky, Helmut, 128n.75

Schlaifer, Robert, 36n.30

Schlatter, Richard, 110n.56

Schopp, Joseph, 145n

Schrödinger, Erwin, 3, 287n.51

Schulze-Delitzsch, Hermann, 98n

science: action as prerogative of scientists, 323–24; behavioral sciences, 45; carrying irreversibility and unpredictability into natural realm, 238; dreams as anticipating, 1–2; earth alienation as hallmark of, 264–65; emphasis shifting from why to how, 295–96; experiment, 150n, 231, 286, 287–88, 295, 312; as history of the universe, 296; hypothesis, 278, 287; instruments in, 295; as making nature, 231–32, 295; mathematics in, 4, 267, 284, 285–87; in modern age's creation, 248, 249–50, 249n; and philosophy, 272, 290, 294, 313; subjectivization of modern,

141, 282; success as criterion of, 278–79; virtues of modern, 278. *See also* natural science

science fiction, 2

scienza nuova, 249n

scribes, 91, 92, 92n.28

sculpture, 82n, 93, 93n.30, 157

secondary qualities, 114n.63, 115

secularization, 253, 320

self: Marxian self-alienation, 89n.21, 162, 210, 254; modern philosophy's concern with the, 254. *See also* introspection

self-sufficiency, 234–35

Seneca, 23, 36n.30, 40n, 218n.53

senses, the: astrophysical world view challenging, 261–62; Cartesian doubt of, 274–79; common sense as uniting, 208–9; in experiencing the world, 114–15, 114n.63; and Galileo's telescope, 260, 260n.11; and mathematization of physics, 287; vision, 114n.63. *See also* common sense

sensualism, 51n.43, 112, 272, 310, 311

servants, 119, 122

servile arts, 91–93

servi publici, 84n.11

shame, 73

sign language, 179

Simon, Yves, 141n

sin: and death, 314; original sin, 310

skholē, 14–15, 82n, 131n.84

slavery: ancient attitude toward, 36n.30; ancient justification of, 83–84; Aristotle on, 83n.9, 84, 84nn. 11, 12, 119n.69; chief function of ancient, 119; Christian view of life affecting, 316; *dēmiourgoi* and slaves distinguished, 81; emancipation contrasted with that of free labor, 217; Euripides on slaves, 84n.10; excellence as lost in, 49n; freed slaves becoming businessmen, 66n.70; Greek word as signifying defeated enemy, 81, 129n.79; labor power in, 88; masters seeing and hearing through their slaves, 120; Periandros's attempt to abolish, 221; Plato on natural slavishness, 36n.30, 316; Plato on treatment of slaves, 34n.37; property owning by slaves, 62, 62n.58; rebellions as rare, 215, 215n.51; in Roman Republic and Empire, 36n.30, 59, 59n.54; scribes as slaves, 92; Sen-

Index

Index

of, 95, 187; and revelation, 178–80,
187; rhetoric, 26, 26n.9; stories resulting from, 97; as superstructure,
33; and thought, 25; in web of relationships, 184. *See also* words
state, the (government): Augustine on
function of, 229n; bureaucracy, 40,
45, 93n.28; contradiction in modern
conception of, 69; economists as hostile to, 109–10, 110n.54; Marx on
withering away of, 45, 60; Marx's
stateless society, 131n.82; Smith on
function of, 220n; violence as monopoly of, 31, 32; wealth accumulation protected by, 72. See also *polis*,
the
state of nature, 32
statistics, 42–43
status, 41, 56
steam engine, 148
Stoicism: abstention from human affairs in, 234; all men as slaves for,
130n.81; attitude toward slavery influenced by, 36n.30; illusion of freedom underlying, 235; modern hedonism contrasted with, 311; pain as
natural experience underlying, 112;
as world alienation, 310
stories: action and speech resulting in,
97; action revealing itself fully only
to the storyteller, 191–92; an agent
revealed in, 184; biography, 97; each
human life telling its story, 184; fictional and real, 186; hero of a story,
184–87, 186n, 194; history as story
of mankind, 184–85; the *polis* as giving permanence to, 198; storytelling
transforming intimate life, 50; and
the web of relationships, 181–88
strength: of action process, 188–89,
233; as bounded, 201–2; power contrasted with, 200; sovereignty and
man's limited, 234; and violence,
140, 203
strong men, 188–89
subject/object split, 312–13
success, as criterion of science, 278–79
suicide, 315, 316
surplus, 88, 108

tame animals, 80, 83n.8
Tartaglia, Niccolò, 249n
teamwork, 161–62, 179, 271n.26
technology: carrying irreversibility and
unpredictability into natural realm,

238; development of modern,
147–50; origins in useless knowledge, 289
telescope, invention of, 248, 257–58,
274, 290
Tertullian, 54n.46, 74, 74n.83
thaumazein, 273, 302, 302n
Thebes, 26n.9
theodicies, 281, 282
theōria: in Aristotle's political philosophy, 14; in experience of the eternal,
20; in Greek political philosophy,
301; Socratic school on, 16; and *thaumazein*, 302; truth as residing in,
278. *See also* contemplation
thing-character of the world, 9, 93–96
thirty-six righteous men, story of, 75
Thomas Aquinas. *See* Aquinas, Thomas
thought: the brain and, 3, 322; cognition distinguished from, 170–71;
contemplation distinguished from,
16; for Descartes, 279n.39; eternity
as center of metaphysical, 20; freedom as condition for, 324; *homo faber*
as inspired by, 171; as inner dialogue, 76, 291; and modern world
view, 285–89; Plato as separating
from action, 223–27; and poetry,
170; as source of works of art,
168–69; and speech, 25; thinking as
laboring, 90; transforming into tangible objects, 76, 90, 95; as unable to
think itself, 236; useless thought,
170
Thucydides, 205–6
Tilgher, Adriano, 305n
tilling of the soil, 138
Tocqueville, Alexis de, 39
tools: and division of labor, 124–25; instrumentality of, 151; labor en-
hanced by, 121–22, 144; machines
contrasted with, 147; machine tools,
148. *See also* machines
torture, 129, 129n.78
totalitarianism, 216
trades, 91
trade unions: division of labor in, 123;
political parties contrasted with,
215–17
tragedy, Greek, 187, 187n.12
Trajan, 130n.81
Trojan War, 197, 198
truth: action as source of, 290; Cartesian doubt of, 274–79; and mathematization of physics, 287; Plato on

Index

Index

wealth: accumulation of as limitless for, 124; as capital, 68, 255; common wealth, 68–69; as condition of citizenship, 65n.68; expropriation in accumulation of, 254–55; government protecting accumulation of, 72; Greek and Roman attitudes toward, 59, 59n.54; growth of, 105, 111; labor as source of, 101; landed wealth, 66; political significance of, 64–65; property distinguished from, 61, 253

Weber, Max: on ancient cities as centers of consumption, 66n.69, 119n.70; on Athens as pensionopolis, 37; on innerworldly asceticism of capitalism, 251, 252n, 254; on loss of certainty of salvation, 277n

web of relationships: defined, 183; and the enacted stories, 181–88; freedom entangling people in, 233–34; realm of human affairs consisting of, 183–84

Weil, Simone, 131n.83, 287n.53

Weizsäcker, Viktor von, 123n, 213n.45

Westermann, William L., 12n.4, 215n.51

Whitehead, Alfred North: on common sense as in retreat, 283; on Galileo and the telescope, 257; on Michelson's interferometer, 295n; on nature as process, 296n.61; on science and organization, 271n.26; on scientific results offending reason, 290; on traditional versus modern science, 267n

will to power, 203, 204n, 245n

Wilson, Edmund, 105n.50

women: Jesus and Paul on creation of, 8n; labor in giving birth, 30; in sexual division of labor, 48n.38; and slaves in same category, 72, 72n.8; species survival as task of, 30; and working class emancipated in modern age, 73

wonder, 273, 302, 302n

words: futility of, 173; immortality of, 19; inserting ourselves in the world with, 176; politics as transacted in, 26

work, 136–74; and action in Greek political philosophy, 301–2; in Aristotle's *bios politikos*, 12, 13; benefactors as doing, 196; defined, 7; as destructive for nature, 100, 139, 153; as having an end, 98, 143; image or model as guiding, 141–42; intellectual work, 90–93;

labor distinguished from, 79–93, 80n.3, 81n.5, 83n.8, 94, 103–4, 138; natality and mortality as connected with, 8–9; as now performed in mode of laboring, 230; a proper location for, 73; rhythm for, 145n; in solitude, 22; and space of appearance, 212; specialization of, 47n, 123, 214n.48; in *vita activa*, 7; world alienation affecting, 301. See also *homo faber*; instruments of work; use objects

work ethos, Protestant, 252n

working class: alienation of, 255; emancipation of labor, 126–35, 217–18, 255; as jobholders like everyone else, 219; labor movement, 212–20; as not always having existed, 66n.70; as wage-earners, 255n; and women emancipated in modern age, 73

working hours, 132, 132n.85

workmanship: labor replacing since industrial revolution, 124; machines contrasted with tools of, 147; in music and poetry, 169; as required only for models, 125; teamwork as destructive of, 161; as unpolitical but not antipolitical, 212

works: good works, 76; greatness of mortals in, 19, 19n.19

works of art. See art

world, the: absence of pain in liberation from, 112–15; alienation from, 6, 209, 248–57, 301, 307, 310; as consisting of use objects, 134; discovery and exploration of the earth, 248, 250–51; durability of, 136–39; natality as miracle that saves, 247; permanence of, 167–74; power in human creation of, 204; "public" as signifying, 52; the senses in experiencing, 114–15, 114n.63; thing-character of, 9, 93–96; trust in reality of, 120; work's products in, 94

worldlessness, 54, 76, 115, 118–19

worldliness: of artists, 323, 323n; fabrication as sustaining, 236; as human condition of work, 7; love as unworldly, 242; of produced things, 96; secularity identified with, 253, 320

Xenophon, 31n, 32n.23, 48n.38, 82n, 182n.7

Zeus Herkeios, 30

Zoroastrianism, 278n.35